Biogenic Amines as Physiological Regulators

Society of General Physiologists' Symposia

HOFFMAN   The Cellular Functions of Membrane Transport
HAYASHI AND SZENT-GYORGYI   Molecular Architecture in Cell Physiology
DAVIS AND WARREN   The Specificity of Cell Surfaces
GOLDSTEIN   Control of Nuclear Activity
CARLSON   Physiological and Biochemical Aspects of Nervous Integration
TOSTESON   The Molecular Basis of Membrane Function
BLUM   Biogenic Amines as Physiological Regulators

PRENTICE-HALL INTERNATIONAL, INC., *London*
PRENTICE-HALL OF AUSTRALIA, PTY., LTD., *Sydney*
PRENTICE-HALL OF CANADA, LTD., *Toronto*
PRENTICE-HALL OF INDIA PRIVATE LTD., *New Delhi*
PRENTICE-HALL OF JAPAN, INC., *Tokyo*

Biogenic Amines as Physiological Regulators

A symposium held under the auspices of
The Society of General Physiologists
at its annual meeting at
The Marine Biological Laboratory,
Woods-Hole, Massachusetts, August 29–
September 1, 1969

J. J. Blum, *Editor*

Prentice-Hall, Inc.
Englewood Cliffs, New Jersey

©1970 by
PRENTICE-HALL, INC.
Englewood Cliffs, New Jersey

All rights reserved. No part of this book may be reproduced in any form or by any means without permission in writing from the publisher.

Current printing (last digit):

10 9 8 7 6 5 4 3 2 1

13-076554-6
Library of Congress Catalog Card Number: 73-126996
Printed in the United States of America

Preface

The Society of General Physiologists has conducted annual symposia since its founding and initiated the practice of publishing these symposia in 1955. The symposium on Biogenic Amines as Physiological Regulators was held at the Marine Biological Laboratory, Woods Hole, Massachusetts, in September 1969 and represented an effort to integrate in a manner useful to general physiologists the already enormous and still rapidly growing literature on the regulatory role of the biogenic amines. Since several recent symposia have adequately dealt with the mechanisms of biosynthesis and storage of these amines, it was decided to concentrate on the comparative aspects of this subject. There are two senses in which the word "comparative" can be used with respect to the biogenic amines. First, although most of the work done with these compounds relates to their roles as regulators of secretion, contractility, metabolism, etc. in mammals, it is clear that lower organisms, too, utilize these potent substances to control some of their physiological processes. The comparative aspects of the control functions of these amines have not been adequately studied and will certainly yield important insights into the evolution of these substances as physiological regulators. For this reason an attempt was made to include papers that dealt with the role of these amines in lower organisms, even at the price of excluding coverage of some of the better understood aspects of biogenic amine function in mammals.

The second sense of the word "comparative" refers to the various types of biogenic amines. Classically, these have been histamine, epinephrine and norepinephrine, and serotonin. Other amines are also found in nature, and some insight may be gained by considering whether these amines have physiological roles as regulatory substances. If they do, additional insight can be gained by contrasting these roles to those of the classical biogenic amines. This, again, necessitated the exclusion of some papers on more familiar subjects, but it is hoped that the gain in breadth of view will more than offset the loss of coverage in depth.

The officers and administration of the Marine Biological Laboratory were very helpful in making this symposium and the annual meeting an enjoyable one for all participants. Financial support was received from the National Institute of Neurological Diseases and Stroke, the Upjohn Company, Merck, Sharpe & Dohme, and Burroughs-Wellcome. The Society is indeed grateful to these institutions and corporations for their generous support. In addition, I am personally grateful to Drs. Norman Kirshner and Saul Schanberg for valuable advice on the organization of the program, to Drs. George Aghajanian, Seymour Cohen, and Betty Twarog for their able chairmanship of sessions of the symposium, and to Mrs. Karen Meadow for the care with which she prepared these manuscripts for publication. Finally, I would like to thank the authors not only for their excellent contributions but also for their cooperation in all other matters pertaining to the organization of the symposium.

J. J. BLUM

Contents

On the Nature of Receptor Sites for Biogenic Amines
    **G. A. Robison, J. W. Dobbs, and
E. W. Sutherland**   3

Factors Affecting Adenyl Cyclase Activity and its
Sensitivity to Biogenic Amines
    Benjamin Weiss   35

Phylogenetic Aspects of the Distribution of
Biogenic Amines
    John H. Welsh   75

Biogenic Amines and Metabolic Control in Tetrahymena
    J. J. Blum   95

Biogenic Amines as Metabolic Regulators in Invertebrates
    Tag E. Mansour   119

Regulation of Cardiac and Skeletal Muscle Glycogen
Metabolism by Biogenic Amines
    Steven E. Mayer   139

Hormonal Control of Glycogen Synthetase Interconversions
    **C. Villar-Palasi, N. D. Goldberg, J. S. Bishop,
F. Q. Nuttall, K. K. Schlender, and J. Larner**   161

Effects of Biogenic Amines on Adipose Tissue Metabolism
    Robert L. Jungas   181

Selected Topics on the Function of Biogenic Amines in
Exocrine Organs with Special Reference to Histamine
in Stomach
    Michael A. Beaven   207

Histamine Formation as Related to Growth and Protein
    Georg Kahlson and Elsa Rosengren   223

Biogenic Amines and Microcirculatory Hemeostasis
  Richard W. Schayer   239

Binding and Metabolism of Gamma-Aminobutyric Acid and Other Physiologically Active Amino Acids in the Brain
  K. A. C. Elliott   253

Do Polyamines Play a Role in the Regulation of Nucleic Acid Metabolism
  A. Raina and J. Janne   275

Norepinephrine and the Circadian Rhythm of Rat Hepatic Tyrosine Transaminase Activity
  Ira B. Black   301

Serotonin and Sleep
  M. Jouvet   321

INDEX   341

Biogenic Amines as Physiological Regulators

On the Nature of Receptor Sites for Biogenic Amines[1]

G. A. Robison, J. W. Dobbs,[2] and E. W. Sutherland[3]

*Departments of Pharmacology and Physiology
Vanderbilt University School of Medicine
Nashville, Tennessee*

The concept of specific drug receptors was introduced in the latter part of the nineteenth century (Langley, 1879; Ehrlich, 1900; see also Gaddum, 1962) and later developed systematically by A.J. Clark (1937). In simplest terms, receptors are what potent drugs such as the biogenic amines interact with to elicit the various responses which they characteristically produce. More appropriately, they can be defined as specific patterns of forces forming a part of some biological

[1]These studies were supported in part by grants from the National Institute of Mental Health (MH-11468) and the National Institutes of Health (HE-08332 and AM-07642) of the U.S. Public Health Service.
[2]Predoctoral Research Fellow of the U.S. Public Health Service.
[3]Career Investigator of the American Heart Association.

system, each of which is complementary to a certain pattern of forces subtended by the drug molecule, such that an interaction between the two patterns may occur (Schueler, 1960). This interaction leads to some change, or stimulus, which in turn leads to the observed response(s).

The existence of such receptors was deduced on the basis of several lines of evidence. Studies on the relationship between chemical structure and biological activity (Schueler, 1960; Ariens, 1966) disclosed that compounds with similar structures tended to produce similar effects. The same response may be produced by compounds with different chemical structures, but in these cases it is often possible to competitively antagonize the response to one set of compounds (by the use of structurally related compounds which have affinity for the receptor but which lack the ability to produce a response) without interfering with the response to the other compounds. Receptor protection experiments, based on the ability of a drug to prevent its own but not other receptors from being alkylated by various site-directed alkylating agents (Furchgott, 1966), have provided additional strong evidence that specific drug receptors do exist.

In only a few cases, such as the case of drugs inhibiting acetylcholinesterase (Wilson, 1967), is the nature of a drug receptor understood in any detail. Even in that case, moreover, our knowledge of the receptor is based largely on a knowledge of the chemicals with which the receptors interact. Such progress that has been made in this area was reviewed not long ago by Ariens (1966). To summarize the results of a large amount of research in a small amount of space, it can be said that we often know a good deal about the drug or hormone, chemically, and also about what these chemicals do in certain complex biological systems. We might think in terms of a black box, with an input (the drug or hormone molecule) and an output (the response), the problem being to understand the nature of the system in more detail.

There has been a tendency on the part of some investigators to define the receptor as the macromolecule or macromolecular complex of which the receptor proper is thought to be a part. As Schueler (1960) emphasized, however, such definitions are not very helpful from the standpoint of general pharmacology (Clark, 1937), the ultimate goal of which is to understand how drugs act at the molecular level. When we contemplate the three-dimensional structure of hemoglobin, for example, certain chemical groupings of which combine to form the receptors for carbon monoxide, we can see how far we are from reaching this ultimate goal, even after the macromolecule involved has been identified. In most cases involving the biogenic amines, the macromolecules or macromolecular complexes with which these agents interact have not even been isolated, not to mention studied with a view to establishing their three-dimensional structure. In many cases, it seems possible that the chemical moieties which combine to form the receptor will be contributed not by one macromolecule but by several. To whatever extent this is the case, the "isolation" of receptors may be a difficult or even impossible task.

It should be emphasized, however, that our inability to isolate receptors does not mean that we are incapable of understanding something of their nature. Let us consider, for example, a different type of pattern, the pattern of sounds which combine to form a symphonic movement. This can be transcribed and recorded, and meaning can be derived from it, but only with difficulty can it be isolated and put in a bottle. A more appropriate example might be the words on this page. Although we hope the reader will find them meaningful, or at least intelligible, and although the chemical structure of the print of which they are composed could be described in some detail, still it is the *pattern* that is important, and this is what is hard to isolate.

We hope that these brief remarks will indicate something of the distance yet to be travelled before an understanding of the chemical nature of the receptors for biogenic amines is achieved. In the meantime, a few steps have been taken along this road by the discovery that some of these agents, and quite a few other hormones besides, owe at least part of their biological activity to their ability to stimulate or inhibit the formation of adenosine 3',5'-phosphate, better known as cyclic AMP. Some of the evidence for this will be presented in the following pages.

Cyclic AMP: Its Formation and Metabolism

Cyclic AMP is now recognized as a versatile regulatory agent which controls the rate of a number of cellular processes. Distributed widely throughout the animal kingdom, its principal established role in multicellular organisms is to act as an intracellular second messenger mediating the effects of a great variety of hormones (Sutherland, et al., 1968). This concept is illustrated in Fig. 1. Different hormones af-

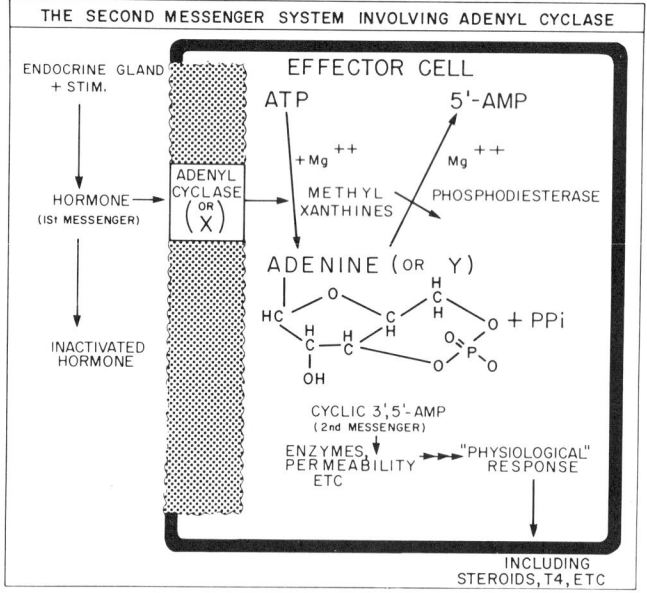

Figure 1

fect the level of cyclic AMP in different tissues, and this change in the level of the nucleotide leads to different effects depending upon the type of cell in which the change occurs.

The level of cyclic AMP is determined by the activity of at least two enzymes. Adenyl cyclase catalyzes the formation of cyclic AMP from ATP (Sutherland, et al., 1962). This reaction, which requires magnesium ions and yields pyrophosphate as the other product (Rall and Sutherland, 1962), can be reversed under some conditions (Greengard, *et al.*, 1969). The forward reaction can be greatly accelerated by the addition of fluoride, at least in particulate preparations from multicellular animals. Where studied, adenyl cyclase has most often been found as a component of the plasma membrane (Davoren and Sutherland, 1963; Rosen and Rosen, 1969), although in white skeletal muscle most of the activity sediments with the microsomal fraction (Rabinowitz, *et al.*, 1965). In some tissues, such as brain (De Robertis, *et al.*, 1967) and cardiac muscle (Entman, *et al.*, 1969), adenyl cyclase may be present in more than one type of membrane. In all of these cases, however, the activity has been restricted to particulate fractions, and this has greatly hindered efforts to solubilize and purify the enzyme. Only in bacteria have efforts to solubilize the enzyme without the aid of detergents been successful (Hirata and Hayaishi, 1967).

Perhaps the most striking feature of adenyl cyclase obtained from the cells of multicellular animals is its sensitivity to hormonal stimulation. As reviewed elsewhere (Robison, *et al.*, 1970), the hormones which are now known to be capable of stimulating adenyl cyclase in the cells of their respective target tissues include such diverse agents as the catecholamines, glucagon, ACTH, luteinizing hormone, TSH, MSH, vasopressin, and parathyroid hormone. These hormones are therefore capable of causing an increase in the intracellular level of cyclic AMP. The catecholamines differ from most of these hormones in that they are effective in a greater variety of cell types, and also in that they are capable of causing a *decrease* in the level of cyclic AMP in some cells. Several other hormones, including insulin, the prostaglandins, and melatonin, are also capable of causing a fall in the intracellular level of cyclic AMP, but only the catecholamines and the prostaglandins are presently known to be capable of changing the level in both directions. This is discussed in more detail in the next section.

The other enzyme known to influence the level of cyclic AMP is a relatively specific phosphodiesterase, which catalyzes the hydrolysis of cyclic AMP to 5'-AMP

(Butcher and Sutherland, 1962). This enzyme, which is found not only in cell membranes but in soluble fractions as well, can be inhibited by several drugs, including the methylxanthines (e.g., caffeine and theophylline). It may also be influenced directly or indirectly, by several hormones (see, for example, Senft, et al., 1968). Imidazole stimulates the phosphodiesterase (Butcher and Sutherland, 1962), and there is now evidence (Goodman, 1968) that some of the effects of histamine may be secondary to this type of action.

Finally, in this section, it should be noted that many other factors may influence the intracellular level of cyclic AMP. These would include some of the permissive or developmental hormones (e.g., growth hormone, thyroxine, and certain steroids), which may at times increase the amount of adenyl cyclase in some cells or enhance its sensitivity to stimulation by other hormones. Changes in the ionic environment may also have an effect, although the effect of a given ionic change may differ considerably from one type of cell to another. The effects of ions on adenyl cyclase are discussed in more detail by Dr. Weiss elsewhere in this book.

The Catecholamines

The catecholamines, epinephrine and norepinephrine, are the biogenic amines which have been most thoroughly studied, probably in general and certainly from the standpoint of their effects on cyclic AMP formation. Before going on to some of the newer data, it may be useful to review briefly the development of current concepts in this area.

The first evidence that these agents could react with two different types of receptors was provided by Sir Henry Dale (1906). He found that an increase in blood pressure occurred when epinephrine was injected into anesthetized cats. After such animals had been given a dose of ergotoxine, however, blood pressure fell in response to the injection of epinephrine. Later other compounds were discovered which could inhibit or reverse the pressor response. Although these drugs could inhibit many of the other excitatory effects of epinephrine, it was established that they did not affect the positive inotropic and chronotropic effects of epinephrine. Several attempts were made to account for these data in terms of a single type of receptor, perhaps the best known of these being that of Cannon and Rosenblueth (1937). They suggested that either of two substances might be formed when epinephrine reacted with its receptors, which one depending on the type of cell involved. One of these hypothetical substances, which they called "sympathin E," was thought to be responsible for the excitatory effects of epinephrine, while the other, "sympathin I," was thought to mediate the inhibitory effects. Although this theory had several attractive features, and was widely accepted for a time, it later became clear that it could not explain all of the facts. Among these were the ability of a single catecholamine to produce both effects in the same cell, as in the presence and absence of ergotoxine, and also the inability of ergotoxine to prevent the "excitatory" effects of epinephrine in the heart.

Ahlquist (1948) later showed that a number of adrenergic responses could be classified into two relatively distinct groups, according to the order of potency of a series of catecholamines in producing them. The responses in one group, which included most of the "excitatory" effects, such as smooth muscle contraction, could be blocked by ergotoxine and other known adrenergic blocking agents. The responses in the other group, which included most of the "inhibitory" effects, such as smooth muscle relaxation, but which also included some of the "excitatory" effects, such as cardiac stimulation, could not be prevented by drugs known at the time. These data supported Dale's earlier conclusion that there were at least two types of adrenergic receptors. Since they did not seem to correlate well with either "excitation" or "inhibition" Ahlquist suggested that they might best be described in terms of the Greek alphabet, at least until more was known about them. *Alpha* receptors were said to mediate the effects which could be blocked by drugs, while *beta* receptors mediated those which could not.

The subsequent introduction of dichloroisoproterenol (DCI), the first of a series of drugs capable of preventing the responses which Ahlquist had suggested were mediated by *beta* receptors (Powell and Slater, 1958), did much to gain acceptance for Ahlquist's method of classifying these receptors (Moran, 1967). This method has survived to the present day with surprisingly little modification. Although Ahlquist originally defined these receptors in terms of the order of potency of agonists, it is now recognized that the relative potencies of these agents may vary considerably from one tissue to another, or for the same tissue, from one species to another. Attempts have even been made to subclassify *beta* receptors (Lands, *et al.*, 1967), although the usefulness of such a procedure seems questionable. It is also recognized that the relative potencies of adrenergic blocking agents vary, and not necessarily with any obvious relation to the relative potencies of agonists (Robison, *et al.*, 1969a). Thus *alpha* and *beta* receptors seem to constitute two different classes or families or receptors, distinguishable partly on the basis of the order of potency of agonists (isoproterenol, a synthetic catecholamine, is generally a more potent *beta* agonist but a weaker *alpha* agonist than the naturally occurring catecholamines) but chiefly on the basis of the type of blocking agent which will prevent the response in question.

It should be emphasized at this point that both types of blocking agents may have many other actions not related to their adrenergic blocking activity. It is nevertheless true, however, that *alpha* adrenergic blocking agents are capable of competitively antagonizing some adrenergic responses without affecting others, while *beta* adrenergic blocking agents competitively block another set of responses in doses which do not interfere with those antagonized by *alpha* blockers. At present, this differential sensitivity to blockade is the most commonly used criterion for distinguishing the two types of receptors. Although there are some grey areas where this has not been established, most of the known physiologically important actions of the catecholamines have been characterized on this basis as being mediated by either *alpha* or *beta* adrenergic receptors.

The introduction of DCI as the first *beta* adrenergic

blocking agent coincided closely in time with the identification of cyclic AMP as the mediator of the hepatic glycogenolytic effect of epinephrine (Sutherland and Rall, 1960). After it was established that epinephrine increased the hepatic level of cyclic AMP by stimulating adenyl cyclase, it was found that this effect could be antagonized by DCI (Murad, *et al.*, 1962). Subsequently, as reviewed in more detail elsewhere (Robison, *et al.*, 1969a), it was found that wherever adenyl cyclase was stimulated by a catecholamine this effect could be competitively antagonized by a *beta* adrenergic blocking agent. In most cases studied, the standard *alpha* blocking agents have been ineffective, although in some tissues, notably the liver, high concentrations of ergotamine may be inhibitory (Murad, *et al.*, 1962).

The selective effect of pronethalol, a newer *beta* adrenergic blocking agent, is illustrated by the data in Fig. 2, from the work of Butcher, *et al.* (1968). Fat cells isolated from rat adipose tissue, unlike their counterparts from human tissue, are sensitive to stimulation by ACTH and glucagon as well as the catecholamines. These hormones all act by stimulating adenyl cyclase, thereby causing an increase in the intracellular level of cyclic AMP. As shown in Fig. 2, pronethalol completely prevents this response to epinephrine in doses which do not prevent or may even enhance the response to the other hormones. This selective type of blockade is also demonstrable at the level of lipolysis, although experiments at this level may be complicated by the tendency of higher concentrations of both *alpha* and *beta* blocking agents to noncompetitively inhibit lipolysis at a step beyond the formation of cyclic AMP (Aulich, *et al.*, 1967).

The experiments summarized to this point suggested that many responses mediated by adrenergic *beta* receptors were in fact mediated by cyclic AMP. Since adenyl cyclase in washed particulate preparations invariably retained the characteristics of the *beta* receptor in intact tissues, it was further suggested that these receptors might constitute an integral part of the adenyl cyclase system, in those tissues in which *beta* receptors occurred (Sutherland, 1965; Robison, *et al.*, 1967). The obvious corollary of this hypothesis was

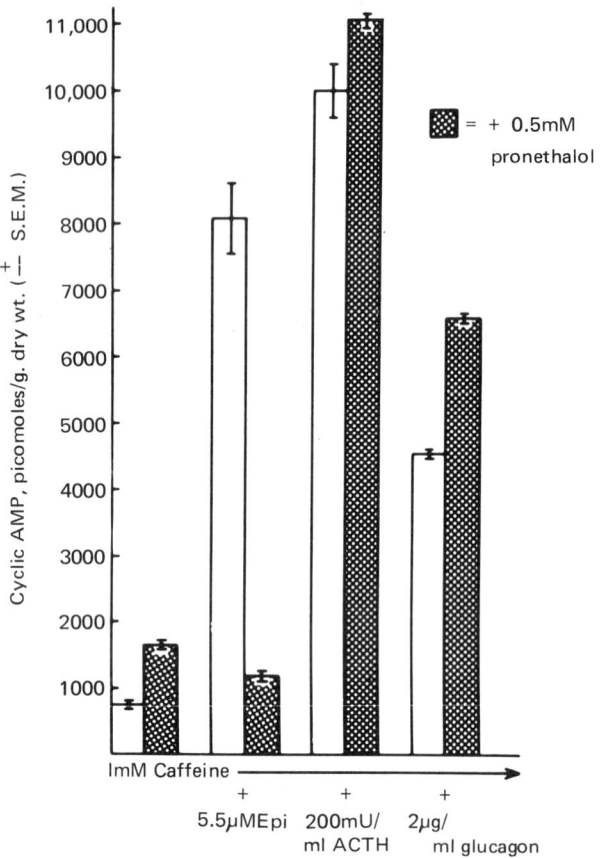

Figure 2. Effect of pronethalol on cyclic AMP levels in isolated fat cells. Cells were incubated for 10 min in the presence of the indicated concentrations of epinephrine, ACTH, and glucagon, in the presence (shaded bars) and absence (open bars) of pronethalol. From Butcher, et al. (1968).

that all *beta* receptor effects without exception might be mediated by cyclic AMP. It was felt by some investigators, however, that this concept might not apply to those functional or mechanical effects upon which the classification of *alpha* and *beta* receptors was initially based.

The first "mechanical" response to be studied with

this hypothesis in mind was the positive inotropic response in the heart. Briefly, it was found that this response to epinephrine in the isolated perfused working rat heart was preceded by a rapid increase in the level of cyclic AMP (Robison, *et al.*, 1965) and that the smallest doses of epinephrine capable of producing an increase in contractile force also produced a significant increase in the level of cyclic AMP (Robison, *et al.*, 1967). Similar results were obtained in the rat heart *in situ* (Namm and Mayer, 1968). Murad, *et al.* (1962) had previously shown that the order of potency of a series of catecholamines in stimulating adenyl cyclase in washed particulate preparations from dog heart was the same as their order of potency as inotropic agents *in vivo*. Rall and West (1963) showed that norepinephrine and theophylline acted synergistically to stimulate rabbit atria *in vitro*, as would be expected if the response was mediated by cyclic AMP. In line with this, Kukovetz and Pöch (1967) found that imidazole, which stimulates the phosphodiesterase *in vitro*, inhibited the inotropic response to catecholamines and methylxanthines. Glucagon stimulates cardiac adenyl cyclase and produces an inotropic effect similar to that produced by the catecholamines (Murad and Vaughan, 1969; Levey and Epstein, 1969). The action of the catecholamines at both levels can be prevented by *beta* adrenergic blocking agents in concentrations which do not prevent the response to glucagon (Murad and Vaughan, 1969), and isopropylmethoxamine, which is less potent than pronethalol in preventing the stimulation of adenyl cyclase *in vitro*, is likewise less potent than pronethalol as an antagonist of the inotropic response (Robison, *et al.*, 1967). Acetylcholine, which inhibits the accumulation of cyclic AMP in broken cell preparations from cardiac muscle, likewise inhibits the positive inotropic response in the intact heart (Meester and Hardman, 1967). It has also been possible, at least under some conditions, to mimic the cardiac effects of the catecholamines with exogenous cyclic AMP both *in vivo* (Levine, *et al.*, 1968) and *in vitro* (Kukovetz, 1968). Although many details remain to be clarified, the currently available evidence appears to support the hypothesis that the positive inotropic response to the catecholamines is mediated by cyclic AMP.

Another "mechanical" *beta* adrenergic response which has been studied with this hypothesis in mind is uterine relaxation (Dobbs and Robison, 1968). Uteri from ovariectomized estrogen-primed rats contract spontaneously and rhythmically when placed in an organ bath, and relax in response to catecholamines. Homogenates and washed particulate preparations from these uteri contain adenyl cyclase activity which can be stimulated by catecholamines and phosphodiesterase activity which can be inhibited by theophylline. Catecholamines act synergistically with theophylline to increase the level of cyclic AMP in intact uteri (Fig. 3), and likewise act synergistically to produce relaxation (Dobbs and Robison, 1968; Levy and Wilkenfeld, 1968). Theophylline produces only a slight increase in the level of cyclic AMP by itself, in the absence of exogenous cacatecholamines, and likewise produces only a slight effect on motility.

The order of potency of the catecholamines in elevating cyclic AMP is the same as their order of potency as relaxing agents (isoproterenol > epinephrine > norepinephrine, as in most of the responses studied by Ahlquist). The d-isomer of epinephrine does not cause an increase in the level of cyclic AMP and likewise does not cause the uterus to relax. The rise in cyclic AMP in response to catecholamines occurs at least as rapidly as relaxation can be measured. In the continued presence of a catecholamine, the level of cyclic AMP rises and falls (Fig. 4), just as contractions disappear and recur, and the level of cyclic AMP correlates at all times, within limits, with the functional state of the organ.

Relaxation in response to the catecholamines can be antagonized competitively by *beta* adrenergic blocking agents, and Fig. 5 illustrates this at the level of cyclic AMP. In these experiments we used propranolol, one of the newer and more potent *beta* blocking agents (Moran, 1967). As shown in the upper panel of Fig. 6, propranolol was capable of completely preventing relaxation in response to isoproterenol. In the absence of propranolol, this dose of isoproterenol would have

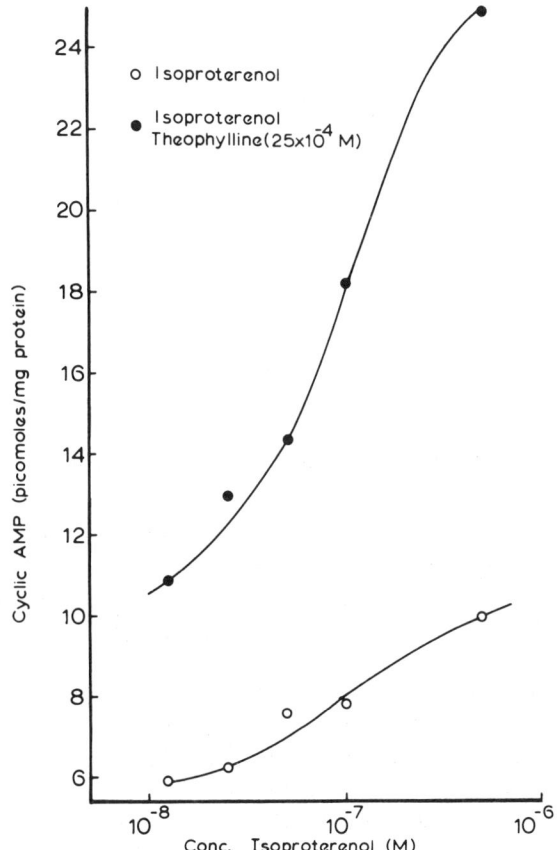

Figure 3. Effect of increasing concentrations of isoproterenol on level of cyclic AMP in isolated rat uteri in the absence (open circles) and presence (closed circles) of 2.5 X 10^{-4} M theophylline. Theophylline was added 5 min and isoproterenol 2 min before freezing. Each determination required 6 uterine horns.

completely abolished uterine motility for at least 30 minutes. By contrast, as shown in the lower panel of this figure, propranolol did not prevent the relaxing effect of the dibutyryl derivative of cyclic AMP (Henion, et al., 1967). Other nucleotides and nucleo-

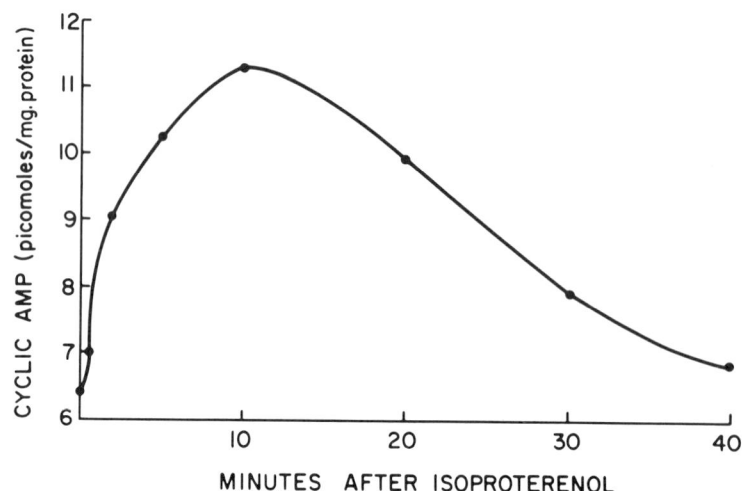

Figure 4. Effect of isoproterenol on uterine level of cyclic AMP. Uteri were incubated in presence of 5×10^{-7} M isoproterenol, and frozen at the indicated times.

sides do not cause the rat uterus to relax, and in higher concentrations may even induce contracture.

It should be noted that no single piece of this evidence is very strong when taken by itself. It could be argued, for example, that the correlations observed between the level of cyclic AMP and the functional state of the uterus are meaningless, since they might simply indicate that the stimulation of adenyl cyclase, on the one hand, and relaxation of the uterus, on the other, are both mediated by adrenergic *beta* receptors. Potentiation by theophylline might also be explained in other ways, and indeed attempts to do so would not be entirely unreasonable. The methylxanthines are among the "dirtiest" drugs in all of pharmacology, i.e., they have many actions which are unrelated to their ability to inhibit phosphodiesterase (Sutherland, *et al.*, 1968). We might note in passing, however, that it would be very difficult to explain this particular effect of theophylline in terms of any of its other known actions. To continue with our argument, the ability of the dibutyryl derivative of cyclic AMP to

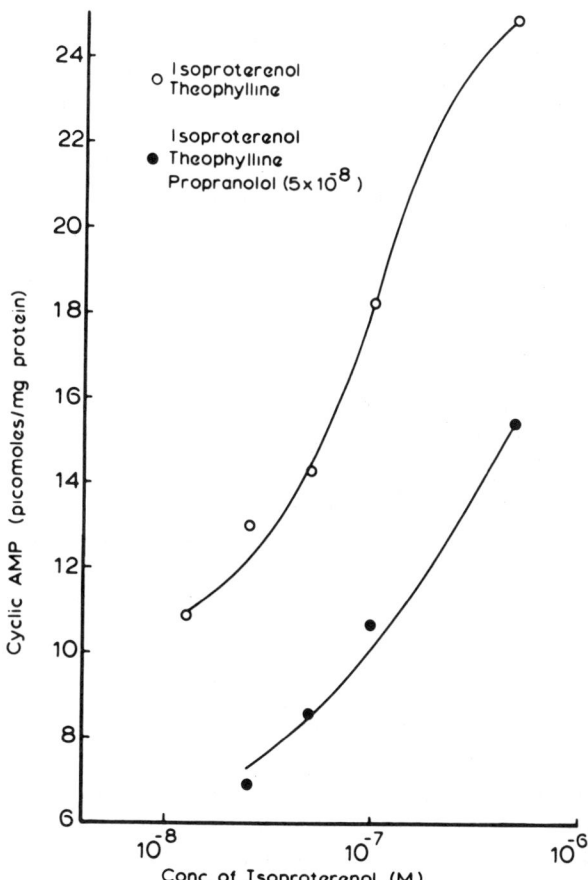

Figure 5. Effect of increasing concentrations of isoproterenol on uterine cyclic AMP levels in absence (open circles) and presence (closed circles) of 5 X 10^{-8} M propranolol. Theophylline (2.5×10^{-4} M) present throughout. Propranolol was added 10 min, theophylline 5 min, and isoproterenol 2 min before freezing.

mimic the relaxing effect of the catecholamines could be understood as nothing but a coincidence, and so on for each of the other points summarized. Taken together, however, the data seem to suggest that uterine relaxation in response to the catecholamines is medi-

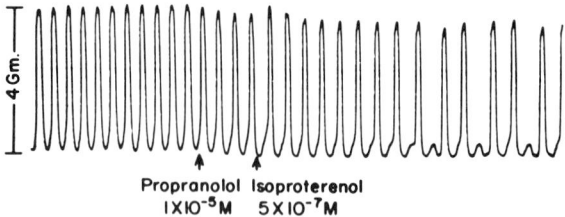

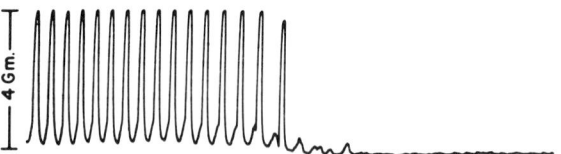

1×10⁻⁵M   1×10⁻⁴M
Propranolol Dibutyryl Cyclic AMP

Fig. 6. Effect of dibutyryl cyclic AMP (lower panel) and lack of effect of isoproterenol (upper panel) on spontaneous contractions of the isolated rat uterus in the presence of propranolol. Tracings are from opposite horns of the same uterus, from an ovariectomized estrogen-pretreated rat. Time marker intervals represent 30 sec. Contractions measured isometrically.

ated by cyclic AMP. When viewed in the light of observations made in other systems, the data also support the hypothesis that the *beta* adrenergic receptor is an integral component of the adenyl cyclase system, at least in those tissues where *beta* receptors occur.

A possible model to illustrate this concept is shown in Fig. 7. In this model, the protein component of the adenyl cyclase system is visualized as being composed of two types of subunits, a regulatory subunit (R)

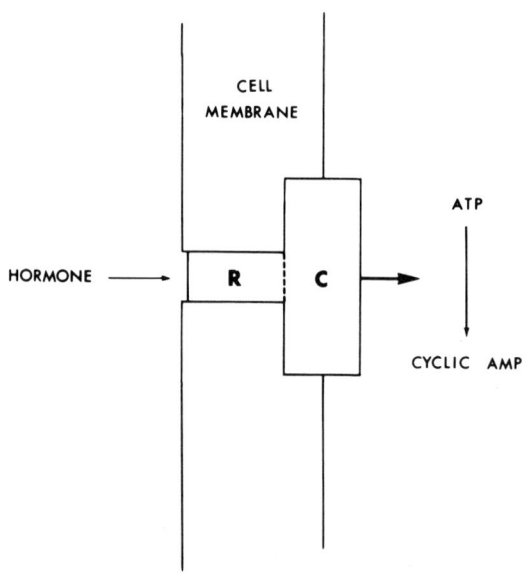

Figure 7. Model of the protein component of the membrane adenyl cyclase system. From Robison, et al. (1967).

facing the extracellular space, and a catalytic subunit (C) in contact with the cytoplasm. The *beta* receptor is visualized as a specific pattern of forces on the surface of the regulatory subunit. Although this pattern may differ slightly from one system to another, it is thought that all *beta* receptors resemble each other more than any one of them resembles the receptors for other hormones which stimulate adenyl cyclase. To account for the ability of two or more structurally different hormones to stimulate the same adenyl cyclase, as in the case of glucagon and the catecholamines in rat adipocytes and also in cardiac muscle from several species, it can be imagined that separate patterns of forces occur on the same regulatory subunit or that more than one type of regulatory subunit unfluences the same catalytic unit. The result of the hormone-receptor interaction is the same in all cases, which is to induce a conformational change which leads to an increase in the catalytic activity of adenyl cyclase.

Whether future research will substantiate or invalidate this hypothesis remains to be seen.

Whether the release of insulin from pancreatic *beta* cells in response to *beta* adrenergic stimulation should be regarded as a "mechanical" effect or a "metabolic" effect will presumably depend upon the previous orientation of the viewer. It is, in any event, another *beta* adrenergic effect which seems to be mediated by cyclic AMP. Under most physiological conditions *alpha* receptors predominate in the pancreas, so that this effect is generally seen only in the presence of *alpha* blocking agents. In the absence of adrenergic blockade (and even more strikingly in the presence of *beta* blockade) epinephrine and norepinephrine cause a fall in the level of cyclic AMP in pancreatic islets and inhibit the release of insulin (Turtle and Kipnis, 1967). It had been suggested previously (Sutherland, 1965; Robison, *et al.*, 1967) that many *alpha* adrenergic effects might be caused by a fall in the level of cyclic AMP, and the results with pancreatic islets represented one of the first experimental demonstrations of this.

Turtle and Kipnis (1967) also showed that epinephrine could reduce the level of cyclic AMP in isolated toad bladders, thus explaining the reduced permeability to water which occurs in this preparation in response to *alpha* adrenergic stimulation (Handler, *et al.*, 1968). The increase in cyclic AMP in response to epinephrine in the presence of phentolamine may have occurred in smooth muscle as well as epithelial cells.

Another example of the ability of *alpha* agonists to lower the level of cyclic AMP is shown in Fig. 8, from a study of cyclic AMP formation in the dorsal skin of *Rana pipiens* (Abe, *et al.*, 1969). In this preparation, melanocyte-stimulating hormone (MSH) causes an increase in the level of cyclic AMP, which leads to melanocyte dispersion and hence skin-darkening. Epinephrine and norepinephrine oppose the action of MSH and cause the skin to lighten. The antagonism between MSH and norepinephrine at the level of cyclic AMP is illustrated in Fig. 8, which also shows that the cyclic AMP lowering effect of norepinephrine can be prevented by phentolamine and dihydroergotamine, two *alpha* adrenergic

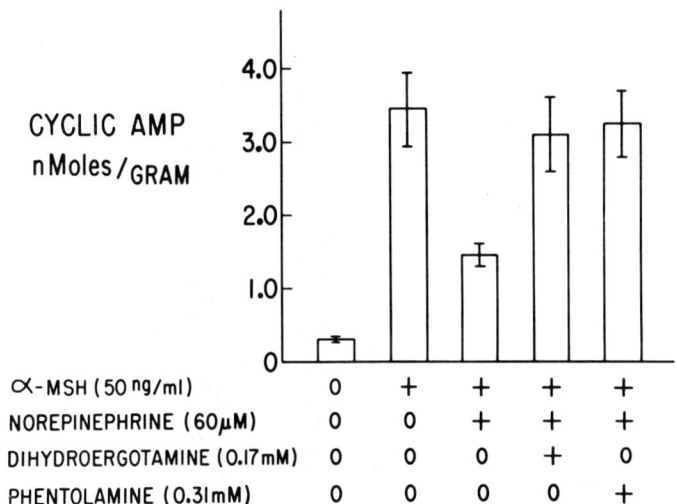

Figure 8. Effects of MSH, norepinephrine, and α-adrenergic blocking agents on the level of cyclic AMP in dorsal frog skin *in vitro*. The frog skins were frozen 30 min after addition of MSH or MSH plus norepinephrine. Phentolamine or dihydroergotamine were added 15 min before epinephrine. From Abe, *et al.* (1969).

blocking agents. Isoproterenol did not mimic MSH in these experiments, and *beta* blocking agents did not alter the response to epinephrine or norepinephrine, indicating that melanocytes from *Rana pipiens* do not contain *beta* receptors. This is in contrast to melanocytes from many other species, which may contain both types of receptors, *alpha* mediating skin lightening and *beta* mediating darkening (Gupta and Bhide, 1967). That epinephrine could produce both effects was in fact one of the first arguments to be raised against the sympathin theory, although Cannon and Rosenblueth (1937) dismissed it on the grounds that it was probably not relevant to mammalian physiology.

The ability of epinephrine to lower the level of cyclic AMP in human blood platelets (Robison, *et al.*, 1969b) is shown in Fig. 9. Prostaglandin E_1 causes an increase in the level of cyclic AMP in these cells and

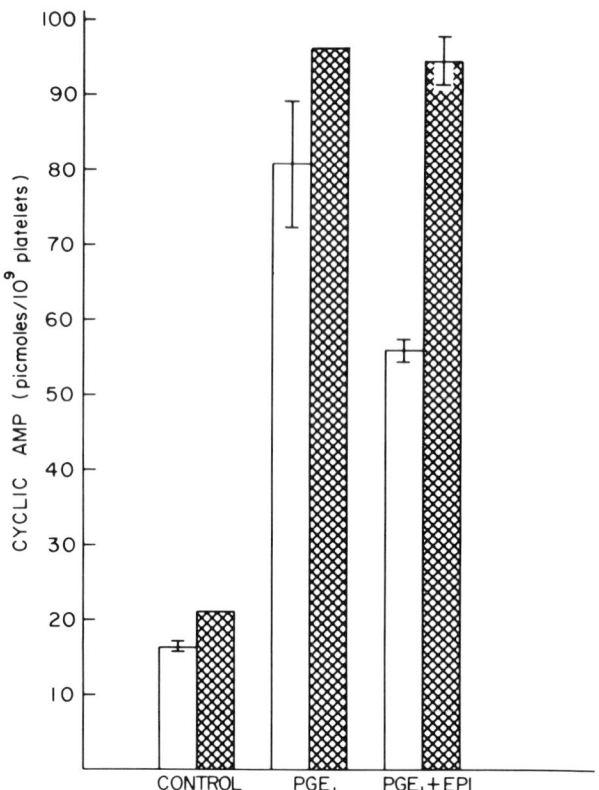

Figure 9. Effects of prostaglandin E_1 (PGE$_1$) and epinephrine (EPI) on level of cyclic AMP in human blood platelets in absence (open bars) and presence (shaded bars) of 5×10^{-4} M phentolamine. Platelets were incubated for 10 min at 37°C in platelet-rich plasma containing 2 mM theophylline. Phentolamine, epinephrine (5×10^{-5} M), and PGE$_1$ (100 ng/ml) were added in that order at intervals of 1 min, and incubation continued for an additional 10 min. The mixtures were then centrifuged for 5 min at 40°C, and the pellets rapidly frozen. From Robison, et al. (1969b).

increase in the level of cyclic AMP in these cells and

opposite effect and stimulates platelet aggregation. Phentolamine prevents the cyclic AMP-lowering effect as well as platelet aggregation, whereas propranolol in similar concentrations had no effect on either parameter. Like frog skin melanocytes, therefore, human blood platelets apparently do not contain adrenergic *beta* receptors.

A different situation prevails in the case of human fat cells. Burns and Langley (1968) had reported several interesting differences between the adipocytes of rats and humans, and have more recently found that human adipocytes contain both *alpha* and *beta* adrenergic receptors. *Beta* receptors usually predominate, so that the normal effect of the catecholamines, in the absence of blocking agents, is to stimulate lipolysis. This effect is greatly enhanced in the presence of *alpha* blocking agents, while in the presence of *beta* blocking agents the catecholamines actually inhibit lipolysis. In collaboration with Burns and Langley, we have recently initiated a study of cyclic AMP formation in these cells. The results of the first experiment in this series are presented in Table 1.

These results are extraordinary for several reasons, one being the almost linear relationship between cyclic AMP and lipolysis over an unusually wide range. In every other system where this has been studied, including rat adipocytes (Butcher, *et al.*, 1968), a relatively small increase in the level of cyclic AMP, on the order of 2 to 5-fold, has been sufficient to produce a maximal effect on cell function. Further increases in the level of cyclic AMP are not associated in these cases with any further effect. By contrast, the data in Table 1 indicate that in human fat cells a 10-fold increase in the level of cyclic AMP is associated with a 10-fold increase in the rate of lipolysis. This linear relationship is not only of great theoretical interest, but may also be extremely helpful experimentally. In Figs. 8 and 9, for example, the lowered levels of cyclic AMP in the presence of catecholamines were still above the range in which cyclic AMP is rate-limiting. These unphysiological conditions were deliberately chosen because of the difficulties involved in accurately measuring changes in cyclic AMP

TABLE 1
Effects of epinephrine and adrenergic
blocking agents on cyclic AMP and
lipolysis in isolated human fat cells

	Cyclic AMP after 30 min (picomoles/ml)	Glycerol released after 4 hr (μmoles/ml)
Controls	14.1	0.276
Epinephrine 5×10^{-5} M	21.2	0.412
Epinephrine + phentolamine (5×10^{-5} M)	67.3	1.486
Epinephrine + propranolol (5×10^{-5} M)	6.8	0.140

Fat cells were prepared and suspended in Krebs bicarbonate buffer containing albumin and glucose, as described previously (Burns and Langley, 1968). For cyclic AMP measurements, aliquots of 10 ml each were incubated for 30 min, then poured into liquid nitrogen. The frozen mass was then homogenized with 10 ml of 0.2 N HCl, and the resulting mixture purified and assayed. Glycerol was measured in separate aliquots of 1 ml each after incubating for 4 hours.

at lower levels. As a consequence, although they enabled us to demonstrate the ability of the catecholamines to reduce the level of cyclic AMP, they prevented us from quantitatively correlating this effect with effects on cell function. In human fat cells, by contrast, the quantitative relationship between the level of cyclic AMP and the change in cell function is obvious.

The data summarized to this point support the hypothesis that at least some *alpha* receptor responses are mediated by a fall in the level of cyclic AMP. Five such responses have been studied with this hypothesis in mind (inhibition of insulin release in rat pancreatic islets, reduced permeability to water in toad bladder, melanocyte condensation in frog skin, platelet aggregation, and inhibition of lipolysis in human fat cells), and each has been associated with a fall in the level of cyclic AMP. Unfortunately, the mechanism by which *alpha* receptors mediate this fall in the level of cyclic AMP has not been carefully studied. Therefore, the exact relationship of *alpha* receptors to the cyclic AMP system, and the related question of whether all *alpha* receptor effects are mediated in the same fashion, remain as subjects for future research. It is tempting to suppose that adrenergic *alpha* receptors are related to adenyl cyclase in much the same fashion postulated for *beta* receptors, except that interaction with a catecholamine leads to an opposite effect on the catalytic activity of the enzyme. On the other hand, it seems possible that *alpha* receptors may resemble the receptors for insulin in rat adipocytes (Kono, 1969). The interaction of insulin with these receptors leads to a change which suppresses the formation of cyclic AMP, on the one hand, and at the same time facilitates glucose transport, on the other (Park, *et al.*, 1968). To the extent that *alpha* adrenergic receptors resemble insulin receptors, in this functional sense, then to that extent it seems likely that some *alpha* receptor effects may be secondary to a fall in the level of cyclic AMP while others may not be. It would appear in any event that some interesting new problems are now open for investigation.

In restrospect, it was probably a mistake on the part of early investigators to define certain responses to the biogenic amines and other drugs as "excitatory" or "inhibitory" without a better knowledge of their real nature. We now know, for example, that cyclic AMP causes smooth muscle to relax while causing cardiac muscle to contract more forcefully, but whether one of these responses should be thought of as more "excitatory" or "inhibitory" than the other is not obvious, when one reflects on the possible mechanisms that might be involved. The sudden application of the brake pedal in a speeding automobile might seem quite excitatory to the person sitting beside the driver, but inhibitory to one viewing the scene from above. While it may seem puzzling that a single compound is capable of producing such seemingly diverse effects, our ignorance of the mechanisms of action of cyclic AMP is so vast, and the differences between the tissues in which it occurs so enormous, that it should perhaps be regarded less as a subject for speculation than as one requiring additional intensive research. In stressing our ignorance of the mechanisms of action of cyclic AMP, it should of course be noted that in some areas progress in this direction has been substantial. One such area is reviewed elsewhere in this volume by Dr. Mayer (p. 139). It is still too soon to know, however, whether a single type of action, such as the phosphorylation of a protein, can account for more than a few of the known effects of cyclic AMP.

A question which does invite speculation is why some cells contain both types of adrenergic receptors. Such an arrangement seems of dubious value, as Cannon and Rosenblueth (1937) noted, even though it has long been recognized as a feature of many forms of smooth muscle. *Alpha* receptors always seem to predominate in the pancreas, such that epinephrine or norepinephrine always seem to inhibit insulin release, unless the opposite effect is unmasked by *alpha* adrenergic blocking agents. Conversely, *beta* receptors seem to predominate in human fat cells, such that the catecholamines always seem to stimulate lipolysis, unless the opposite effect is unmasked by *beta* adrenergic blocking agents. Still, one wonders if this is always the case. Although inhibition of insulin release and stimulation of lipolysis in response to sympathetic stimulation have obvious survival value, and fit the traditional view of the catecholamines as emergency hormones (Sutherland and Robison, 1969), it is possible to imagine conditions under which the opposite effects might be of value. There is one well-known example of the ratio of *alpha* to *beta* receptor predominance changing in response to a physiological change, and this occurs in the uterus in response to changes in the estrogen/progesterone balance. The physiological significance of this is far from clear, especially since the direction of the response seems to differ in different species (Robison, *et al.*, 1967), but it does raise the possibility that similar changes might occur in other tissues. Our own studies in this area are in a very preliminary stage, and we still do not know whether physiologically meaningful changes in the estrogen/progesterone balance alter the uterine response to catecholamines by changing the amount or sensitivity of adenyl cyclase (or phosphodiesterase), by changing the nature or ratio of the adrenergic receptor population, by altering the sensitivity of one or more systems to cyclic AMP, or by some combination of these various possible effects. These and other possibilites are being considered in our present research in this area.

Finally, in this section, we should confess that we have made a deliberate attempt to fit some of our present knowledge about cyclic AMP into earlier concepts, such as the concept of *alpha* and *beta* receptors. This has seemed worth doing, if only because so much previous research was based on these concepts, but it is possible that we may ultimately have to admit defeat. To cite but one troublesome example, we have tended to think of the adrenergic receptors in the rat liver (which appear to be similar to those in the human liver) as *beta* receptors, because, like all other *beta* receptors we have studied, they mediate the stimulation of adenyl cyclase by catecholamines. However, as Arnold and McAuliff (1968) have pointed out, and as we ourselves have noted, these receptors resemble *alpha* receptors in their surface characteristics at least as much as they resemble *beta* receptors. For example, agents such as propranolol, which are potent inhibitors of adenyl cyclase stimulation (by catecholamines) in most tissues, are relatively weak in the rat liver. Isoproterenol, which in all other tissues examined has been at least equipotent and usually more potent than the naturally occurring catecholamines, may be less potent in the rat liver. Our proposed solution to this conceptual problem, to redefine *beta* receptors as those receptors which mediate the stimulation of adenyl cyclase by catecholamines, has seemed useful, but admittedly stretches the original definition pretty much beyond recognition. Possibly in the future adrenergic receptors will have to classified in some other manner.

In summary, it is now clear that many adrenergic responses are mediated by an increase in the level of cyclic AMP, secondary to the stimulation of adenyl cyclase, while certain other adrenergic responses are mediated by a fall in the level of cyclic AMP, secondary to an unknown action, possibly a direct inhibition of adenyl cyclase. A majority of the responses mediated by an increase in the level of cyclic AMP involve receptors which share certain characteristics (generally associated in the past with adrenergic *beta* receptors), while those mediated by a decrease involve receptors with different characteristics (generally associated in the past with adrenergic *alpha* receptors). Whether or not all of the physiologically important effects of the catecholamines are mediated by one or the other of these divergent effects on the level of cyclic AMP is presently unclear.

Other Biogenic Amines

The response to the other biogenic amines, including histamine, serotonin, and melatonin, have been studied much less extensively from the standpoint of the participation of cyclic AMP than have the responses to the catecholamines. Although the stated purpose of this article was to discuss receptors, and not necessarily cyclic AMP, we feel that we can more usefully contribute by staying as close as possible to our own area of competence. Instead of attempting to discuss these agents in detail, therefore, we will refer the reader to standard textbooks for background material and to other papers in this volume for some of the more recent advances.

Certainly many of the responses to these agents resemble one or another of the responses to the catecholamines, and it is therefore natural to think that many of them may be mediated by an increase or a decrease in the level of cyclic AMP. We believe that this is probably the case, but hard data to support this view are not overly abundant.

Superficial similarities are not reliable guides to mechanism. For example, in view of the evidence summarized previously that the uterine relaxing effect of the catecholamines was mediated by cyclic AMP (Dobbs and Robison, 1968), it seemed likely that oxytocin might stimulate the uterus, at least in part, by causing a fall in the level of cyclic AMP. Upon investigation, however, it was found that oxytocin antagonized the relaxing effect of the catecholamines (and of the dibutyryl derivative of cyclic AMP) without changing the level of cyclic AMP. Similarly we thought acetylcholine might cause a fall in the level of cyclic AMP in intestinal smooth muscle. This seemed especially likely in view of the previously demonstrated ability of choline esters to inhibit the accumulation of cyclic AMP in preparations from cardiac muscle (Murad, *et al.*, 1962). Instead, in collaboration with Hurwitz and Joiner, we found that doses of acetylcholine which increased the tension of the longitudinal smooth muscle of the guinea pig ileum did not produce a detectable change in the level of cyclic AMP. The mechanisms by which oxytocin and acetylcholine stimulate smooth muscle to contract are therefore still obscure, or at least are not well understood by us.

There is at least one effect of serotonin which seems clearly to be mediated by cyclic AMP, and this is the stimulation of phosphofructokinase in the liver fluke. Studies in this area were initiated some years ago by Mansour and his colleagues, and Dr. Mansour will review these studies in a following paper (p. 119). The similarities between the effects of serotonin and exogenous cylic AMP in stimulating fluid secretion by insect salivary glands (Berridge and Patel, 1968) and sodium transport across frog skin (Sayoc and Little, 1967; Baba, *et al.*, 1967) suggest but do not prove that these resopnses also are mediated by cyclic AMP. It seems probable that many of the effects of serotonin in molluscs, where serotonin seems to play a role analogous to that played by the catecholamines in higher species, are mediated by cyclic AMP. However, experiments to test this hypothesis have not to our knowledge been carried out. Studies in mammals have likewise been few and far between. Serotonin had little or no effect on the formation of cyclic AMP in chopped brain tissues, under conditions where norepinephrine produced a substantial increase (Palmer, *et al.*, 1969), but the range of these conditions was very limited.

Turning now to melatonin, we have one example where an effect of this agent seems clearly to be mediated by a *fall* in the level of cyclic AMP. This is the antagonism of MSH in dorsal frog skin (Abe, et al., 1969), illustrated in Fig. 10. A difference between this ef-

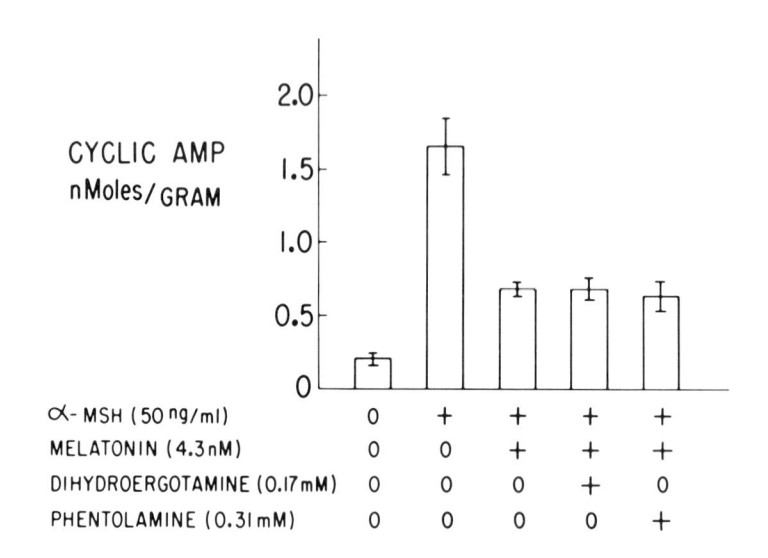

Figure 10. Effects of MSH and melatonin (and lack of effect of α-adrenergic blocking agents) on the level of cyclic AMP in dorsal frog skin *in vitro*. The frog skins were frozen 30 min after addition of MSH or MSH plus melatonin. Phentolamine or dihydroergotamine were added 15 min before melatonin. From Abe, et al. (1969).

fect and the apparently similar effect of norepinephrine was that the effect of melatonin could not be prevented by *alpha* adrenergic blocking agents, even when present in very high concentrations. This was interpreted to mean that melatonin and the catecholamines probably interacted with separate receptors, even though these receptors might be part of the same system. Whether any or all of the effects of melatonin in mammals are mediated by a fall in the level of cyclic AMP is unknown.

Histamine may produce several of its effects in mammalian tissues by altering the accumulation of cyclic AMP, and may act by several different mechanisms. Some of the effects of histamine, including the stimulation of gastric secretion in the frog (Harris and Alonso, 1965) and of cardiac contractility in the rat (Dean, 1968), are enhanced by theophylline, suggesting that histamine may at times stimulate adenyl cyclase. In line with this, Kakiuchi and Rall (1968) observed an increase in the level of cyclic AMP in rabbit brain slices in response to histamine. On the other hand, we found that histamine by itself did not increase the level of cyclic AMP in rat brain slices but did potentiate the effect norepinephrine (Palmer, *et al.*, 1969), suggesting a possible inhibitory effect on phosphodiesterase. Conversely, Goodman (1968) interpreted his results with histamine in rat adipose tissue as indicating a stimulatory effect on phosphodiesterase, and he has now demonstrated this effect in cell-free preparations (personal communication). If this occurs in some forms of smooth muscle, it could explain many of the known effects of histamine. However, a fall in the level of cyclic AMP in response to histamine in an intact tissue has thus far not been demonstrated. Our present picture of the role of cyclic AMP in the response to histamine is not a very clear one.

It seems likely that the possible role of cyclic AMP in the responses to these and other biogenic amines will receive an increasing amount of attention in the years to come. The results of this research may not tell us much about the nature of the receptors for these agents, but they will probably be helpful in telling us where to look for more information. We feel that studies with the catecholamines have already been helpful in this regard, and we hope that in this article we have been able to communicate some of the reasons for this view.

REFERENCES

Abe, K., G.A. Robison, G.W. Liddle, R.W. Butcher, W.E. Nicholson, and C.E. Baird. 1969. *Endocrinology* 85: 674.

Ahlquist, R.P. 1948. *Am. J. Physiol.* 153: 586.

Ariens, E.J. 1966. *Advances in Drug Research* 3: 235.
Arnold, A. and J.P. McAuliff. 1968. *Experientia* 24: 674.
Aulich, A., K. Stock and E. Westermann. 1967. *Life Sciences* 6: 929.
Baba, W.I., A.J. Smith and M.M. Townshend. 1967. *Quart. J. Exptl. Physiol.* 52: 416.
Berridge, M.J. and N.G. Patel. 1968. *Science* 162: 462.
Burns, T.W. and P. Langley. 1968. *J. Lab. Clin. Med.* 72: 813.
Butcher, R.W. and E.W. Sutherland. 1962. *J. Biol. Chem.* 237: 1244.
Butcher, R.W., C.E. Baird and E.W. Sutherland. 1968. *J. Biol. Chem.* 243: 1705.
Cannon, W.B. and A. Rosenblueth. 1937. *Autonomic Neuroeffector Systems* (New York: MacMillan).
Clark, A.J. 1937. In *Heffter's Handbuch der Experimentellen Pharmakologie*, Vol. 4, ed. W. Heubner and J. Schuller, (Berlin: Springer-Verlag).
Dale, H.H. 1906. *J. Physiol.* 34: 163.
Davoren, P.R. and E.W. Sutherland. 1963. *J. Biol. Chem.* 238: 3016.
Dean, P.M. 1968. *Brit. J. Pharmacol.* 32: 65.
DeRobertis, E., G.R.D.L. Ainaiz, M. Alberici, R.W. Butcher and E.W. Sutherland. 1967. *J. Biol. Chem.* 242: 3487.
Dobbs, J.W. and G.A. Robison. 1968. *Federation Proc.* 27: 352.
Ehrlich, P. 1900. *Proc. Royal Soc.* 66: 424.
Entman, M.L., G.S. Levey and S.E. Epstein. 1969. *Biochem. Biophys. Res. Comm.* 35: 728.
Furchgott, R.F. 1966. *Advances in Drug Research* 3: 21.
Gaddum, J.H. 1962. *Ann. Rev. Pharmacol.* 2: 1.
Goodman, H.M. 1968. *Nature* 219: 1053.
Greengard, P., O. Hayaishi and S.P. Colowick. 1969. *Federation Proc.* 28: 467.
Gupta, I. and N.K. Bhide. 1967. *J. Pharm. Pharmacol.* 19: 768.
Handler, J.S., R. Bensinger and J. Orloff. 1968. *Am. J. Physiol.* 215: 1024.
Harris, J.B. and D. Alonso. 1965. *Federation Proc.* 24: 1368.
Henion, W.F., E.W. Sutherland and T. Posternak. 1967. *Biochim. Biophys. Acta* 148: 106.

Hirata, M. and O. Hayaishi. 1967. *Biochim. Biophys. Acta* 149: 1.
Kakiuchi, A. and T.W. Rall. 1968. *Molec. Pharmacol.* 4: 367.
Kono, T. 1969. *J. Biol. Chem.* 244: 1772.
Kukovetz, W.R. 1968. *Arch. Pharmak. Exptl. Path.* 260: 163.
Kukovetz, W.R. and G. Pöch. 1967. *J. Pharmacol. Exptl. Therap.* 156: 514.
Lands, A.M., A. Arnold, J.P. McAuliff, F.P. Luduena and T.G. Brown. 1967. *Nature* 214: 597.
Langley, J.N. 1878. *J. Physiol.* 1: 339.
Levey, G.S. and S.E. Epstein. 1969. *Circulation Research* 24: 151.
Levine, R.A., L.M. Dixon and R.B. Franklin. 1968. *Clin. Pharmacol. Therap.* 9: 168.
Levy, B. and B.E. Wilkenfeld. 1968. *Brit. J. Pharmacol.* 34: 604.
Meester, W.D. and H.F. Hardman. 1967. *J. Pharmacol. Exptl. Therap.* 158: 241.
Moran, N.C. 1967. *Ann. N. Y. Acad. Sci.* 139: 649.
Murad, F. and M. Vaughan. 1969. *Biochem. Pharmacol.* 18: 1129.
Murad, F., Y.M. Chi, T.W. Rall and E.W. Sutherland. 1962. *J. Biol. Chem.* 237: 1233.
Namm, D.H. and S.E. Mayer. 1968. *Molec. Pharmacol.* 4: 61.
Palmer, E.C., F. Sulser and G.A. Robison. 1969. *The Pharmacologist* 11: 258.
Park, C.R., O.B. Crofford, and T. Kono. 1968. *J. Gen. Physiol.* 52: 296s.
Powell, C.E. and I.H. Slater. 1958. *J. Pharmacol. Exptl. Therap.* 122: 480.
Rabinowitz, M., L. De Salles, J. Meisler and L. Lorand. 1965. *Biochim. Biophys. Acta* 97: 29.
Rall, T.W. and E.W. Sutherland. 1962. *J. Biol. Chem.* 237: 1228.
Rall, T.W. and T.C. West. 1963. *J. Pharmacol. Exptl. Therap.* 139: 269.
Robison, G.A., R.W. Butcher, I. Øye, H.E. Morgan and E.W. Sutherland. 1965. *Molec. Pharmacol.* 1: 168.
Robison, G.A., R.W. Butcher and E.W. Sutherland. 1967. *Ann. N. Y. Acad. Sci.* 139: 703.

Robison, G.A., R.W. Butcher and E.W. Sutherland. 1969a. In *Fundamental Concepts of Drug-Receptor Interactions*, ed. D.J. Triggle, (New York: Academic Press).
Robison, G.A., A. Arnold and R.C. Hartmann. 1969b. *Pharmacol. Res. Comm.* 1: in press.
Robison, G.A., R.W. Butcher and E.W. Sutherland. 1970. *Cyclic AMP.* (New York: Academic Press).
Rosen, O.M. and S.M. Rosen. 1969. *Arch. Biochem. Biophys.* 131: 449.
Sayoc, E.F. and J.M. Little. 1967. *J. Pharmacol. Exptl. Therap.* 155: 352.
Schueler, F.W. 1960. *Chemobiodynamics and Drug Design.* (New York: Blakiston).
Senft, G., G. Schultz, K. Munske and M. Hoffmann. 1968. *Diabetologia* 4: 322.
Sutherland, E.W. 1965. In *Pharmacology of Cholinergic and Adrenergic Transmission*, ed. G.B. Koelle, W.W. Douglas and A. Carlsson, (New York: Pergamon).
Sutherland, E.W. and T.W. Rall. 1960. *Pharmacol. Rev.* 12: 265.
Sutherland, E.W. and G.A. Robison. 1969. *Diabetes* 18: in press.
Sutherland, E.W., T.W. Rall and T. Menon. 1962. *J. Biol. Chem.* 237: 1220.
Sutherland, E.W., G.A. Robison and R.W. Butcher. 1968. *Circulation* 37: 279.
Turtle, J.R. and D.M. Kipnis. 1967. *Biochem. Biophys. Res. Comm.* 28: 797.
Wilson, I.B. 1967. *Ann. N. Y. Acad. Sci.* 144: 664.

Factors Affecting Adenyl Cyclase Activity and its Sensitivity to Biogenic Amines

Benjamin Weiss

Laboratory of Preclinical Pharmacology
National Institute of Mental Health
Washington, D. C.

INTRODUCTION

Investigation into the biological role of cyclic nucleotides is presently in a period of logarithmic growth. The pioneering studies of Sutherland and his co-workers have now been expanded into virtually every field of biology. It is difficult to write a discussion of adenyl cyclase without emphasizing the contributions made by Sutherland's group to this vital area of research, for they not only initiated the study of cyclic nucleotides and the enzymes responsible for their formation and destruction but have also figured prominently in stimulating and developing new concepts of hormone-receptor interactions.

To date, the cyclic nucleotide most studied is cyclic 3',5'-adenosine monophosphate (cyclic AMP), but other closely related compounds, such as cyclic 3',5'-guanosine monophosphate, will surely increase in stature in the near future. Cyclic AMP affects virtually every species and tissue examined and in most cases mimics the response of specific biologically active agents. Thus, it is found in species from bacteria to mammals, it has physiological effects on tissues as diverse as smooth muscle and neurons, and it produces biochemical actions ranging from glycogenolysis to protein synthesis.

It is not the purpose of this review to consider the relevance of cyclic AMP to biology; this topic has been reviewed recently by several authors (see Haugaard and Hess, 1965; Sutherland and Robison, 1966; Robison, *et al.*, 1967, 1968; Sutherland, *et al.*, 1968; Costa and Weiss, 1968, Weiss and Kidman, 1969). Nor will I discuss the importance of cyclic AMP phosphodiesterase, the enzyme that hydrolyzes cyclic AMP; the possible role of this enzyme as a physiological regulator of cyclic AMP levels has also been examined recently (see Weiss and Kidman, 1969). But rather, I will concentrate my attention on the enzyme system that catalyzes the formation of cyclic AMP from adenosine triphosphate (ATP); for it is the activity of adenyl cyclase that is exquisitely controlled by hormones, and it is at this step that the intracellular concentrations of cyclic AMP are probably regulated.

Adenyl cyclase is activated by a wide variety of hormones and neurohormones, most prominently, the catecholamines and other biogenic amines and the polypeptide hormones. Indeed, so many of the actions of these hormones are explicable by their stimulation of adenyl cyclase—with the subsequent increase in tissue concentrations of cyclic AMP—that adenyl cyclase may be thought of as a receptor for these compounds (see Robison, *et al.*, 1967, 1968, this book; Sutherland, *et al.*, 1968; Weiss and Kidman, 1969). In this paper, I will review published data and present new evidence indicating that adenyl cyclase is not only acutely activated by a variety of agents, but its activity and its responsiveness to biogenic amines can also be chronically altered by several exogenous and endogenous factors. Moreover, I will try whenever possible to correlate these biochemical changes in adenyl cyclase activity with physiological responses.

In the context of this dicussion the biogenic amines will be considered as specialized hormones (neurohormones) released from nerve endings exerting a local effect on postjunctional receptor sites, in contrast to other hormones which are secreted into the blood producing more generalized effects at sites distal to their point of synthesis.

DISTRIBUTION OF ADENYL CYCLASE

Species Distribution

In 1962, Sutherland, *et al.* showed that adenyl cyclase is present in mammals, birds, insects, fish, and worms. Adenyl cyclase activity has since been described in bacterial systems (Hirata and Hayaishi, 1965, 1967), in sea urchin eggs (Castaneda and Tyler, 1968), and in frog erythrocytes (Rosen and Rosen, 1968, 1969). The widespread prevalence of this enzyme in the animal kingdom may help explain why such a variety of compounds can activate it. Investigations into the phylogenetic development of adenyl cyclase may show that different adenyl cyclases are formed at different evolutionary stages and that hormonal sensitivity evolves as a result of the changing requirements of the organism. This important study has yet to be made.

Gross Distribution

Adenyl cyclase is found in every species examined and in essentially all types of tissue. Again it was Sutherland and his co-workers (Sutherland, *et al.*, 1962) who first reported the distribution and relative activity of adenyl cyclase in various tissues. Since then they and others have described in more detail the distribution and properties of adenyl cyclase in adipose tissue, adrenal gland, erythrocytes, platelets, brain, heart, liver, ovary, parotid gland, pineal gland, skeletal muscle, and testes (see Table 1 and below for references). The highest adenyl cyclase activity is in brain (Sutherland, *et al.*, 1962), being especially high in cerebral cortex, cerebellum and pineal gland

TABLE 1

Biological agents that alter tissue concentrations of cyclic 3',5'-AMP

Tissue	Agent	Effect	Reference
Adipose Tissue	Epinephrine	Increase	Butcher, et al., 1965; Butcher and Baird, 1968
Fat cells	Prostaglandin E₁	Decrease	Butcher and Baird, 1968
Fat cells	Epinephrine	Increase	Turtle and Kipnis, 1967a, b
Fat cells	Epinephrine	Increase	Butcher, et al., 1968
Fat cells	Insulin	Decrease	Butcher, et al., 1968
Fat cells	ACTH	Increase	Butcher, et al., 1968
Fat cells	Glucagon	Increase	Butcher, et al., 1968
Fat cells	Thyroid stimulating hormone	Increase	Butcher, et al., 1968
Fat cells	Luteinizing hormone	Increase	Butcher, et al., 1968
Fat cells	Norepinephrine	Increase	Butcher, et al., 1968
Fat cells	Isoproterenol	Increase	Butcher, et al., 1968
Fat cells	Nicotinic acid	Decrease	Butcher, et al., 1968
Adrenal	Vasopressin	Increase	Brown, et al., 1963
Adrenal	ACTH	Increase	Haynes, 1958
Adrenal	ACTH	Increase	Grahme-Smith, et al., 1967
Brain			
Cerebellum	Norepinephrine	Increase	Kakiuchi and Rall, 1968a
Cerebellum	Histamine	Increase	Kakiuchi and Rall, 1968a
Cerebellum	Serotonin	Increase	Kakiuchi and Rall, 1968a

Cerebr. cort.	Norepinephrine	Increase	Kakiuchi and Rall, 1968b
Cerebr. cort.	Histamine	Increase	Kakiuchi and Rall, 1968b
E. coli		Decrease	Makman and Sutherland, 1965
Corpus luteum	Luteinizing hormone	Increase	Marsh, *et al.*, 1966
Erythrocytes	Epinephrine	Increase	Davoren and Sutherland, 1963b; Oye and Sutherland, 1966
Erythrocytes	Norepinephrine	Increase	Davoren and Sutherland, 1963b
Eryhtrocytes	Isoproterenol	Increase	Davoren and Sutherland, 1963b
Heart	Norepinephrine	Increase	Murad, *et al.*, 1962
Heart	Epinephrine	Increase	Murad, *et al.*, 1962
Heart	Epinephrine	Increase	Robison, *et al.*, 1965
Heart	Epinephrine	Increase	Williamson, 1966
Heart	Isoproterenol	Increase	LaRaia, *et al.*, 1968
Kidney	Vasopressin	Increase	Brown, *et al.*, 1963
Liver	Epinephrine	Increase	Rall and Sutherland, 1958
Liver	Glucagon	Increase	Rall and Sutherland, 1958
Liver	Catecholamines	Increase	Murad, *et al.*, 1962
Liver	Insulin	Decrease	Jefferson, *et al.*, 1968
Liver Fluke	Serotonin	Increase	Mansour, *et al.*, 1960
Lung	Prostaglandin E$_1$	Increase	Butcher and Baird, 1968
Pancreas	Epinephrine	Decrease	Turtle and Kipnis, 1967b
Pancreas	Glucagon	Increase	Turtle and Kipnis, 1967b

TABLE 1 (cont.)

Tissue	Agent	Effect	Reference
Skeletal muscle	Prostaglandin E$_1$	Increase	Butcher and Baird, 1968
Skeletal muscle	Epinephrine	Increase	Posner, et al., 1965
Skeletal muscle	Epinephrine	Increase	Ludholm, et al., 1967
Skeletal muscle	Epinephrine	Increase	Craig, et al., 1969
Skeletal muscle	Insulin	Increase	Goldberg, et al., 1967
Smooth muscle	Epinephrine	Increase	Bueding, et al., 1966
Spleen	Prostaglandin E$_1$	Increase	Butcher and Baird, 1968
Thyroid	Thyroid stimulating hormone	Increase	Kaneko, et al., 1969
Thyroid	Thyroid stimulating hormone	Increase	Gilman and Rall, 1968
Thyroid	Epinephrine	Increase	Gilman and Rall, 1968
Thyroid	Prostaglandin E$_1$	Increase	Keneko, et al., 1969
Urine	Parathyroid hormone	Increase	Chase and Aurbach, 1967
Uterus	Estradiol	Increase	Szego and Davis, 1967

(Weiss and Costa, 1968a). It is of particular interest, therefore, to find that cyclic AMP has both electrophysiological (Siggins, et al., 1969) and biochemical effects (Miyamoto, et al., 1969) on brain.

Although a variety of tissues possess adenyl cyclase activity, there are profound differences with which the adenyl cyclases respond to hormones. The specificity with which biological agents activate the adenyl cyclase of each particular tissue is essential to the concept of the enzyme being an intermediary in hormonal responses. This topic will be amplified in a subsequent section.

Subcellular Location

To implicate adenyl cyclase as a receptor for neurohormones the enzyme must be readily accessible to its agonist. Since the most likely site at which biogenic amines act is on the plasma membrane, it was important to find that in mammals and birds, adenyl cyclase is totally particulate (Sutherland, et al., 1962) and is associated with the plasma membrane fraction (Davoren and Sutherland, 1963a; Oye and Sutherland, 1966). More recent studies have confirmed the particulate nature and membraneous location for adenyl cyclase in skeletal muscle (Rabinowitz, et al., 1965), brain (deRobertis, et al., 1967; Weiss and Costa, 1968a), pineal gland (Weiss and Costa, 1968a), kidney (Chase and Aurbach, 1968; Dousa and Rychlik, 1968a), fat cells (Rodbell, 1967), myometrium and mammary gland (Dousa and Rychlik, 1968a).

Moreover, to be consistent with the view that the enzyme serves as a receptor for the biogenic amines, adenyl cyclase must be located post-junctional to the nerve endings rather than within the nerves themselves. The presence of adenyl cyclase in the synaptic fraction of brain was shown by deRobertis, et al. (1967), and the post-junctional location of adenyl cyclase was demonstrated in the pineal gland; after total sympathetic denervation of the gland, adenyl cyclase was still present, and the enzyme was still activated by catecholamines (Weiss and Costa, 1967; Weiss, 1969a).

FACTORS INFLUENCING ADENYL CYCLASE ACTIVITY

I will divide the factors influencing adenyl cyclase activity into two general categories; those which acutely activate or inhibit adenyl cyclase and those which chronically alter the amount of adenyl cyclase, possibly by governing its synthesis.

Acute Effects

Ionic environment. Rall and Sutherland (1958) were the first to show that the biosynthesis of cyclic AMP requires magnesium and is stimulated by fluoride. Sutherland, *et al.* (1962) subsequently showed that manganese could at least partially replace magnesium; barium had little effect on adenyl cyclase, but zinc was strongly inhibitory. Of the monovalent cations, neither sodium nor potassium influenced enzyme activity.

These findings were corroborated and expanded in more recent work. Fluoride has now been shown to activate adenyl cyclase from many tissue sources including adipose tissue (Rodbell, 1967; Williams, *et al.*, 1968) adrenal gland (Taunton, *et al.*, 1967), pineal gland (Weiss, 1969b), and turkey (Oye and Sutherland, 1966) and frog erythrocytes (Rosen and Rosen, 1969). Streeto and Reddy (1967) reported an optimum magnesium concentration for adrenal adenyl cyclase of about 10^{-2} M which is somewhat higher than that found in pineal gland (about 3×10^{-3} M; see Fig. 3). That calcium (Weiss, 1969b; Rosen and Rosen, 1969; Taunton, *et al.*, 1969) and zinc inhibit adenyl cyclase has also been substantiated as has the lack of effect of sodium and potassium (Williams, *et al.*, 1968; Rosen and Rosen, 1969). Taunton, *et al.* (1969), on the other hand, showed that high concentrations of sodium and potassium inhibited adrenal adenyl cyclase activity.

Studies of several tissues revealed that the hydrogen ion concentration for maximum adenyl cyclase activity falls within a narrow range. The pH optimum for adrenal (Streeto and Reddy, 1967; Taunton, *et al.*, 1969), pineal (Weiss, 1969b), and frog erythrocyte (Rosen and Rosen, 1969) adenyl cyclase is between pH

7.5 and 8, results comparable to those found earlier by Sutherland, *et al.* (1962).

The ionic environment not only influences the activity of adenyl cyclase *per se* but also affects the stimulation of the enzyme by hormones. For example, maximum adenyl cyclase activity occurs at a lower pH if fluoride or the catecholamines are present. This interesting observation was first noted by Sutherland, *et al.* (1962) in adenyl cyclase of heart and has since been confirmed in studies of the pineal enzyme (Weiss, 1969b). Moreover, Bar and Hechter (1969a) showed that calcium was essential for activating adipose tissue adenyl cyclase by adrenocorticotropic hormone (ACTH); this may explain the absolute requirement of calcium for the lipolytic action of ACTH (Lopez, *et al.*, 1959). Thus several divalent and monovalent cations influence both the "basal" adenyl cyclase activity and the stimulatory effects of hormones on enzyme activity.

Activation of adenyl cyclase by hormones and neurohormones. A hormone-induced increase in the tissue concentration of cyclic AMP may be caused by any one of several mechanisms: the hormone may (a) inhibit the enzymatic destruction of cyclic AMP, (b) reduce the rate of efflux of cyclic AMP from the tissue, or (c) stimulate the rate of formation of cyclic AMP by activating adenyl cyclase. That epinephrine, for example, causes an increased concentration of cyclic AMP in liver does not prove that the catecholamine activates adenyl cyclase.

Assaying cyclic AMP concentrations in tissue and determining adenyl cyclase activity offer complementary approaches to the same basic question. Measuring intracellular levels of cyclic AMP after hormonal treatment provides a sound physiological basis for involving the cyclic nucleotide as a mediator of hormone actions. Studies into hormone-induced activation of adenyl cyclase, on the other hand, may furnish information on the specific biochemical mechanisms responsible for their effects and may indicate the means whereby the concentration of cyclic AMP can be controlled.

Table 1 lists several compounds which increase

tissue concentrations of cyclic AMP. As can be seen, a wide assortment of tissues, species, and hormones are represented. Outstanding among the stimulatory agents are the biogenic amines and the polypeptide hormones. The actions of the inhibitory compounds are more complex; insulin and prostaglandin E_1 have been shown to decrease the concentration of cyclic AMP under certain experimental conditions, but in most cases, prostaglandin E_1 increases tissue levels of the cyclic nucleotide, and even insulin, which inhibits the accumulation of cyclic AMP in adipose tissue and liver, increases the concentrations of cyclic AMP in skeletal muscle.

The recent development of a simple and sensitive means for determining adenyl cyclase activity (Krishna, et al., 1968a) has enabled enzyme measurements in discrete areas and has aided in the rapid expansion this field currently enjoys. Table 2 lists those compounds that have been shown, by direct enzyme assay, to activate adenyl cyclase. In many, but not all instances, there is good correlation between increases in cyclic AMP concentrations and activation of adenyl cyclase. A notable exception is brain (excluding the pineal gland) where several compounds increased tissue concentrations of cyclic AMP (Kakiuchi and Rall, 1968a) but only epinephrine has been shown to increase adenyl cyclase activity (Klainer, et al., 1962).

It is essential that hormones exert great specificity in altering the concentration of cyclic AMP if one is to assign any physiological significance to this phenomenon. Selective stimulation of adenyl cyclase or selective accumulation of cyclic AMP in tissue has now been demonstrated in many preparations. In general, only those compounds having physiological or pharmacological actions in a given tissue influence the adenyl cyclase-cyclic AMP system in that tissue. For example, among several compounds examined only ACTH increased cyclic AMP levels in adrenal slices (Haynes, 1958). More recently, it was shown that those analogues of ACTH which induced steroidogenesis increased adrenal concentrations of cyclic AMP, whereas pharmacologically inert analogues exhibited no effect on cyclic AMP levels (Grahme-Smith, et al., 1967). Such specificity has also been shown for luteinizing hormone on corpus luteum (Marsh, *et al.*, 1966), for glucagon on liver (Pohl, *et al.*, 1969), for gonadotropins on testes (Murad, *et al.*, 1969), for serotonin on *Fasciola hepatica* (Mansour, *et al.*, 1960, and this book), for thyroid stimulating hormone on thyroid (Pastan and Katzen, 1967; Gilman and Rall, 1968) and for pharmacologically active catecholamines on pineal gland (Weiss and Costa, 1968b). Moreover, vasopressin activates adenyl cyclase of the renal medulla whereas parathyroid hormone activates the renal cortex adenyl cyclase (Chase and Aurbach, 1968). These findings may be correlated with the physiological evidence showing that these are the primary sites at which hormones act in effecting changes in water permeability (Berliner and Bennett, 1967) and cellular transfer of ions (Pitts, *et al.*, 1958; Duarte and Watson, 1967). Adipose tissue, on the other hand, is not selective. Several structurally unrelated hormones induce a lipolytic response, and these same compounds increase cyclic AMP levels in adipose tissue (Butcher, *et al.*, 1968).

To summarize, several polypeptide hormones and biogenic amines raise the concentration of cyclic AMP in tissue. In most cases, this increase is due to activation of adenyl cyclase, the enzyme that catalyzes the formation of the cyclic nucleotide. And in most cases, the stimulation of adenyl cyclase activity by hormones is selective, indicating either that there is more than one type of adenyl cyclase, that there are specific sites on the enzyme system with which the hormones interact, or that there is a membranous material surrounding the active site on the enzyme system which is selectively permeable to the different hormones. The correct explanation must await further purification and solubilization of the adenyl cyclase system.

<u>Activation of adenyl cyclase by sodium fluoride</u>. In contrast to the specificity with which hormones activate adenyl cyclase, sodium fluoride enhances adenyl cyclase activity from essentially every tissue examined. This was shown in liver by Rall and Sutherland in 1958 and has since been confirmed in many tissues by them and numerous other investigators.

TABLE 2

Biological agents that alter adenyl cyclase activity

Tissue	Agent	Effect	Reference
Adipose tissue	Insulin	Inhibition	Jungas, 1966
Adipose tissue	ACTH	Activation	Rodbell, 1967
Adipose tissue	Thyroxin	Induction	Brodie, et al., 1966 and Krishna, et al., 1968b
Adipose tissue	Nucleotides	Inhibition	Cryer, et al., 1969
Adipose tissue	ACTH	Activation	Bar and Hechter, 1969
Adrenal gland	ACTH	Activation	Taunton, et al., 1967, 1969
Brain	Epinephrine	Activation	Klainer, et al., 1962
Erythrocytes	Catecholamines	Activation	Davoren and Sutherland, 1963b
Erythrocytes	Serotonin	Inhibition	Davoren and Sutherland, 1963b
Erythrocytes	Epinephrine	Activation	Oye and Sutherland, 1966
Heart	Catecholamines	Activation	Sobel, et al., 1968
Heart	Catecholamines	Activation	Entman, et al., 1969
Heart	Catecholamines	Activation	Murad and Vaughan, 1969

Heart	Thyroid hormone	Activation	Levey and Epstein, 1968
Heart	Glucagon	Activation	Murad and Vaughan, 1969
Liver	Glucagon	Activation	Pohl, et al., 1969
Liver	Glucagon	Activation	Becker and Bitensky, 1969
Liver	Epinephrine	Activation	Becker and Bitensky, 1969
Ovary	Luteinizing hormone	Activation	Dorrington and Baggett, 1969
Ovary	Chorionic gonadotropin	Activation	Dorrington and Baggett, 1969
Parotid gland	Isoproterenol	Activation	Malamud, 1969
Pineal gland	Catecholamines	Activation	Weiss and Costa, 1967, 1968b, and Weiss, 1969a, 1969b
Pineal gland	Purines	Inhibition	Weiss, 1968a
Platelets	Prostaglandin E_1	Activation	Wolfe and Shulman, 1969
Platelets	Prostaglandin E_1	Activation	Zieve and Greenough, 1969
Platelets	Catecholamines	Inhibition	Zieve and Greenough, 1969
Kidney	Parathyroid hormone	Activation	Dousa and Rychlik, 1968b
Renal cortex	Parathyroid hormone	Activation	Chase and Aurbach, 1968
Renal medulla	Vasopressin	Activation	Chase and Aurbach, 1968
Skeletal muscle	Parathyroid hormone	Activation	Chase, et al., 1969
Testes	Gonadotropins	Activation	Murad, et al., 1969
Thyroid	Thyroid stimulating hormone	Activation	Pastan and Katzen, 1967

The mechanism by which fluoride activates adenyl cyclase is still a matter of controversy. However, some information is available suggesting that fluoride and hormones act at different sites. In 1966, Oye and Sutherland published data showing that epinephrine increased the accumulation of cyclic AMP only when incubated with intact avian erythrocytes, whereas fluoride stimulated adenyl cyclase activity in homogenates but not in whole cells indicating that the two compounds have different sites or different mechanisms of action. More recently, I have shown that in the pineal gland fluoride activates adenyl cyclase by a mechanism unrelated to inhibition of ATPase or phosphodiesterase and at a site on the adenyl cyclase system distinctly different from that of the catecholamines (Weiss, 1969b). Accordingly, the activation of adenyl cyclase by maximally effective concentrations of sodium fluoride was increased further by the addition of norepinephrine. Moreover, *beta* adrenergic blocking agents completely inhibited the activation by norepinephrine, but either they had no effect or actually enhanced the stimulatory response to sodium fluoride. These data show that in the same tissue there are different sites with which compounds can act to increase adenyl cyclase activity, But whether these sites are on the same enzyme system or on a different adenyl cyclase is still unknown.

Inhibitors of adenyl cyclase activity. Compounds which inhibit adenyl cyclase activity may be divided into two major groups: (a) those which reduce the "basal" enzyme activity; that is, the adenyl cyclase activity measured in the absence of any activator, and (b) those which do not inhibit the basal activity but antagonize the activation of the enzyme. The sites of action of these two groups of compounds may be discussed using the model proposed by Robison, *et al.* (1967) which depicts the adenyl cyclase molecule as consisting of two subunits, a regulatory subunit facing the extracellular fluid and a catalytic subunit in contact with the interior of the cell (see Fig. 1). The first group of compounds, those which reduce the basal enzyme activity, may act non-specifically on the catalytic subunit of the adenyl cyclase system. These compounds

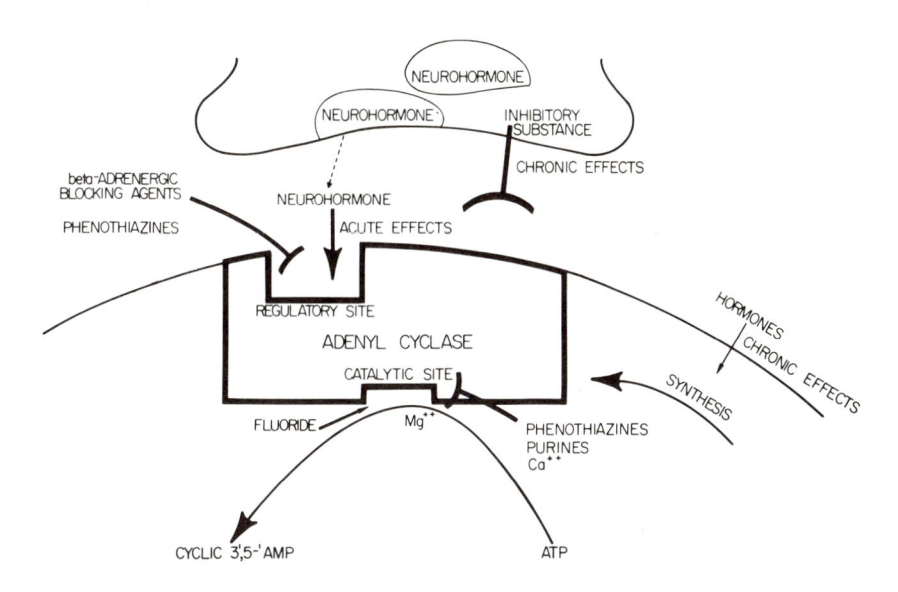

Figure 1. Factors affecting adenyl cyclase activity. Heavy arrows indicate enzyme activation. Inverted arrows represent enzyme inhibition. (See text for discussion.)

would be expected to reduce adenyl cyclase activity whether or not an activator was present and without regard to the type of activator. The second group of compounds may react with the regulatory unit thus blocking hormonal activation of the enzyme. In this case, since the catalytic subunit is still free, the basal adenyl cyclase activity would be unaffected.

Adrenergic blocking agents. As early as 1962, Murad *et al.* showed that adrenergic blocking agents can inhibit the catecholamine-induced enhancement of the accumulation of cyclic AMP in dog heart. More recent work indicated that *beta* adrenergic blocking agents are considerably more effective than *alpha* adrenergic blocking agents in antagonizing the activation of adenyl cyclase by norepinephrine. For example, the *beta* adrenergic blocking drugs, dichloroisoproterenol (DCI) and propranolol, inhibited the norepinephrine-induced activation of pineal adenyl cyclase by about 90%, but they had no effect on the basal enzyme activity (Weiss and Costa, 1968; Weiss, 1969b). Moreover, Butcher, Sutherland, and their co-workers showed that DCI (Butcher, *et al.*, 1965) and pronethalol (Butcher and Sutherland, 1967) antagonized the epinephrine-induced rise in cyclic AMP concentrations in adipose tissue without altering the normal levels of the cyclic nucleotide. DCI also blocked the rise in cerebellar cyclic AMP caused by norepinephrine without reducing the control levels (Kakiuchi and Rall, 1968a).

Prostaglandins. The inhibitory effects of prostaglandin E_1 on cyclic AMP levels in isolated fat cells resemble in some respects those seen with the *beta* adrenergic blocking agents in that prostaglandin E_1 antagonized the hormone-induced rise in cyclic AMP without reducing the basal levels (Butcher and Baird, 1968). However, an inhibitory action of prostaglandin E_1 on adenyl cyclase has not yet been demonstrated.

Purine derivatives. Purine derivatives antagonize the lipolytic response to catecholamines (Davies, 1968) presumably by inhibiting adenyl cyclase since they do not block the lipolytic actions of cyclic AMP itself (Weiss, *et al.*, 1966). Since the purine derivatives are structurally similar to ATP, one could easily envision that they would compete with ATP for the catalytic site of the adenyl cyclase system and thus constitute a second type of inhibitor. By acting at the catalytic site, these compounds should reduce enzyme activity whether the enzyme is activated or not. Adenine derivatives such as adenosine, 5'-adenosine monophosphate, and adenosine diphosphate (ADP) have been reported to inhibit adenyl cyclase activity of pineal gland (Weiss, 1968a) and of adipose tissue (Cryer, *et al.*, 1969). In addition, ADP inhibits the basal enzyme activity and the norepinephrine- and sodium fluoride-induced stimulation of adenyl cyclase by about the same extent (Table 3). Upon closer examination of these responses, certain anomalies become evident. For example, Fig. 2 shows that ADP inhibits adenyl cyclase in a non-competitive rather than in a competitive manner. Therefore, it is not competing with ATP for the catalytic site. Moreover, this inhibition is unrelated

TABLE 3

Inhibition of the accumulation of cyclic 3',5'-AMP-^{14}C in rat pineal gland by adenosine diphosphate

Additions	Apparent cyclic 3',5'-AMP formed (μμmoles/mg protein/min) Control	ADP (10^{-3} M)	% Inhibition
None	97 ± 2	43 ± 2	66
Norepinephrine (10^{-4} M)	309 ± 4	141 ± 4	64
NaF (10^{-2} M)	423 ± 7	178 ± 5	68

Adenyl cyclase activity was determined in pineal homogenates from the rate of conversion of ATP-^{14}C to cyclic 3',5'-AMP-^{14}C. The incubation mixture consisted of a pineal gland homogenate (about 100 μg protein) 0.5 μc ATP-8-^{14}C (0.1 μmole), "carrier" cyclic 3',5'-AMP (0.3 μmole), MgSO$_4$ (0.3 μmole) and the compounds under study in a total volume of 100 μl of Tris-HCl buffer (0.05 M), pH 7.4. Incubations were conducted in air at 30° C. After 10 min cyclic 3',5'-AMP-^{14}C was isolated according to the method of Krishna, *et al.* (1969a) as previously described (Weiss and Costa, 1967, 1968b). Each value is the mean of four experiments ± S.E.

to the binding of ADP to magnesium (Fig. 3); increasing magnesium concentrations could not completely overcome the effects of ADP on the accumulation of radiolabeled cyclic AMP.

Ordinarily, one uses the rate of accumulation of radioactive cyclic AMP from radiolabeled ATP as a determination of adenyl cyclase activity. But the adenine nucleotides themselves may be converted to ATP in the tissue preparation, reducing the specific activity of ATP. Therefore, there may be only an apparent reduction of cyclic AMP formation. In the pineal gland, for example, ADP does significantly decrease the specific

52   Adenyl Cyclase Activity

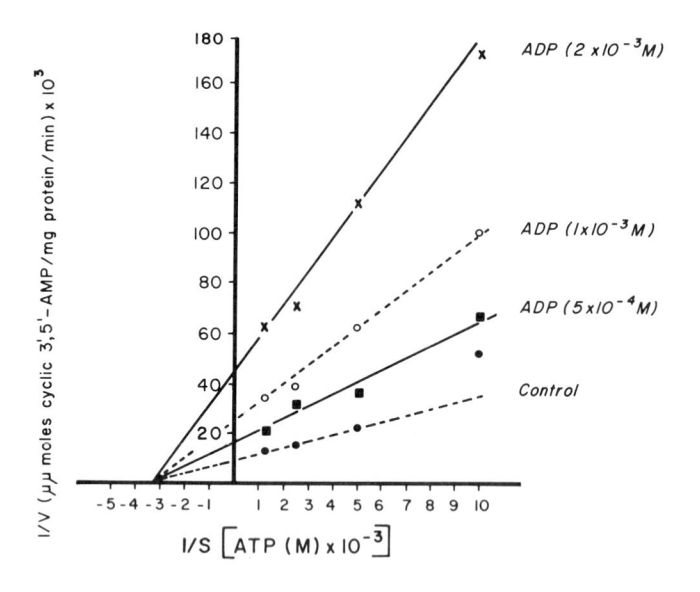

Figure 2. Effect of adenosine diphosphate on the formation of radioactive cyclic 3',5'-AMP from ATP-8-^{14}C in bovine pineal gland. Incubation and assay procedures were conducted as outlined in the legend to Table 3 with the exception that the concentration of ATP was varied.

activity of radiolabeled ATP (Weiss, unpublished).

Possibly, purine derivatives inhibit adenyl cyclase by acting at some allosteric site on the enzyme system, but one must be extremely cautious in interpreting results with purine derivatives, and no definite conclusions should be drawn until more careful studies are made taking into account the changes in specific activity of ATP.

Phenothiazines. Inhibition of adenyl cyclase activity by other agents is also complex and not readily amenable to classification. The phenothiazines are a case in point. Weiss and Kidman (1969) reported that chlorpromazine blocks the norepinephrine-induced activation of pineal adenyl cyclase. More recent experiments showed that chlorpromazine not only prevents the activation by norepinephrine but reduces the basal

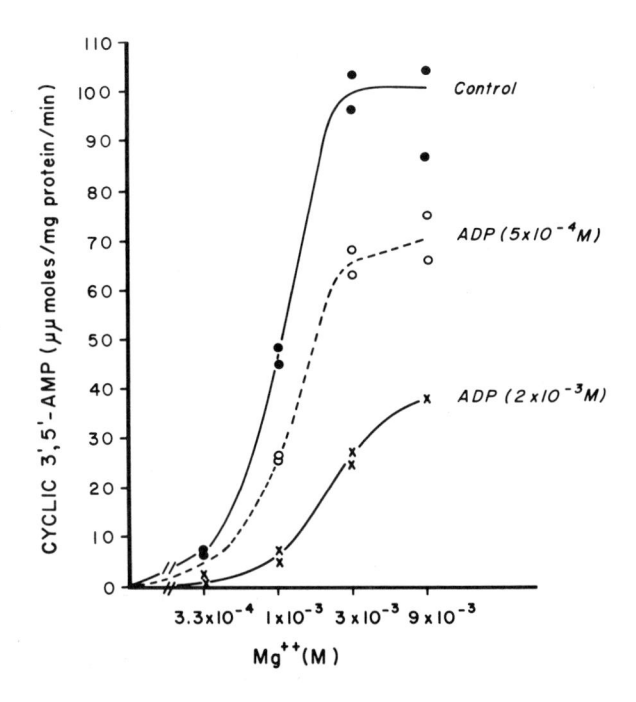

Figure 3. Effect of magnesium on the ADP-induced inhibition of the formation of cyclic 3',5'-AMP-^{14}C from ATP-8-^{14}C in bovine pineal glands. Incubation and assay procedures were conducted as outlined in the legend to Table 3 with the exception that the magnesium concentration was varied.

adenyl cyclase activity and partially inhibits the effects of sodium fluoride as well (Table 4). Since norepinephrine and sodium fluoride act at different sites on the pineal adenyl cyclase system (Weiss, 1969b), the phenothiazine at these high concentrations apparently was affecting the total enzyme rather than a site specific for the catecholamine. That is, it may be acting at the catalytic rather than at the regulatory site or at both the catalytic and regulatory sites (see Fig. 2). Studies using lower concentrations of the phenothiazines might reveal greater specificity for blocking catecholamines.

TABLE 4

Inhibition by chlorpromazine of rat pineal adenyl cyclase activity

	Cyclic 3',5'-AMP formed (µµmoles/mg protein/min)	
Additions	Control	Chlorpromazine (10^{-3} M)
None	192 ± 14	31 ± 3
Norepinephrine (10^{-4} M)	394 ± 70	30 ± 4
NaF (10^{-2} M)	889 ± 159	363 ± 39

Adenyl cyclase activity of rat pineal gland homogenates was determined as described in the legend to Table 3. Each value is the mean of five experiments ± S.E.

That phenothiazines act at several sites on the adenyl cyclase system (or on different adenyl cyclases) gains further support from experiments reported by Kakiuchi and Rall (1968a). They showed that chlorpromazine blocks both the histamine- and norepinephrine-induced increase in cyclic AMP levels in cerebellum although these two amines apparently act at different sites.

This non-selective blockade of adenyl cyclase activity by chlorpromazine does not necessarily mean that inhibition of adenyl cyclase has no pharmacological significance. For it is well established that certain phenothiazines block not only the actions of catecholamines (Courvoisier, et al., 1953; Bradley, et al., 1966) but also those of serotonin (see Gyermek, 1966) and histamine (McKeon, 1963a, b). Moreover, a study of the dose-response curves for chlorpromazine, trifluorperazine, and their sulfoxide metabolites showed that the potency of these phenothiazines on the central nervous system (see Janssen, et al., 1965) was directly related to their effectiveness in inhibiting pineal adenyl cyclase (Fig. 4). Thus, trifluorperazine was more potent than chlorpromazine, and their sulfoxides, which are relatively inactive metabolites

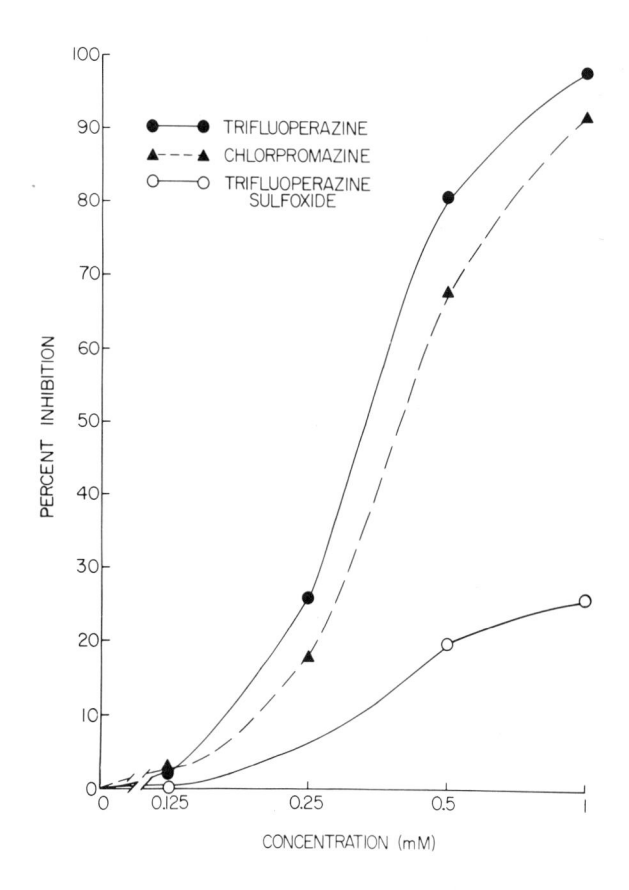

Figure 4. Inhibition of pineal adenyl cyclase activity by phenothiazine derivatives. Incubation and assay procedures were conducted as outlined in the legend to Table 3 in the presence of l-norep-nephrine (10^{-4} M).

(Salzman, et al., 1955), had little effect on the enzyme system. These results support the notion that inhibition of adenyl cyclase may explain some of the central actions of phenothiazines. Further support for this concept comes from experiments showing that cyclic AMP has effects on cerebellar neurons similar to those of norepinephrine and is probably responsible for the action of the catecholamine on these neurons (Siggins, et al., 1969); chlorpromazine has previously been reported to antagonize the neuronal response to norepinephrine in brain stem (Bradley, *et al.*, 1966). It is pertinent to note here that phenothiazines inhibit phosphodiesterase of brain (Honda and Imamura, 1968). If this were operative on the pineal gland then the system would be still more complex. Nevertheless, inhibition of phosphodiesterase by phenothiazines would not alter our conclusions concerning their effects on adenyl cyclase since any changes in the destruction of cyclic AMP during incubation are corrected for when enzyme activity is computed.

Insulin. The inhibitory action of insulin on the accumulation of cyclic AMP has been demonstrated in adipose tissue (Jungas, 1966; Butcher, *et al.*, 1968) and liver (Jefferson, *et al.*, 1968), but whether this is due to an action on adenyl cyclase, phosphodiesterase or on cellular transport is still a matter of considerable controversy.

Chronic effects. The studies described so far dealt only with the acute effects of biologically active compounds on adenyl cyclase; that is, the activation or inhibition of the enzyme system seen immediately upon the addition of the agent to the enzyme preparation. I wish now to discuss those factors which may chronically alter either the formation of adenyl cyclase or the sensitivity of adenyl cyclase to hormonal activation.

Ontogenetic development. Rosen and Rosen (1968) were the first to show a difference in the sensitivity of adenyl cyclase to hormones depending upon the stage of development of the animal. Adenyl cyclase of tadpole hemolysates was unresponsive to the catecholamines whereas that from adult frogs was stimulated by these amines. In contrast, fluoride activated adenyl cyclase from both tadpole and frog.

Castaneda and Tyler (1968) reported that adenyl cyclase activity of sea urchin eggs increased within ten minutes after fertilization. However, they did not show whether hormonal activation was altered.

Studies of the ontogenesis of mammalian pineal adenyl cyclase showed that the enzyme of adult rats is more responsive to the stimulatory effects of norepinephrine than that of immature animals (Table 5). But

TABLE 5

Ontogenetic development of adenyl cyclase activity of rat pineal gland

Cyclic 3',5'-AMP formed
(μμmoles/mg protein/min)

Age	Control	Norepinephrine (10^{-4} M)
2 days	97 ± 16	155 ± 33
60 days	201 ± 38	377 ± 37

Pineal adenyl cyclase activity of rats either two days or 60 days of age was determined in the absence and presence of norepinephrine as described in the legend to Table 3. Each value is the mean of five experiments ± S.E.

since the basal enzyme activity also increases with age it is difficult to say whether there is a greater sensitivity toward norepinephrine in particular or merely more enzyme.

Environmental lighting. Light affects the activity of several enzymes in the pineal gland. For example, continuous exposure of rats to light decreases pineal hydroxyindole-O-methyltransferase (HIOMT) activity (Wurtman, et al., 1963; Weiss, 1968b). Conversely, tyrosine hydroxylase activity is higher during darkness than during light (McGeer and McGeer, 1966). Since the effects of HIOMT are apparently mediated by the sympathetic nervous system (Wurtman, et al., 1964; Brownstein and Heller, 1968), it was of interest to determine whether light could alter the activity of adenyl cyclase, the proposed receptor for adrenergic neurohormones.

Studies of the circadian variation of adenyl cyclase in the rat pineal gland showed that enzyme activity in-

Adenyl Cyclase Activity

creased when the animals were kept in light and decreased in darkness. The responsiveness of adenyl cyclase to norepinephrine, however, was not appreciably affected; the catecholamines increased enzyme activity approximately two-fold regardless of the lighting condition (Fig. 5). In contrast, subjecting rats to con-

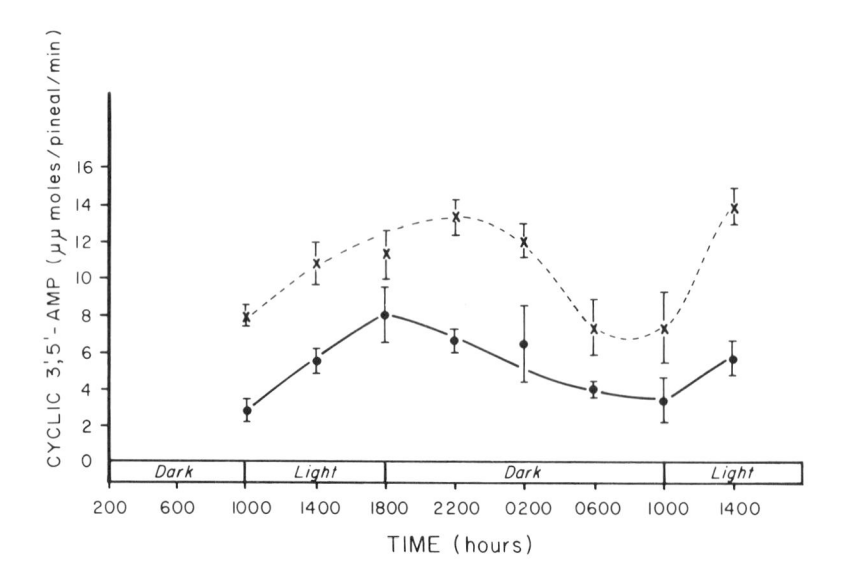

Figure 5. Effect of alternating light and dark periods on adenyl cyclase activity of rat pineal glands. Rats were kept in alternating periods of 16 hr dark:8 hr light. After the first dark period (at 10:00 hr) a group of rats were sacrificed, the pineals removed and frozen. Sixteen hr after the last group was sacrificed, the pineal glands were allowed to thaw, the glands were homogenized and adenyl cyclase was assayed in the absence (o——o) and in the presence (x----x) of l-norepinephrine (10^{-4} M) as described in the legend to Table 3. Adenyl cyclase activity was significantly lower ($P < 0.05$) at the end of the long dark period (10:00 hr) than that at the end of the long light period (18:00 hr). Each point represents the mean of 4 experiments. Brackets indicate the S.E.

tinuous light or darkness for prolonged periods did significantly alter the activation of adenyl cyclase by norepinephrine (Weiss, 1969a). Pineal adenyl cyclase of rats maintained in constant light for 10 consecutive days was more responsive to norepinephrine than that of rats maintained in darkness or in alternating light:dark periods. That these effects of light were mediated by the sympathetic nervous system was shown by experiments in which the sympathetic innervation to the gland was interrupted. Bilateral removal of the superior cervical ganglia, which completely denervates the gland (Kappers, 1960), abolished the usual increased responsiveness of adenyl cyclase to norepinephrine (Weiss, 1969a).

Modifying other environmental parameters has not as yet produced significant changes in the pineal adenyl cyclase system nor have there been any reports on the effects of the environment on adenyl cyclase of other organs. Conceivably, many environmental factors, cold exposure for example, may chronically alter certain hormone-sensitive adenyl cyclases from specific organs. This is an area ripe for exploitation.

Neuronal activity. In 1967, Weiss and Costa reported that denervating the rat pineal gland increased the responsiveness of pineal adenyl cyclase to norepinephrine; this was the first indication that neuronal activity may chronically influence the hormone-sensitive adenyl cyclase system. More recent results (Weiss, 1969a) showed that maximal as well as submaximal concentrations of norepinephrine were potentiated by denervation and that the activation of adenyl cyclase caused by sodium fluoride was also enhanced. These findings suggest that denervation caused a general alteration of the enzyme system rather than a greater affinity of norepinephrine for adenyl cyclase.

The chronic nature of the effects of denervation are shown in Fig. 6. An increased responsiveness to norepinephrine was not evident two weeks after bilateral superior cervical ganglionectomy but became manifest only when four or more weeks had elapsed between ganglionectomy and assay of the enzyme.

These studies should be viewed in light of experi-

Adenyl Cyclase Activity

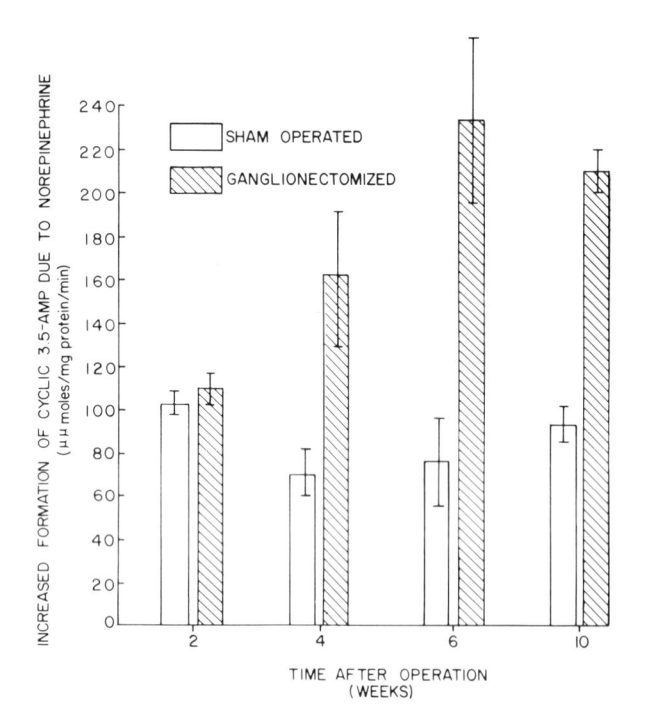

Figure 6. Responsiveness of rat pineal adenyl cyclase to norepinephrine after varying periods of superior cervical ganglionectomy. Bilateral superior cervical ganglionectomy or sham-operations were performed from two to ten weeks prior to sacrifice. Pineal adenyl cyclase activity was determined in the absence and presence of norepinephrine (10^{-4} M) as outlined in the legend to Table 3. Each point is the mean of 3 to 5 experiments. Brackets indicate the standard error. Data taken from Weiss, 1969a.

ments, conducted over the past 60 years (see Trendelenburg, 1966), showing that sympathetically innervated structures are hypersensitive to the catecholamines after chronic denervation. Several parallels between the two phenomena can be drawn: (a) Both the chronic physiological response, the supersensitivity of the cat nictitating membrane to norepinephrine (Langer and

Trendelenburg, 1968), and the chronic biochemical response, the increased responsiveness of pineal adenyl cyclase (Weiss and Costa, 1967; Weiss, 1969a) are clearly post-junctional. (b) Chronic decentralization of the superior cervical ganglia produces effects similar to denervation on the nictitating membrane (Flemming, 1963) and on adenyl cyclase (Weiss and Kidman, 1969). (c) As in sympathetically denervated tissues, the increased responsiveness of pineal adenyl cyclase to catecholamines appears only after the nerve terminals have degenerated.

Since chronic denervation of the pineal gland and long term exposure to light induced similar changes in the pineal adenyl cyclase system—in both conditions there was an increased activation of the enzyme system by norepinephrine and sodium fluoride, and neither denervation nor light caused any significant alteration in the basal adenyl cyclase activity—it is tempting to speculate that the two phenomena are related. But if this is true, one must conclude that continuous exposure to light results in a decreased neuronal activity in the pineal gland. This apparently untenable position gained support, however, from the recent experiments of Taylor and Wilson (1969) who showed that light exposure decreased nerve activity in rat pineal gland four-fold and that the reduced activity continued as long as the light remained on.

If there is in fact more hormone-sensitive adenyl cyclase after chronic denervation and exposure to continuous light, and if this is due to decreased neuronal activity then perhaps there are substances ordinarily released from the nerve ending which inhibit the synthesis of adenyl cyclase. The validity of this supposition might be best tested in isolated tissue culture where the suspected inhibitory factors could be readily isolated.

To study further the neural factors influencing adenyl cyclase activity, experiments were conducted to determine if the pineal gland requires sympathetic innervation to develop a hormone-sensitive adenyl cyclase. Sympathetic neurons apparently do not start invading the interior of the pineal gland until one or two days after birth, and innervation does not become complete until approximately three weeks (Machado, *et al.*, 1968). Therefore, superior cervical ganglionectomy of newborn rats should prevent the sympathetic innervation of the gland. These operations were performed on one-day-old male rats. Nine weeks after surgery, the pineal glands were removed and adenyl cyclase activity was determined. As can be seen from Table 6, norepinephrine and sodium

TABLE 6

Effect of superior cervical ganglionectomy of newborn rats on norepinephrine and sodium fluoride-induced stimulation of pineal adenyl cyclase activity

Treatment	Cyclic 3',5'-AMP formed (μμmoles/mg protein/min)			Increase due to	
	Control	Norepi-nephrine	NaF	Norepi-nephrine	NaF
Sham-operated	94 ± 21	191 ± 42	504 ± 72	97 ± 46	410 ± 66
Ganglion-ectomized	74 ± 12	245 ± 47	665 ± 118	171 ± 32	591 ± 109

Sham operations or bilateral superior cervical ganglionectomies were performed on male rats one day old. Nine weeks after surgery the pineal glands were removed and adenyl cyclase activity was determined in the absence and presence of l-norepinephrine (10^{-4} M) and NaF (10^{-2} M) as described in the legend to Table 3. Each value represents the mean of five experiments ± S.E.

fluoride increased adenyl cyclase activity in both groups of rats. And like the previously described experiments, denervation caused greater responsiveness to both stimulants.

Administering the antiserum to nerve growth factor (see Levi-Montalcini and Booker, 1960) also prevents the sympathetic innervation of the gland. Rats were treated with the antiserum two days after birth, and enzyme activity was determined at ten weeks of age. Table 7 shows that in the immunosympathectomized groups of animals the stimulatory effects of norepinephrine and sodium fluoride were still evident and even slightly increased.

TABLE 7

Effect of immunosympathectomy on adenyl cyclase activity of rat pineal gland

Treatment	Cyclic 3',5'-AMP formed ($\mu\mu$moles/mg protein/min)			Increase due to	
	Control	Norepi-nephrine	NaF	Norepi-nephrine	NaF
None	149 ± 12	279 ± 33	649 ± 56	130 ± 30	500 ± 48
Immuno-sympathec-tomized	181 ± 24	346 ± 39	717 ± 84	165 ± 19	536 ± 66

Male Sprague-Dawley rats, two days old, were administered the antiserum to nerve growth factor, 2000 units a day subcutaneously once a day for four consecutive days. After 70 days, the pineal glands were removed and adenyl cyclase activity was determined in the absence and presence of l-norepinephrine (10^{-4} M) or NaF (10^{-2}M) as described in the legend to Table 3. Each value represents the mean of six experiments ± S.E.

These results show that reducing neuronal activity to the pineal gland, whether initiated at birth or in adulthood, causes greater activation of adenyl cyclase by the sympathetic neurotransmitter. These results also show that the development of adenyl cyclase can proceed without sympathetic innervation.

Endocrine factors. Considered from the teleological standpoint, the endocrine system must play a prominent role in chronically influencing adenyl cyclase. For it is endocrine system that is largely responsible for regulating the metabolic activity of cells and must govern not only the activities of enzymes but the actual amounts of enzyme capable of being activated.

It is well known that the thyroid state governs the lipolytic response to catecholamines; greater lipolysis is produced in hyperthyroidism and less lipolysis in hypothyroidism (Debons and Schwartz, 1961; Deykin and Vaughan, 1963; Felt, et al., 1963). The studies of Brodie, et al. (1966) showed that this phenomenon might be explained by a change in adenyl cyclase. Administering thyroxin to rats for five days increased adenyl cyclase activity in adipose tissue two-fold. More recently, Krishna, *et al.* (1968b), studying the rat epididymal fat pad, showed that hypothyroid rats contained less adenyl cyclase, and hyperthyroid rats contained more adenyl cyclase than did euthyroid animals.

As part of a continuing program designed to determine the factors affecting the long term responsiveness of adenyl cyclase to neurotransmitters, I have studied the effect of various endocrine deficiencies on adenyl cyclase activity of rat pineal gland. This gland is unique in that of all hormones examined, only the pharmacologically active catecholamines specifically and consistently stimulate the pineal adenyl cyclase system (Weiss and Costa, 1968b). Moreover, the pineal gland is interesting from the functional point of view since it affects gonadal activity, probably via the pineal hormone, melatonin (Reiter, 1967; Wurtman, *et al.*, 1967).

The first indication that sex hormones might alter pineal adenyl cyclase activity came from studies comparing the enzyme activity from male and female rats. As may be seen from Table 8, the basal adenyl cyclase

TABLE 8

Comparison of pineal adenyl cyclase activity of male and female rats

Sex	Control	Cyclic 3',5'-AMP formed ($\mu\mu$moles/mg protein/min) Norepinephrine (10^{-4} M)	NaF (10^{-2} M)
Male	149 ± 15	406 ± 53	796 ± 115
Female	169 ± 22	221 ± 43	643 ± 105

Adenyl cyclase activity of male and female rat pineal gland homogenates was determined in the absence and presence of norepinephrine and sodium fluoride as described in the legend to Table 3. Each value is the mean of five experiments ± S.E.

activity and the enzyme activity measured in the presence of sodium fluoride was not significantly different in the two groups of rats. However, adenyl cyclase from male rat pineal glands was more responsive to norepinephrine than that from female rats. This demonstrated once more the existence of different sites at which norepinephrine and fluoride act. Apparently, in the pineal gland, only the norepinephrine-sensitive adenyl cyclase is governed by sex hormones.

Pineal adenyl cyclase activity of female rats in various stages of their estrus cycle was also studied. Figure 7 shows that in proestrus there was less adenyl

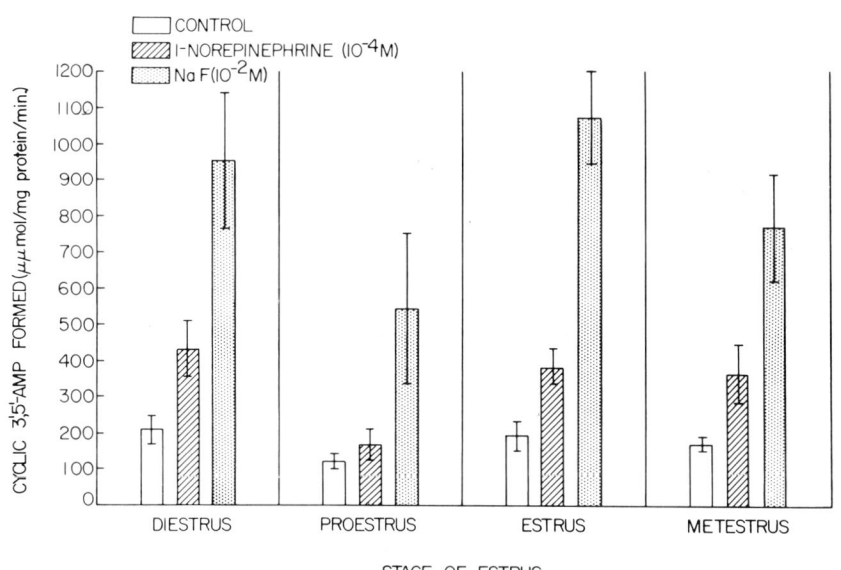

Figure 7. Influence of the estrus cycle on adenyl cyclase activity of rat pineal gland. The stage of the estrus cycle was determined over a two-week period by vaginal smear. Pineal adenyl cyclase activity was determined in the absence and presence of norepinephrine and sodium fluoride as described in the legend to Table 3. (Crayton and Weiss, unpublished.)

cyclase activity measured either in the absence or presence of norepinephrine or sodium fluoride. The greatest effect was on the response to norepinephrine which significantly enhanced pineal adenyl cyclase activity from rats in all stages except proestrus. Whether the estrogens and progestins or whether the male sex hormones are involved in these responses is currently being investigated.

Regardless of which chemicals are involved, these results show that a reciprocal relationship exists between the pineal gland and its target tissue. Pineal activity influences the gonads, and gondal factors may govern pineal adenyl cyclase. A more complete picture of this interrelationship would emerge if one could demonstrate an action of cyclic AMP on the formation of melatonin.

Preliminary studies of other endocrine deficiencies, such as thyroidectomy, parathyroidectomy, adrenalectomy and hypophysectomy, have shown no significant changes in pineal adenyl cyclase. But, obviously, if one is to examine the role of the endocrine system on adenyl cyclase, the specific target structures for each hormone must be studied. Moreover, one must also pay particular attention to the specific hormone activator of that adenyl cyclase system. A more pertinent question would be—what are the chronic effects of the follicle stimulatory hormone on the luteinizing hormone-sensitive adenyl cyclase of corpus luteum?

The main conclusion I would like to draw from these studies is that certain hormones may alter the amount of specific adenyl cyclases which in turn are activated by other hormones. Thus, by acutely activating adenyl cyclase, hormones allow the organism to meet its immediate requirements, and by inducing the formation of more adenyl cyclase, the endocrine system enables the organism to chronically alter its capacity to respond to these hormones.

Fasting. The only report on the influence of fasting on adenyl cyclase activity is that of Brodie, *et al.* (1969) which shows that rat epididymal fat pads from fasted rats contained about two times the amount of adenyl cyclase than did fat pads from fed animals.

These data might explain why norepinephrine induces greater lipolysis in adipose tissue from fasted rats (Brodie, *et al.*, 1969). The mechanism by which these changes are induced is still unclear. In fact, these workers suggest that the observed effect is not due to the dietary condition *per se* but rather to an alteration of the hormonal balance which in turn causes the changes in adenyl cyclase. Detailed studies into the mechanism involved may place this finding more properly under the heading of "endocrine factors".

SUMMARY AND CONCLUSIONS

There are two general mechanisms by which the organism can regulate the activity of adenyl cyclase and therefore the intracellular concentrations of cyclic AMP. The first of these is by acutely activating the enzyme by changing the immediate chemical environment of the adenyl cyclase system. The chemicals involved may be certain divalent cations, which have been shown to profoundly influence enzyme activity or they may be specific hormones or neurohormones, which selectively activate adenyl cyclase from discrete target organs. Recently, evidence has also been presented suggesting that various endogenously formed substances such as insulin, prostaglandin, and purine derivatives may inhibit adenyl cyclase activity. The interplay of all these biologically active compounds enables the organism to cope with the short term demands placed upon it.

In addition, the quantity of adenyl cyclase capable of being activated may fluctuate allowing diurnal or even seasonal adjustment of the animal to its environment. The amount of adenyl cyclase may be governed by the ontogenetic development of the organism, by environmental lighting, by neuronal activity, and by the endocrine or hormonal state. Thus, the organism may regulate the formation of cyclic AMP by acutely activating adenyl cyclase or by chronically altering the synthesis or degradation of adenyl cyclase.

REFERENCES

Bar, H. and O. Hechter. 1969. *Biochem. Biophys. Res. Comm.* 35: 681.
Becker, F.F. and M.W. Bitensky. 1969. *Proc. Soc. Exptl. Biol. Med.* 130: 983.
Berliner, R.W. and C.M. Bennett. 1967. *Am. J. Med.* 42: 777.
Bradley, P.B., J.H. Wolstencroft, L. Hosli and G.L. Avanzino. 1966. *Nature* 212: 1425.
Brodie, B.B., G. Krishna and S. Hynie. 1969. *Biochem. Pharmacol.* 18: 1129.
Brodie, B.B., J.I. Davies, S. Hynie, G. Krishna and B. Weiss. 1966. *Pharmacol. Rev.* 18: 273.
Brown, E., D.L. Clarke, V. Roux and G.H. Sherman. 1963. *J. Biol. Chem.* 238: PC-852.
Brownstein, M.J. and A. Heller. 1968. *Science* 162: 367.
Bueding, E., R.W. Butcher, J. Hawkins, A.R. Timms and E.W. Sutherland, Jr. 1966. *Biochim. Biophys. Acta* 115: 173.
Butcher, R.W. and E.W. Sutherland. 1967. *Ann. N.Y. Acad. Sci.* 139: 849.
Butcher, R.W., C.E. Baird and E.W. Sutherland. 1968. *J. Biol. Chem.* 243: 1705.
Butcher, R.W. and C.E. Baird. 1968. *J. Biol. Chem.* 243: 1713.
Butcher, R.W., R.J. Ho, H.C. Meng and E.W. Sutherland. 1965. *J. Biol. Chem.* 240: 4515.
Castaneda, M. and A. Tyler. 1968. *Biochem. Biophys. Res. Comm.* 33: 782.
Chase, L.R., S.A. Fedak and G.D. Aurbach. 1969. *Endocrinol.* 84: 761.
Chase, L.R. and G.D. Aurbach. 1967. *Proc. Nat. Acad. Sci.* 58: 518.
Chase, L.R. and G.D. Aurbach. 1968. *Science* 159: 545.
Costa, E. and B. Weiss. 1968. In *Psychopharmacology—A Review of Progress, 1957-1967*, (Washington, D.C.: Superintendent of Documents, U.S. Government Printing Office) p. 39.
Courvoiser, S., J. Fournel, R. Ducrot, M. Kolsky and P. Koetschet. 1953. *Arch. Intern. Pharmacodyn.* 92: 305.
Craig, J.W., T.W. Rall and J. Larner. 1969. *Biochim.*

Biophys. Acta 177: 213
Cryer, P.E., L. Jarett and D.M. Kipnis. 1969. *Biochim. Biophys. Acta* 177: 586.
Davies, J.I. 1968. *Nature* 218: 349.
Davoren, P.R. and E.W. Sutherland. 1963a. *J. Biol. Chem.* 238: 3016.
Davoren, P.R. and E.W. Sutherland. 1963b. *J. Biol. Chem.* 238: 3009.
Debons, A.F. and I.L. Schwartz. 1961. *J. Lipid Res.* 2: 86.
DeRobertis, E., G. Rodriguez DeLores Arnaiz, M. Alberici, R.W. Butcher and E.W. Sutherland. 1967. *J. Biol. Chem.* 242: 3487.
Deykin, D. and M. Vaughan. 1963. *J. Lipid Res.* 4: 200.
Dorrington, J.H. and B. Baggett. 1969. *Endocrinol.* 84: 989.
Dousa, T. and I. Rychlik. 1968a. *Life Sci.* 7: 1039.
Dousa, T. and I. Rychlik. 1968b. *Biochim. Biophys. Acta* 158: 484.
Duarte, C.G. and J.F. Watson. 1967. *Am. J. Physiol.* 212: 1355.
Entman, M.L., G.S. Levey and S.E. Epstein. 1969. *Biochem. Biophys. Res. Comm.* 35: 728.
Felt, V., V. Vrbensky and P. Benes. 1963. *Experientia* 19: 315.
Fleming, W.W. 1963. *J. Pharmacol. Exp. Ther.* 141: 173.
Gilman, A.G. and T.W. Rall. 1968. *J. Biol. Chem.* 243: 5867.
Goldberg, N.D., C. Villar-Palasi, H. Sasko and J. Larner. 1967. *Biochim. Biophys. Acta* 148: 665.
Grahame-Smith, D.G., R.W. Butcher, R.L. Ney and E.W. Sutherland. 1967. *J. Biol. Chem.* 242: 5535.
Gyermek, L. 1966. In *Handbuch der experimentellen Pharmakologie*, Vol. 19, O. Eichler and A. Farah, eds. (New York: Springer-Verlag) p. 471.
Haugaard, N. and M.E. Hess. 1965. *Pharmacol. Rev.* 17: 27.
Haynes, R.C., Jr. 1958. *J. Biol. Chem.* 233: 1220.
Hirata, M. and O. Hayaishi. 1967. *Biochim. Biophys. Acta* 149: 1.
Hirata, M. and O. Hayaishi. 1965. *Biochem. Biophys. Res. Comm.* 21: 361.
Honda, F. and H. Imamura. 1968. *Biochim. Biophys. Acta*

161: 267.
Janssen, P.A.J., C.J.E. Niemegeers and K.H.L. Schellekens. 1965. *Arznimittel-Forsch.* 15: 1196.
Jefferson, L.S., J.H. Exton, R.W. Butcher, E.W. Sutherland and C.R. Park. 1968. *J. Biol. Chem.* 243: 1031.
Jungas, R.L. 1966. *Proc. Nat. Acad. Sci.* 56: 757.
Kakiuchi, S. and T.W. Rall. 1968a. *Mol. Pharmacol.* 4: 367.
Kakiuchi, S. and T.W. Rall. 1968b. *Mol. Pharmacol.* 4: 379.
Kaneko, T., U. Zor and J.B. Field. 1969. *Science* 163: 1062.
Kappers, J.A. 1960. *Z. Zellforschung. mikrosk. Anat.* 52: 163.
Klainer, L.M., Y.-M. Chi, S.L. Freidberg, T.W. Rall and E.W. Sutherland. 1962. *J. Biol. Chem.* 237: 1239.
Krishna, G., B. Weiss and B.B. Brodie. 1968a. *J. Pharmacol. Exp. Ther.* 163: 379.
Krishna, G., S. Hynie and B.B. Brodie. 1968b. *Proc. Nat. Acad. Sci.* 59: 884.
Langer, S.Z. and U. Trendelenburg. 1968. *J. Pharmacol. Exp. Ther.* 163: 290.
LaRaia, P.J., J. Craig and W.J. Reddy. 1968. *Amer. J. Physiol.* 215: 968.
Levey, G.S. and S.E. Epstein. 1968. *Biochem. Biophys. Res. Comm.* 33: 990.
Levi-Montalcini, R. and B. Booker. 1960. *Proc. Nat. Acad. Sci.* 46: 384.
Lopez, E., J.E. White and F.L. Engel. 1959. *J. Biol. Chem.* 234: 2254.
Lundholm, L., T. Rall and N. Vemos. 1967. *Acta. Physiol. Scand.* 70: 127.
Machado, A.B.M., Machado, C.R.S. and L.E. Wragg. 1968. *Experientia* 24: 464.
Makman, R.S. and E.W. Sutherland. 1965. *J. Biol. Chem.* 240: 1309.
Malamud, D. 1969. *Biochem. Biophys. Res. Comm.* 35: 754.
Mansour, T.E., E.W. Sutherland, T.W. Rall and E. Bueding. 1960. *J. Biol. Chem.* 235: 466.
Marsh, J.M., R.W. Butcher, K. Savard and E.W. Sutherland. 1966. *J. Biol. Chem.* 241: 5436.
McGeer, E.G. and P.L. McGeer. 1966. *Science* 153: 73.
McKeon, W.B., Jr. 1963a. *Arch. Intern. Pharmacodyn.*

145: 396.
McKeon, W.B., Jr. 1963b. *Arch. Intern. Pharmacodyn.* 146: 374.
Miyamoto, E., J.F. Kuo and P. Greengard. 1969. *Science* 165: 63.
Murad, F. and M. Vaughan. 1969. *Biochem. Pharmacol.* 18: 1053.
Murad, F., B.S. Strauch and M. Vaughan. 1969. *Biochim. Biophys. Acta* 177: 591.
Murad, F., Y.-M. Chi, T.W. Rall and E.W. Sutherland. 1962. *J. Biol. Chem.* 237: 1233.
Øye, I. and E.W. Sutherland. 1966. *Biochim. Biophys. Acta* 127: 347.
Pastan, I. and R. Katzen. 1967. *Biochem. Biophys. Res. Comm.* 29: 792.
Pitts, R.F., R.S. Gurd, R.H. Kessler and K. Hierholzer. 1958. *Am. J. Physiol.* 194: 125.
Pohl, S.L., L. Birnbaumer and M. Rodbell. 1969. *Science* 164: 566.
Posner, J.B., R. Stern and E.G. Krebs. 1965. *J. Biol. Chem.* 240: 982.
Rabinowitz, M., L. DeSalles, R.J. Meisler and L. Lorand. 1965. *Biochim. Biophys. Acta* 97: 29.
Rall, T.W. and E.W. Sutherland. 1958. *J. Biol. Chem.* 232: 1065.
Reiter, R.J. 1967. *Neuroendocrinol.* 2: 138.
Robison, G.A., R.W. Butcher, I. Øye, H.E. Morgan and E.W. Sutherland. 1965. *Mol. Pharmacol.* 1: 168.
Robison, G.A., R.W. Butcher and E.W. Sutherland. 1968. *Ann. Rev. Biochem.* 37: 149.
Robison, G.A., R.W. Butcher and E.W. Sutherland. 1967. *Ann. N.Y. Acad. Sci.* 139: 703.
Rodbell, M. 1967. *J. Biol. Chem.* 242: 5744.
Rosen, O.M. and S.M. Rosen. 1968. *Biochem. Biophys. Res. Comm.* 31: 82.
Rosen, O.M. and S.M. Rosen. 1969. *Arch. Biochem. Biophys.* 131: 449.
Salzman, N.P., N.C. Moran and B.B. Brodie. 1955. *Nature* 176: 1122.
Siggins, G.R., B.J. Hoffer and F.E. Bloom. 1969. *Science* 165: 1018.
Sobel, B.E., P.J. Dempsey and T. Cooper. 1968. *Biochem. Biophys. Res. Comm.* 33: 758.

Streeto, J.M. and W.J. Reddy. 1967. *Anal. Biochem.* 21: 416.
Sutherland, E.W., G.A. Robison and R.W. Butcher. 1968. *Circulation* 37: 279.
Sutherland, E.W. and G.A. Robison. 1966. *Pharmacol. Rev.* 18: 145.
Sutherland, E.W., T.W. Rall and T. Menon. 1962. *J. Biol. Chem.* 237: 1220.
Szego, C.M. and J.S. Davis. 1967. *Proc. Nat. Acad. Sci.* 58: 1711.
Taunton, O.D., J. Roth and I. Pastan. 1967. *Biochem. Biophys. Res. Comm.* 29: 1.
Taunton, O.D., J. Roth and I. Pastan. 1969. *J. Biol. Chem.* 244: 247.
Taylor, A.N. and R.W. Wilson. 1969. *Anat. Rec.* 163: 327.
Trendelenburg, U. 1966. *Pharmacol. Rev.* 18: 629.
Turtle, J.R. and D.M. Kipnis. 1967a. *Biochemistry* 6: 3970.
Turtle, J.R. and D.M. Kipnis. 1967b. *Biochem. Biophys. Res. Comm.* 28: 797.
Weiss, B. 1968a. *Fed. Proc.* 27: 752,
Weiss, B. 1968b. In *Advances in Pharmacology*, Vol. 6A, E. Costa, S. Garattini, M. Sandler and P. Shore, eds. (N.Y.: Academic Press), p. 152.
Weiss, B. 1969a. *J. Pharmacol. Exp. Ther.* 168: 146.
Weiss, B. 1969b. *J. Pharmacol. Exp. Ther.* 166: 330.
Weiss, B. and E. Costa. 1967. *Science* 156: 1750.
Weiss, B. and E. Costa. 1968a. *Biochem. Pharmacol.* 17: 2107.
Weiss, B. and E. Costa. 1968b. *J. Pharmacol. Exp. Ther.* 161: 310.
Weiss, B., J.I. Davies and B.B. Brodie. 1966. *Biochem. Pharmacol.* 15: 1553.
Weiss, B. and A.D. Kidman. 1969. In *Advances in Biochemical Psychopharmacology*, Vol. 1, E. Costa and P. Greengard, eds. (New York: Raven Press), p. 131.
Williams, R.H., S.A. Walsh and J.W. Ensinck. 1968. *Proc. Soc. Exp. Biol. Med.* 128: 279.
Williamson, J.R. 1966. *Mol. Pharmacol.* 2: 206.
Wolfe, S.M. and N.R. Shulman. 1969. *Biochem. Biophys. Res. Comm.* 35: 265.
Wurtman, R.J., J. Axelrod and L.S. Phillips. 1963.

Science 142: 1071.
Wurtman, R.J., J. Axelrod and J.E. Fischer. 1964. *Science* 143: 1328.
Wurtman, R.J., J. Axelrod, E.W. Chu, A. Heller and R. Y. Moore. 1967. *Endocrinol.* 81: 509.
Zieve, P.D. and W.B. Greenough, 3rd. 1969. *Biochem. Biophys. Res. Comm.* 35: 462.

Phylogenetic Aspects of the Distribution of Biogenic Amines

John H. Welsh

Biological Laboratories
Harvard University
Cambridge, Massachusetts

INTRODUCTION

The term *biogenic amine* as used by some biologists and biochemists (e.g. Ackerman, 1962) includes a great variety of amines that have been isolated from plants and animals. The functions of many of these are uncertain or unknown. Since the theme of this symposium is the involvement of biogenic amines in metabolic regulation, and since the catecholamines (CA) and 5-hydroxytryptamine (5-HT, serotonin) have been most extensively studied in this connection, this consideration of the phylogenetic distribution of biogenic amines will be largely confined to these compounds.

---

[*]The preparation of this paper was supported by Grant No. NB-00623 from the United States Public Health Service.

Until rather recently, most studies on the occurrence and quantitative distribution of CA and 5-HT have been done by means of bioassays of extracts of whole small organisms or of isolated organs, tissues and secretory products (e.g., venoms) of large animals. These studies revealed the widespread occurrence of CA and of 5-HT in most animal phyla and in certain plants. Bioassays do not readily identify these monoamines and not until chromatography and spectrofluorometry came into general use (after 1955) was it possible to make more certain their identification and to obtain more exact quantitative estimates of their levels in a given tissue. Since the early 1960s an improved histochemical fluorescence method, the Falck and Hillarp method (see Falck and Owman, 1965 for details) has been available that permits the localization and identification of CA and 5-HT in individual cells. This has made it possible to determine the exact sites of these monoamines in entire small animals or in organs or tissues.

The several methods that are now available make it possible to locate monoamine-containing cells; to identify the monoamine by microspectrofluorometry and other procedures; to estimate the quantity of the monoamine by appropriate fluorescence measurements; and, by means of combined histochemistry and electron microscopy, to recognize the amine-containing organelles of a given cell. These are important advances toward a better understanding of the functional role of the biogenic monoamines.

In view of limited space, no attempt will be made to summarize all of the existing data on the distribution of CA and 5-HT in the various phyla. Instead, certain of the quantitative estimates of levels in nervous systems and some non-nervous sources will be chosen to show the wide ranges in values that have been obtained. This will be followed by some examples of the application of the Falck and Hillarp histochemical fluorescence method showing the distribution of CA- and 5-HT-containing cells in representative animals. Finally, the recent study by Rude, Coggeshall and Van Orden (1969) on the Retzius cells of the leech, *Hirudo medicinalis*, will be presented as an example of the information that can be obtained by the application of a variety of techniques to a given monoamine-containing cell type.

Since our own studies have been largely restricted to the invertebrates a failure to deal adequately with the extensive literature on the vertebrates will, perhaps, be forgiven.

I. THE QUANTITATIVE DISTRIBUTION OF BIOGENIC AMINES

In this section the widespread distribution of certain of the biogenic amines will be indicated with examples to show the quantitative variation between representatives of phyla and classes and between certain tissues, cells, and cellular products.

5-Hydroxytryptamine

Early studies on a physiologically active factor, released by the clotting of mammalian blood, lead to the use of the term "serotonin" or "thrombocytin." This substance was eventually isolated and shown to be 5-hydroxytryptamine (see Page, 1954). Independently, Erspamer and coworkers were concerned with an active product of enterochromaffin cells that they called *enteramine*. This was also shown to be identical with 5-HT (see Erspamer, 1954). Eventually, in addition to the presence of 5-HT in blood platelets and enterochromaffin cells, it was found in certain plants and fruits (e.g., bananas); in mast cells (especially of rat and mouse); in certain glandular tissues such as the posterior salivary glands of *Octopus*, the hypobranchial glands of the snail *Murex*, and the pineal gland; in certain protozoans (e.g., *Tetrahymena*); in the venoms of many species; and in the nervous system of representatives of most animal phyla. The most complete summary of the quantitative distribution of 5-HT is that by Erspamer (1966).

5-HT in non-nervous tissues and products. The functional roles of 5-HT are not yet fully understood.

However, many physiological and pharmacological actions have been reported (see Erspamer, 1966). Among them are actions on systemic blood pressure, vascular and cellular permeability, chromatophores, smooth muscle other than vascular, histamine release, and certain enzymes.

In the vertebrates, where peripheral 5-HT-containing nerves are absent, there are several non-nervous sources of 5-HT. For example, 5-HT from the enterochromaffin cells of the gut almost certainly plays a role in regulating intestinal peristalsis and permeability. 5-HT released from blood platelets may act on blood vessels. 5-HT from mast cells—at least in the rat and mouse—is involved in the inflammatory reaction. Various sources of 5-HT could serve as regulators of certain enzymes either directly or through changes in cellular permeability.

In the invertebrates, some of the non-nervous sources of 5-HT are the following: coelenteric tissue of the sea anemone, *Calliactis parasitica*; excretory organs of crustaceans and molluscs; and many venom producing structures. These latter structures include the posterior salivary glands of some species of octopus; the hypobranchial glands of some muricid molluscs; and venom glands of many social wasps, some spiders, some scorpions, and, possibly, the nematocysts of coelenterates.

Some venoms have yielded the highest levels of 5-HT found in any natural sources (see Table 1). The most plausible role of this 5-HT is that of a chemical weapon of defense (Welsh, 1964), since it is one of the most active known pain producers when applied to cutaneous nerve endings (Keele and Armstrong, 1964).

The 5-HT found in certain tissues and organs of invertebrates (e.g., Welsh and Moorhead, 1960) derives from peripheral 5-HT-containing nerves since, unlike the vertebrates, these are present in many invertebrate species.

5-HT in nervous systems. The quantitative distribution of this monoamine in nervous systems has been reviewed by Erspamer (1966) and Welsh (1968). While it has been found in nervous systems of representatives

TABLE 1

Some high levels of 5-hydroxytryptamine in venoms

Leiurus quinquestriatus (scorpion)	up to 4 mg/g dry venom	Adam & Weiss, 1956
Phoneutria fera (Brazilian spider)	1.5–2.7 mg/g dry venom	Welsh & Batty, 1963
Vespa crabro (hornet)	up to 19 mg/g dry venom sacs	Bhoola et al., 1961
Synoeca surinama (S.A. wasp)	13 mg/g dry sting apparatus	Welsh & Batty, 1963

of all of the major phyla the amounts vary considerably as may be seen in Table 2. Annelids and molluscs (especially the bivalves) have relatively high levels of 5-HT in their nervous systems, while arthropods, echinoderms, and tunicates have very low levels and it may even be absent from some species. The vertebrate nervous system has levels of 5-HT which are intermediate between those of the so-called lower and higher invertebrate phyla.

The quantitative distribution of 5-HT in various parts of the brain and spinal cord is much the same in all classes of vertebrates. Highest levels (1-3 µg/g) have been found in midbrain and hypothalamus and lowest levels (none to 0.3 µg/g) in cerebellar tissue.

While there is good evidence that 5-HT functions as a neuro-transmitter in certain of the invertebrates (especially the molluscs; e.g., Twarog, 1954; Loveland, 1963; Gerschenfeld and Stefani, 1965), the inaccessibility of 5-HT synapses in the vertebrates makes difficult a direct approach to observations on its transmitter role.

TABLE 2

5-Hydroxytryptamine in nervous systems*

Phylum and Class	Tissue	Number of species	Range µg/g wet tissue
ANNELIDS	nerve cords	(5)	3.1-10.4
MOLLUSCS			
Bivalves	ganglia	(7)	10-60
Gastropods	ganglia	(8)	2-10.6
Cephalopods	ganglia	(4)	0.7-4
ARTHROPODS			
Crustaceans	ganglia	(10)	<0.02-0.2
Crustaceans	pericardial organs	(2)	2.6-4
Insects	ventral cord	(1)	<0.02
Insects	heads	(3)	0.05-0.47
ECHINODERMS	radial nerves	(1)	none or trace
TUNICATES	ganglia	(3)	none or trace
VERTEBRATES			
Fishes	whole brains	(8)	0.15-0.48
Amphibians	whole brains	(13)	0.57-3.7
[*Bufo americanus*	whole brains		9.1]
Reptiles	whole brains	(8)	1.7-3.53
Birds	whole brains	(4)	0.4-1.2
Mammals	whole brains	(6)	0.13-1.0

*Data from various sources. See Erspamer (1966), Welsh (1968), and Welsh and Moorhead (1960) for most references.

Catecholamines
―――――――――――

The primary catecholamines—dopamine (DA) and norepinephrine (NE)—have been isolated from nervous systems of a wide variety of invertebrates but, unlike 5-HT, they appear not to occur in any non-nervous tissues.

Table 3 summarizes some of the results from recent studies on invertebrate nervous systems where procedures such as thin layer chromatography and conversion to fluorescing derivatives have been used for identification and quantitation of these amines.

TABLE 3

Catecholamines in nervous systems of some invertebrates

Phylum and Genus	DA* µg/g wet tissue	NE*	Reference
ANNELIDS			
Lumbricus	2.4	1.2	Rude (1969)
Lumbricus	trace	1.0	Myhrberg (1967)
MOLLUSCS			
10 genera		none detected	Sweeney (1963)
Anodonta	12-47	—	
Helix	7.25	—	Dahl et al. (1966)
Buccinum	17.30	—	
Spisula	40-50	5-6	Cottrell (1967)
Eledone	8-13	2-5	
Helix	5.5	—	Kerkut et al. (1966)
ARTHROPODS			
Crustaceans			
Hyas	<0.5	<0.5	
Carcinus	0.5-1.0	<0.5	Cottrell (1967)
Carcinus	6.9-7.8	—	Kerkut et al. (1966)
ECHINODERMS			
Echinus	2.5-7.0	1.5-3.5	Cottrell (1967)
Asterias	3-8	0.5-2.0	

*DA, dopamine; NE, norepinephrine.

Not shown in Table 3 are some recent results (King, Goldstone, and Welsh, unpublished) showing that extracts of the planarian, *Dugesia tigrina*, contain approximately equal amounts of DA, NE and epinephrine (E). In other recent studies only Myhrberg (1967) has reported finding E in an invertebrate nervous system. It has also been suggested (e.g., Cottrell, 1967) that in some other species it may be present but below a detectable level with the method used.

Table 3 shows that DA is the dominant CA in most invertebrate nervous systems thus far studied. Especially high levels occur in molluscan nervous systems, but in two crustaceans none was detected in *Hyas* ganglia and only 0.5-1.0 µg/g wet tissue in *Carcinus*. The significance of large amounts of DA and 5-HT in molluscan ganglia and very much smaller amounts of both in crustacean nervous systems is not yet apparent.

Pharmacological studies on a molluscan heart (Greenberg, 1960), on a molluscan ganglion (Sweeney, 1965) and on individual molluscan neurons (Kerkut and Walker, 1961) provide partial evidence that dopamine plays a neurotransmitter role in this invertebrate group.

In the vertebrates, NE is the neurotransmitter of adrenergic neurons of the autonomic system while both NE and E are present (in different ratios depending on species) in the adrenal glands of mammals (see v. Euler, 1956). Data on the quantitative distribution of DA and NE in different parts of the central nervous system of the dog, cat and man have been summarized by McLennan (1963) and in the rat brain their relative amounts in different areas have been estimated from the intensity of their fluorescence by Fuxe (1965).

Histamine

This amine is a natural constituent of many animal and plant species. Such studies as have been made on invertebrates (e.g., Mettrick and Telford, 1965) show that it is widely distributed among species and tissues. However, even within a given phylum the amounts that have been found may vary from none in some species up to levels of 10-60 µg per gram of whole animal in others.

Some venoms contain unusually high levels of histamine. Dry venom of the honeybee contains 10 mg/g (Reinert, 1936); dry venom sacs of the wasp, *Vespa vulgaris*, 16 mg/g (Jaques and Schachter, 1954); dry venom sacs of the hornet, *Vespa crabro*, from 14-30 mg/g

(Bhoola, *et al.*, 1961).

In the vertebrates, histamine occurs in a wide variety of tissues and body fluids. Extensive tables of values may be found in the review by Vugman and Rocha e Silva (1966). The occurrence of histamine in mast cells and mast cell tumors has been reviewed by Riley and West (1966).

II. HISTOCHEMICAL FLUORESCENCE STUDIES

Quantitative estimates of the levels of a biogenic amine in entire organisms or tissues usually fail to indicate the cellular sites of the amine. While early histochemical methods (e.g., the chromaffin reaction) located amine-containing cells, they failed, usually, to identify the amine and also often failed to allow the visualization of cellular details such as fine axons and their terminals. With the advent of the Falck and Hillarp histochemical fluorescence method, it became possible to locate monoamine-containing cells and usually to identify the specific amine by the fluorescence of its reaction product. This method has now been used in studies (especially of nervous systems) of several representative invertebrates and vertebrates. However, the difficulty of adequate freeze-drying of tissues of marine animals has, thus far, limited its successful use to fresh-water or terrestrial animals or to euryhaline species that may be maintained at low salinities for extended periods before the freeze-drying of their tissues.

Brief summaries of some results from the use of the Falck and Hillarp method follow. It is of interest to note the correspondence between the numbers and types of fluorescing cells and the amounts of monoamines previously found by means of bioassay or spectrofluorometric assay.

<u>Coelenterates</u>. Fluorescing sensory (or, possibly, sensory-motor) cells have been described in the tentacular ectoderm of the sea anemones, *Metridium senile*, and *Telia felina* (Dahl, *et al.*, 1963). Their charac-

teristic fluorescence appears to be due to the presence of primary catecholamines.

Flatworms. The nervous system of the planarian, *Phagocata oregonensis*, contains large numbers of both 5-HT- and CA-containing neurons (Welsh and Williams, 1969). These are present in the brain, ventral cords and peripheral networks (Fig. 1). Sensory cells of both types (5-HT and CA) appear to be present. Other planarians (*Dugesia* and *Procotyla*) resemble *Phagocata*. These latter species have been shown to contain acetylcholine esterase-reactive neurons (Lentz, 1968). It is of interest that the several principal chemical classes of neurons are present in the nervous systems of such primitive animals as planarians.

Annelids. The results of studies by Rude (1966) and Myhrberg (1967) on the distribution of monoamine-containing cells in the earthworm, *Lumbricus terrestris*, are in close agreement. Large numbers of both 5-HT and CA cells are present in the cerebral ganglia. Each

Figure 1. (Opposite page.) Diagram of nervous system of the planarian, *Phagocata oregonensis*, as seen by induced fluorescence. Stippled areas of brain and ventral cords indicate blue fluorescing (CA-containing) fiber tracts and neuropiles. The four pairs of large blue fluorescing cells of the brain (I, II, III, IV) and the blue fluorescing nerve net of the pharynx are shown in black. Yellow fluorescing (5-HT-containing) nerve cells are shown in white. Ventral cord commisures and lateral nerves contain both blue and yellow fluorescing nerve fibers. The extensive branching of lateral nerves and the peripheral plexus of both types of fluorescing neurons have been omitted from this schema. See Welsh and Williams (1969) for further details.
    ant. br. comm. = anterior brain commisure
    post. br. comm. = posterior brain commisure
    v. cord = ventral cord
    v. cord comm. = ventral cord commisure
    lat. n. = lateral nerve
    epid. = epidermis
(This figure reproduced by permission of the Journal of Comparative Neurology, The Wistar Institute Press.)

ganglion in the ventral cord contains four bilaterally disposed groups of 5-HT cells but only one pair of CA cells. However, CA-containing fiber tracts are present in the cord. These appear to be the central processes of CA sensory cells that are present in the ectoderm.

The central nervous system of the leech, *Hirudo medicinalis*, differs from that of *Lumbricus* in having a smaller number of 5-HT cells (Ehinger, *et al.*, 1968;

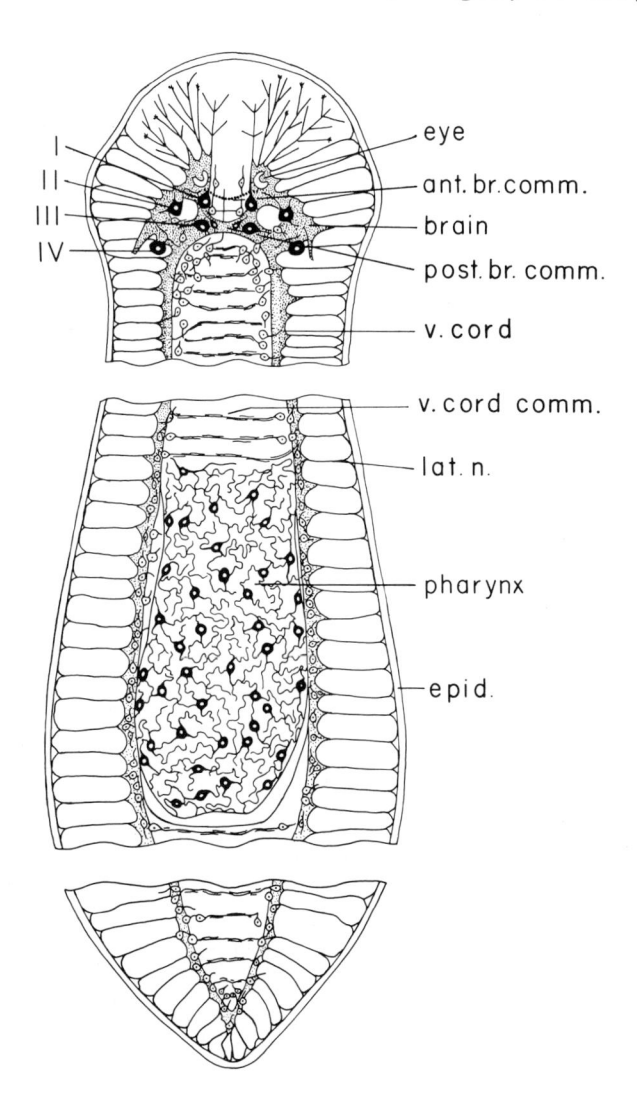

Rude, 1969). Except at anterior and posterior ends where there is fusion of ganglia, each ganglion contains only seven large, 5-HT-containing nerve cells. A single pair of CA cells are located in the anterior roots of these ganglia. Neuropiles of dense networks of 5-HT and CA processes are present in the ganglia.

Molluscs. Ganglia of several species of gastropod and bivalve molluscs have been studied using the Falck and Hillarp technique. Consistent with the finding of large amounts of 5-HT and DA in these classes of molluscs has been the finding of large numbers of fluorescing cells of both types as well as extensive neuropiles.

Among the genera studied have been the gastropods, *Helix* (Dahl, et al., 1968) and *Strophocheilus* (Jaeger, et al., unpublished); and the bivalves, *Anodonta* (Dahl, et al., 1966) and *Sphaerium* (Sweeney, 1968).

In both *Helix* and *Strophocheilus* there appear to be some nerve cells that contain *both* 5-HT and CA (Sedden, et al., 1968; Jaeger, et al., unpublished). These cells deserve further study, especially by means of microspectrofluorometry.

Sensory cells with a fluorescence characteristic of primary catecholamines have been seen in molluscs (Sweeney, 1968; Jaeger, et al., unpublished).

Arthropods. The more anterior ganglia of the crayfish, *Astacus*, have been examined by Elofsson, et al. (1966). Groups of primary CA-containing cells were found as well as fiber tracts and neuropiles. Only one pair of 5-HT cells was found in the brain and at least one such cell in the subesophageal ganglion. This correlates with the low levels or apparent absence of 5-HT in the CNS of crayfish and other crustaceans (see Section I).

The pericardial organs of crabs are nerve plexuses that lie in the pericardial cavity. They consist of axons some of which branch extensively to form an outer layer of nerve endings which release substances into the hemolymph. The fluorescence of certain of these axons (e.g., Fig. 2) indicates the presence of 5-HT in some and CA in others (Cooke and Goldstone, personal

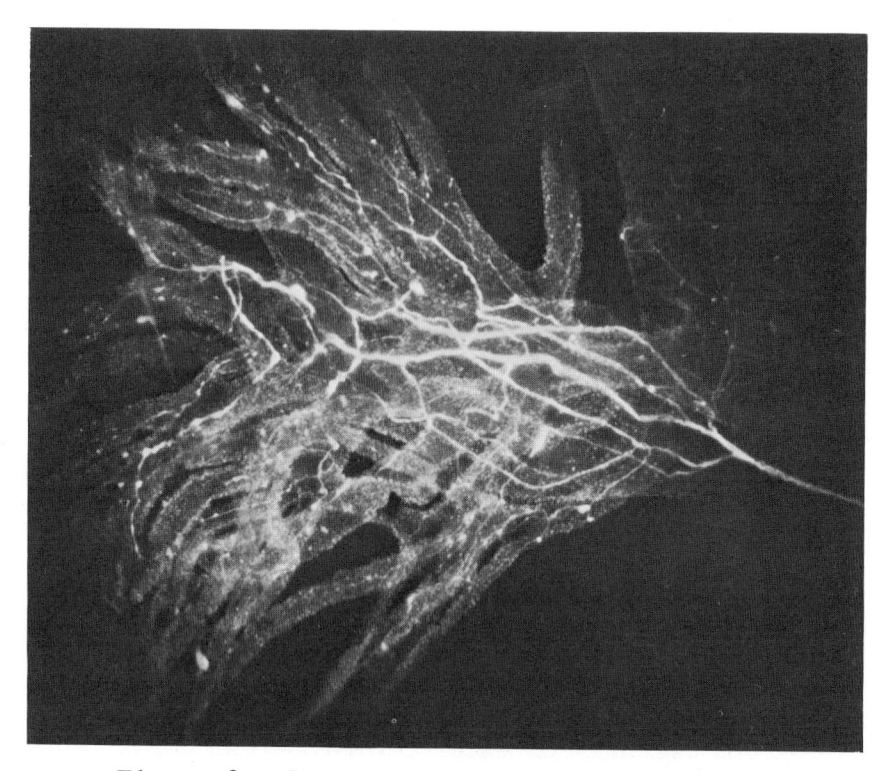

Figure 2. Posterior region of one of the paired pericardial organs of the crab, *Cardisoma guanhumi*, showing a fluorescing catecholamine-containing axon and its many terminal branches. The outer layer of axon endings produces the granular fluorescence. X 60. (Photograph supplied by Ian Cooke and M. Goldstone.)

communication). The pericardial organs of *Cancer borealis* were shown by Maynard and Welsh (1959) to contain 2.6-4.0 µg 5-HT/g wet tissue. This is more than was found elsewhere in the nervous system of this or any other species of crab (Welsh and Moorhead, 1960).

Frontali (1968) has described the distribution of fluorescing cells and neuropiles in the brain of the cockroach, *Periplaneta americana*. They appear to contain dopamine and norepinephrine. No 5-HT-containing neurons were found.

Vertebrates. Application of the Falck and Hillarp fluorescence technique in studies of the distribution of amine-containing neurons in the vertebrate nervous system has yielded much valuable information. For example, the locations of 5-HT-, DA- and NE-containing nerve cells in the mammalian brain have been mapped and their axons traced (e.g., Andén, et al., 1966; Fuxe, 1965; Dahlström and Fuxe, 1965).

The distribution and destinations of CA-containing nerve fibers of the autonomic systems of various vertebrates have been extensively investigated (see Falck and Owman, 1966).

III. THE STUDY OF AN IDENTIFIABLE CELL

We may consider the matter of the site of origin of a monoamine and the nature of its storage organelles in a given type of cell in concluding this brief summary of the distribution of biogenic amines. Such studies have been made on the adrenal medulla and other chromaffin tissues; on enterochromaffin cells; on mast cells and blood platelets; on the pineal gland; and on adrenergic neurons of the sympathetic system. These studies on vertebrates are too numerous to review. Far less attention has been given to the identification and localization of a biogenic amine in a recognizable invertebrate cell. For this reason a recent study of the Retzius cells of the leech, *Hirudo medicinalis*, by Rude, Coggeshall, and van Orden (1969) will be briefly recounted.

The chromaffin cells of *Hirudo* have long been thought to contain epinephrine. Based on histochemical fluorescence studies, however, Kerkut, et al. (1967) concluded that these cells contain 5-HT and not epinephrine. Rude, et al. (1969) have confirmed this in a study that might well serve as a model for future work aimed at the identification and location of a biogenic amine in a given type of cell. However, few monoamine-containing cells are as suitable as are Retzius "colossal" cells for such detailed analysis.

The main points of the study by Rude, et al. are the

following:

1. The cell bodies of colossal neurons can easily be distinguished from all other neurons in the ganglion because of their size (larger than the rest of the cell bodies in the ganglion) and location (ventral surface of the anterior ventral packet). They can also be readily dissected from the ganglion.
2. When colossal neurons are treated by the formaldehyde-condensation technique for the histochemical demonstration of monoamines, they develop an intense yellow fluorescence in two concentric layers, one about the nucleus and the other in the peripheral cytoplasm (Fig. 3). In addition

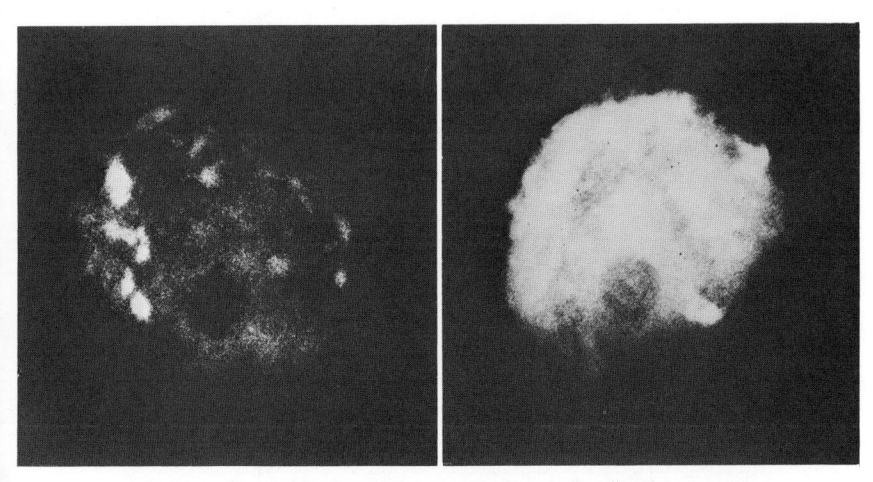

Figure 3. Photomicrographs of whole nerve cell bodies dissected from ganglia of the leech, *Hirudo medicinalis* ("colossal cells of Retzius"). X 600. *A*. Dried untreated cell body with several autofluorescent small glial cells adhering to its surface. *B*. A cell body after exposure to formaldehyde vapor, showing the development of an intense formaldehyde-induced fluorescence in the cytoplasm. (Photographs supplied by Sonia Rude.)

a weaker yellow fluorescence develops between the two bands, beyond the peripheral band, and in the axon hillock.
3. 5-Hydroxytryptamine is the monoamine responsible for the yellow, formaldehyde-induced fluorescence as demonstrated by three methods:

    a. Microspectrofluorometry—after treatment with the fluorescence histochemical technique for monoamines, the colossal cells and 5-HT develop identical peaks of excitation and emission. There is no evidence for the presence of CA by this technique.
    b. Chromatography—when extracts of colossal cells are chromatographed a single spot develops which runs to the same level as 5-HT.
    c. Spectrofluorometry—5-HT can be extracted from colossal cell bodies and measured spectrofluorometrically. By this technique an average colossal neuron cell body contains 6 mM 5-HT.

4. When colossal neurons are examined with the electron microscope, the only organelles in the cell having the same distribution as the yellow formaldehyde-induced fluorescence are dense-cored granules (Fig. 4). Since the yellow fluorescence has been equated with 5-HT, the dense-cored granules probably contain 5-HT.
5. Further strength for this argument comes from:

    a. the use of chromaffin reaction modified for the electron microscope. When applied to the colossal neurons, the cores of the dense-cored granules are the only organelles that stain.
    b. electron microscopic, autoradiographic studies showing that after a short pulse of 5-HT or 5-hydroxytryptophane, the dense-cored granules are the only organelles that label.

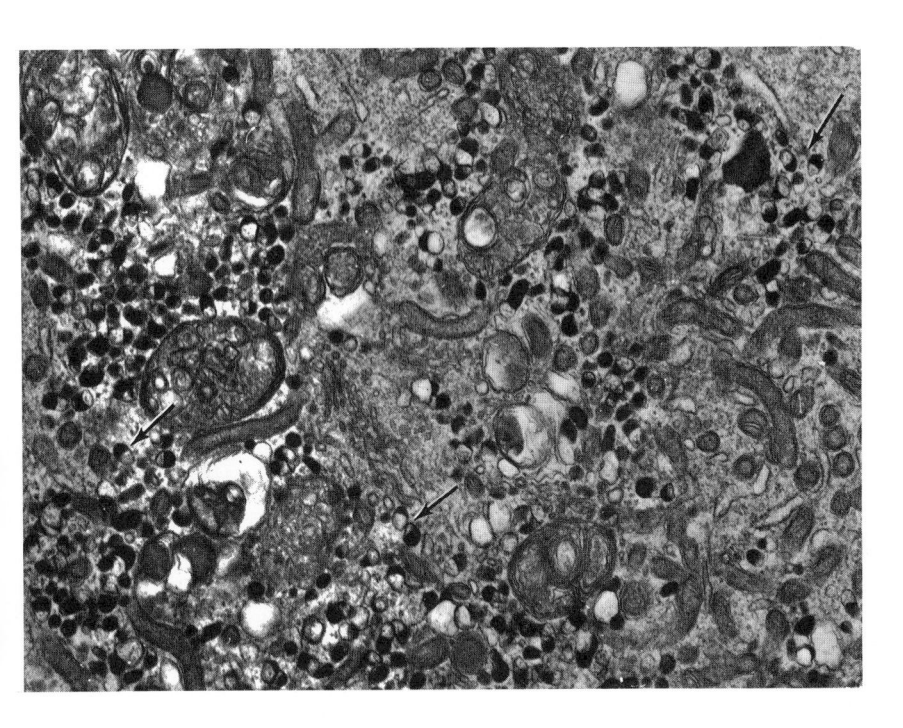

Figure 4. Electron micrograph of a region of a colossal cell showing numerous 5-HT storage granules. The electron dense cores often appear in an eccentric position (see arrows). These granules have an average diameter of 1000 Å. (Photographs supplied by Sonia Rude. For further details see Rude, *et al.*, 1969).

CONCLUSIONS

Some conclusions that may be drawn from this brief review of the phylogenetic distribution of certain biogenic amines and their cellular sites are the following:

1. 5-HT and CA in the invertebrates are largely restricted to nervous systems although 5-HT and histamine are present in some venoms in relatively enormous amounts.
2. In the vertebrates there are several non-nervous

sites of 5-HT such as blood platelets, mast cells and chromaffin tissue.
3. By means of a histochemical fluorescence method the cellular distribution of 5-HT and CA can be determined. Neurons containing these amines are present in such primitive animals as planarians.
4. The numbers of 5-HT- or CA-containing neurons vary greatly in different phyla. Some molluscan ganglia have more cells of both types per unit volume of tissue and higher levels of amines than any other group.

REFERENCES

Ackermann, D. 1962. *Berichte der Physikalisch-Medizinischen Gesellschaft Würzburg.* 70: 1.
Adam, K.R. and C. Weiss. 1956. *Nature* (London) 178: 421.
Andén, N.-E., A. Dahlström, K. Fuxe, K. Larsson, L. Olson, and U. Ungerstedt. 1966. *Acta Physiol. Scand.* 67: 313.
Bhoola, K.D., J.D. Calle, and M. Schachter. 1961. *J. Physiol.* (London) 159: 167.
Cooke, I.M. and M. Goldstone. Personal communication.
Cottrell, G.A. 1967. *Br. J. Pharmacol.* 29: 63.
Dahl, E., B. Falck, C. von Mecklenburg, and H. Myhrberg. 1963. *Quart. J. Micr. Sci.* 104: 531.
Dahl, E., B. Falck, C. von Mecklenberg, H. Myhrberg, and E. Rosengren. 1966. *Z. Zellforsch.* 71: 489.
Dahlström, A. and K. Fuxe. 1965. *Acta Physiol. Scand.* 64 (Suppl. 247): 1.
Ehinger, B., B. Falck, and H.E. Myhrberg. 1968. *Histochemie* 15: 140.
Elofsson, R., T. Kauri, S.-O. Nielsen, and J.-O. Stromberg. 1966. *Z. Zellforsch.* 74: 464.
Erspamer, V. 1954. *Pharmacol. Rev.* 6: 425.
Erspamer, V. 1966. *Handbook Exp. Pharmacol.* 19: 132.
v. Euler, U.S. 1956. *Noradrenaline.* (Springfield, Ill.: C.C. Thomas.)
Falck, B. and C. Owman. 1965. *Acta Univ. Lund.* Sec. II, No. 7.
Falck, B. and C. Owman. 1966. In *Mechanisms of Release of Biogenic Amines.* (Oxford: Pergamon Press.)

Frontali, N. 1968. *J. Insect Physiol.* 14: 881.
Fuxe, K. 1965. *Acta Physiol. Scand.* 64:(Suppl. 247): 37.
Gerschenfeld, H.M. and E. Stefani. 1965. *Nature* (London) 205: 1216.
Greenberg, M.J. 1960. *Br. J. Pharmacol.* 15: 365.
Jaeger, C.P., E. Jaeger, and J.H. Welsh. Unpublished.
Jaques, R. and M. Schachter. 1954. *Br. J. Pharmacol.* 9: 53.
Keele, C.A. and D. Armstrong. 1964. *Substances Producing Pain and Itch.* Williams and Wilkins.
Kerkut, G.A., C.B. Sedden, and R.J. Walker. 1967. *Comp. Biochem. Physiol.* 21: 687.
Kerkut, G.A. and R.J. Walker. 1961. *Comp. Biochem. Physiol.* 3: 143.
Lentz, T.L. 1968. *Comp. Biochem. Physiol.* 27: 715.
Loveland, R.E. 1963. *Comp. Biochem. Physiol.* 9: 95.
Maynard, D.M. and J.H. Welsh. 1959. *J. Physiol.* (London) 149: 215.
McLennan, H. 1963. *Synaptic Transmission.* (Philadelphia: W.B. Saunders.)
Mettrick, D.F. and J.M. Telford. 1965. *Comp. Biochem. Physiol.* 16: 547.
Myhrberg, H.E. 1967. *Z. Zellforsch.* 81: 311.
Page, I.H. 1954. *Physiol. Rev.* 34: 563.
Reinert, M. 1936. Bee venom. In *Festschrift Emile Barell.* (Basel: F. Reinhardt.)
Riley, J.F. and G.B. West. 1966. *Handbook Exp. Pharmacol.* XVIII/1: 116.
Rude, Sonia. 1966. *J. Comp. Neurol.* 128: 397.
Rude, Sonia. 1969. *J. Comp. Neurol.* 136: 349.
Rude, Sonia R.E. Coggeshall, and L.S. van Orden, 3rd. 1969. *J. Cell. Biol.* 41: 832.
Sedden, C.B., R.J. Walker, and G.A. Kerkut. 1968. *Symp. Zool. Soc. London* No. 22: 19.
Sweeney, D. 1963. *Science* 139: 1051.
Sweeney, D. 1965. *Am. Zoologist* 5: 671.
Sweeney, D. 1968. *Comp. Biochem. Physiol.* 25: 601.
Twarog, Betty M. 1954. *J. Cell. Comp. Physiol.* 44: 141.
Vugman, I. and M. Rocha e Silva. 1966. *Handbook Exp. Pharmacol.* XVIII/1: 81.
Welsh, J.H. 1964. *Ann. Rev. Pharmacol.* 4: 293.

Welsh, J.H. 1968. *Advances in Pharmacol.* 6A: 171.
Welsh, J.H. and C.S. Batty. 1963. *Toxicol.* 1: 165.
Welsh, J.H. and M. Moorhead. 1960. *J. Neurochem.* 6: 146.
Welsh, J.H. and Lois D. Williams. 1969. *J. Comp. Neurol.* (in press).

Biogenic Amines and Metabolic Control in Tetrahymena

J. J. Blum

Dept. of Physiology and Pharmacology
Duke University Medical Center
Durham, North Carolina

In her penetrating analysis of the origins of mitosis Sagan (1967) has argued that eukaryotic cells developed from the blue green algae as a result of the internalization of bacteria which became the organelles we now recognize as mitochondria, kinetosomes or centrioles, and chloroplasts. The origins of the metazoa, then, must be sought in the protozoa and, in particular, in those protozoa which do not contain chloroplasts. Whittacker (1969) in his recent revision of the phylogenetic tree places ciliates close to the line of evolution of the metazoa, while Hadzi (1963), arguing on purely taxonomic grounds, suggested that the metazoa evolved directly from the ciliates. Because it can be

---
*The work reviewed in this paper was supported by research grants from the National Institutes of Health (5 R01 HD 01269) and the National Science Foundation (GB 5617) and by a research Career Development Award (5 K3 GM 2341) from the National Institutes of Health.

grown conveniently in defined media *Tetrahymena pyriformis* has been studied more extensively than any other ciliate and, indeed, more than most protozoa. The literature on *Tetrahymena* contains several observations which lend credence to the notion that the metabolism of this ciliate may be more closely related to the metabolism of the metazoa than that of other protozoa. A few such observations are:

1. There is an excellent correlation of the relative abundance of amino acids of crude mammalian proteins to that of whole *Tetrahymena* and a striking resemblance in the pattern of excretion of conjugated amino acids (Wu and Hogg, 1956);
2. Ten of the eleven amino acids required by *Tetrahymena* are also required by mammals (see Kidder, 1967 for references);
3. Certain non-essential amino acids can spare the quantitative needs of certain essential amino acids in a pattern very similar to that found in vertebrates (Genghof, 1949; Hogg and Elliot, 1951);
4. *Tetrahymena* is more animal like in its folic acid requirements than any other unicellular organism studied (Kidder and Dewey, 1947);
5. *Tetrahymena* and *Paramecium* contain hemoglobin and are the only microorganisms so far shown to contain this pigment (see Ryley, 1967 for references);
6. The chemical and physical properties of the glycogen of *Tetrahymena* are closely similar to those of liver glycogen (Ryley, 1952; Barber, *et al.*, 1965);
7. Tetrahymenids have developed the phenylalanine hydroxylating system (typically an animal system) for the synthesis of tyrosine (Kidder, 1967).

The purpose of listing these examples is not to deny the many ways in which the metabolism of *Tetrahymena* differs from that of animal cells, nor to disagree with Holz's (1966) view that *Tetrahymena* is *sui generis*, but merely to suggest that these cells may be particularly useful for the study of the phylogenetic origins of metazoan metabolic control processes.

In view of these statements, the discoveries of Janakidevi, Dewey and Kidder (1966a; 1966b) that *Tetrahymena* contained epinephrine, norepinephrine, and serotonin, led us to test the effect of reserpine on *Tetrahymena*. We found (Blum, *et al.*, 1966) that reserpine inhibited the growth of these cells and depleted the catecholamine content. These observations suggested that *Tetrahymena* might contain other components of the catecholamine system found in metazoa. We therefore examined the effects of a variety of drugs known to effect the uptake, metabolism, or storage of catecholamines and serotonin in the higher metazoa. For brevity we shall refer to such drugs as adrenergic or serotoninergic drugs, without implying however that these drugs have only this action on the metazoa or, *a fortiori*, in protozoa. It was found that many of these drugs inhibited the growth of *Tetrahymena* and altered cell glycogen content (Blum, 1967). In this review we shall concentrate on the effects of only a few of these drugs.

Effects of reserpine on growth and glycogen content. *Tetrahymena* are very sensitive to reserpine. Growth is scarcely inhibited at 10^{-5} M reserpine, but is completely blocked at 4 X 10^{-5} M. In the experiment shown in Fig. 1 cells in a culture in early stationary phase were transferred to fresh proteose-peptone medium and reserpine was added at zero time to the experimental flask. The control cells began rapid exponential growth and their glycogen content decreased to about 0.4 mg/10^6 cells. As the culture entered stationary phase the glycogen content increased to about 1.2 mg/10^6 cells and, finally, after about 50 hr in stationary phase, the cells rapidly utilized their glycogen reserves. The cells exposed to reserpine also grew exponentially, but at a slower rate. The initial decrease in glycogen content occurred at about the same rate and to the same extent as the control cells, but there was no subsequent increase in glycogen content. Similar results were obtained with cells growing in medium supplemented with glucose (Blum, 1967). Thus whether glucose was being synthesized *de novo* from

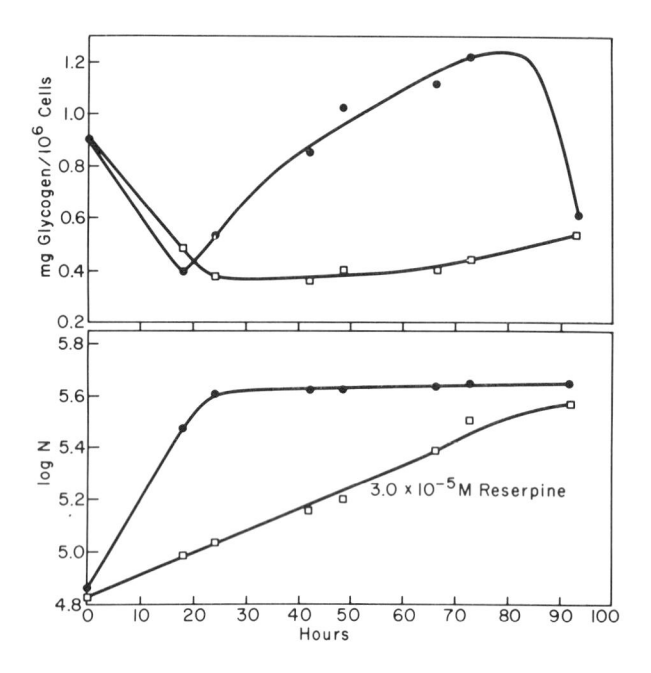

Figure 1. Effect of reserpine on growth and glycogen content of *Tetrahymena*. Cells were grown in a medium consisting of 1% proteose-peptone and 0.05% liver extract in 0.02 M KH_2PO_4, pH 6.5 Cultures were grown at 25° in 1 liter Erlenmeyer flasks containing about 45 ml. Reprinted from Blum (1967).

amino acids or was added to the culture medium, reserpine decreased cell glycogen content. Recently we have shown that the specific activity of glycogen in washed cells that were supplied with ^{14}C-labeled glucose, acetate, or pyruvate was less in cells exposed to reserpine than in the controls (Wexler and Blum, to be published), thus indicating that a decrease in synthesis probably contributes to the reduction in cell glycogen content.

Effect of α- and β-adrenergic blocking agents. Propanolol, one of the most potent β-blocking agents known, inhibits the growth of *Tetrahymena*, the inhibition increasing with increasing time of exposure to the drug (Blum, 1967). In the experiment shown in Fig. 2 a low

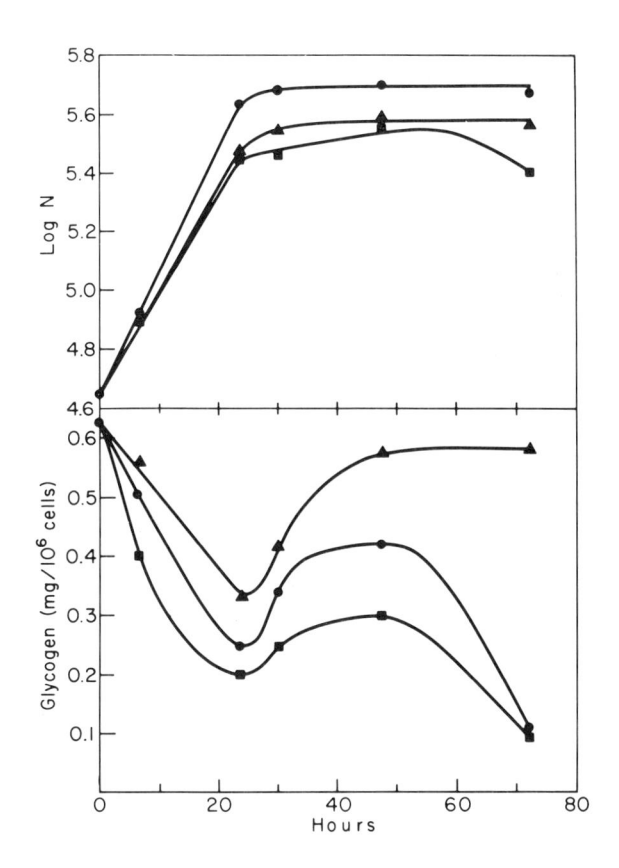

Figure 2. Effects of propanolol and of theophylline on growth and glycogen content of *Tetrahymena*. Cells were grown in the proteose-peptone-liver extract medium described in the legend to Fig. 1. ●----●, control cells; ■----■, 7.1 X 10^{-5} M propanolol; ▲----▲, mM theophylline.

concentration of propanolol was used and growth inhibition was small, but glycogen content was below that of the control cells throughout the life of the culture.

Dichlorisoproterenol (DCI) is β-blocking agent which prevents the epinephrine induced conversion of phosphorylase *b* to phosphorylase *a* in skeletal muscle

(Helmreich and Cori, 1966) and prevents the stimulation of heart adenyl cyclase by catecholamines (Murad, et al., 1962). At 0.1 mM, DCI had little effect on the growth of Tetrahymena, but interfered with glycogen synthesis (Blum, 1967).

Dibenzyline, which presumably interacts irreversibly with α-adrenergic receptor sites, slightly inhibited the growth of Tetrahymena at 0.1 mM, and considerably potentiated the growth inhibitory effect of reserpine (Blum, 1967). Iwata, et al. (1967) have reported that very low concentrations of the α-adrenergic blocking agents ergotamine and dibenamine promoted the growth of Tetrahymena.

Thus Tetrahymena is sensitive to both α- and β-adrenergic blocking agents and in the case of DCI, at least, a concentration which scarcely inhibits growth decreases cell glycogen content. Sensitivity to these agents does not, of course, imply that Tetrahymena has α- or β-adrenergic receptor sites, since the action of these drugs could be on adrenergic sites which cannot be characterized as α or β types or could be on growth controlling sites irrelevant to an adrenergic system.

Effect of triiodothyronine. It is well known that hyperthyroid animals are hypersensitive to apinephrine and that hypothyroid aninals are hyposensitive to epinephrine. Haugaard and Hess (1966) showed that thyroxine increases the percent phosphorylase a of rat heart, and, more recently, Levey and Epstein (1968) demonstrated that triiodothyronine (T_3) increased the adenyl cyclase of cat heart homogenates. It was already known (Wingo and Cameron, 1952) that thyroxine at concentrations of 10 mg/liter or more inhibited the growth and increased the respiratory rate of Tetrahymena but such uncoupling effects are relatively unspecific. We found that 10^{-5} M T_3 did not inhibit growth but reduced cell glycogen content (Blum, 1967). In the experiment shown in Fig. 3 neither T_3 nor T_3 plus DCI inhibited growth but, even in the presence of glucose, T_3 markedly inhibited net glycogen synthesis while T_3 plus DCI completely prevented glycogen accumulation after about six hr exposure to the drugs.

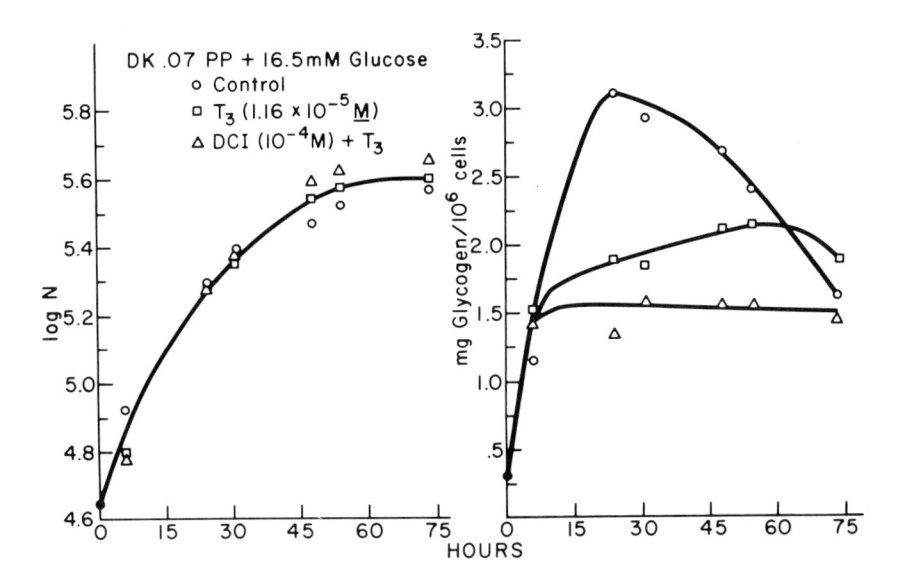

Figure 3. Effects of DCI and triiodothyronine on growth and glycogen content of *Tetrahymena*. Cells were grown in a synthetic medium supplemented with 0.07% proteose-peptone and 16.5 mM glucose. Reprinted from Blum (1967).

Effects of theophylline on growth and glycogen content. Caffeine and theophylline are known to inhibit the specific phosphodiesterase which converts 3',5'-cyclic AMP to 5'-AMP. Theophylline is slightly more potent an inhibitor of the beef heart enzyme than caffeine (Butcher and Sutherland, 1961) and is commonly used to enhance metabolic effects mediated by cyclic AMP. Although it is usually assumed that the *in vivo* effect of theophylline or caffeine results from its inhibition of cyclic AMP phosphodiesterase, this is not necessarily the case since caffeine inhibits calcium uptake by sarcoplasmic reticulum, but it has not been possible to demonstrate an effect of cyclic AMP in this system (Weber, 1968). We found that 1.3 mM caffeine slightly inhibited the growth of *Tetrahymena* and potentiated the growth-inhibitory effect of reserpine (Blum, 1967). At 4 mM caffeine growth was completely inhibited. Theophylline was more potent than caffeine as an inhibitor of growth. In contradistinction to any of the drugs we have so far tested, theophylline increased cell glycogen content (Fig. 2).

Phosphodiesterase activity. Since caffeine and theophylline affected growth and metabolism, it was important to establish whether *Tetrahymena* possessed a methyl xanthine-inhibitable phosphodiesterase. We found that sonic lysates of *Tetrahymena* catalyzed the conversion of 3',5'-cyclic AMP to 5'-AMP (Blum, 1970). Most of the activity remained in the supernatant after centrifugation for 1 hr at 27,000 X g whether the cells were disrupted by ultrasound or by the presumably much gentler method of grinding with glass beads. The level of activity in whole sonic lysates was about 0.7 μmole/mg protein·hr, which is comparable to whole rat brain homogenates (Cheung and Salganicoff, 1967) and *E. coli* (Brana and Chytil, 1966). Both caffeine and theophylline inhibited the cyclic AMP phosphodiesterase activity of *Tetrahymena* (Fig. 4). If it is assumed that

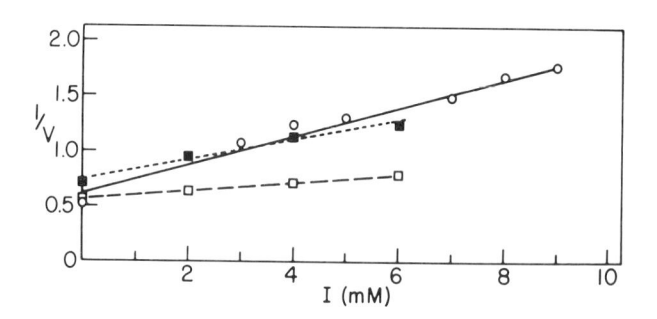

Figure 4. Inhibition of phosphodiesterase activity by caffeine and theophylline. Activity was assayed at 30° and was expressed as μM/mg protein·hr. The reciprocal of the activity is plotted on the ordinate and the inhibitor concentration (I) used is shown on the abscissa. The concentrations of cyclic AMP and the inhibitor used were: O———O, mM cyclic AMP, theophylline; ▫— — —▫, mM cyclic AMP, caffeine; ■----■, 0.30 mM cyclic AMP, caffeine.

the inhibition is competitive, then the ratio of slope to intercept for each line in Fig. 4 is $K_m/K_i(K_m + S)$, where K_m is the Michaelis-Menten constant of the enzyme and K_i is the dissociation constant for the competitive inhibitor. From the data shown in Fig. 4 the following values were computed: K_m, 0.52 mM; K_i (caffeine), 5.3 mM; K_i (theophylline), 1.7 mM. Thus theophylline is a stronger inhibitor than caffeine of the phosphodiesterase activity as well as of growth. It must be stressed, however, that the presence of a theophylline-inhibitable phosphodiesterase does not prove that the growth inhibitory effect of theophylline or its effect on glycogen content are mediated via inhibition of this enzyme, especially since neither the presence of cyclic AMP nor of adenyl cyclase has yet been demonstrated in *Tetrahymena*.

Peroxisomes and gluconeogensis in *Tetrahymena*. *Tetrahymena* resemble liver cells in that they have a high capacity for gluconeogenesis and contain peroxisomes. These organelles, found originally in liver, kidney, and *Tetrahymena* (de Duve and Baudhuin, 1966) are now known to be present in several other protozoa, in germinating fatty seedlings, and in green leaves (de Duve, 1969). Their physiological role is uncertain, but since they contain enzymes which could give rise to pyruvate and to other precursors of glucose and are found in tissues with high glyconeogenetic capacity it has been suggested that they play a role in gluconeogenesis (de Duve and Baudhuin, 1966). Such a role is made more probable by the demonstration that isocitrate lyase and malate synthase are localized in the peroxisomes of *Tetrahymena* (Muller, *et al*., 1968). These enzymes circumvent the oxidative decarboxylation steps of the tricarboxylic acid cycle and permit the net formation of malate from acetyl Co A, thus permitting gluconeogenesis from endogenous lipid (Hogg and Kornberg, 1953) as well as from other sources of acetyl Co A. Indeed, *Tetrahymena* can synthesize up to a quarter of its dry weight as glycogen during growth on proteose-peptone media (Manners and Ryley, 1952). The involvement of peroxisomes in gluconeogenesis is further indicated by the effect of growth conditions on isocitrate lyase activity. Transfer of *Tetrahymena* from aerobic to partially anaerobic conditions, for example, activates gluconeogenesis and increases isocitrate lyase activity, while addition of glucose to the medium stops gluconeogenesis and decreases isocitrate lyase activity (Levy and Hunt, 1967). With these observations as a background, we investigated the effects of several adrenergic drugs on the isocitrate lyase, catalase, and D-amino acid oxidase activities of *Tetrahymena* (Blum, 1968). None of the drugs shown in Table 1 altered the D-amino acid oxidase activity, and only reserpine (at fairly high concentrations) decreased the catalase activity. Isocitrate lyase activity, however, was reduced by reserpine, DCI, propanolol, and triiodothyronine. It should be stressed that these effects are not simply related to growth inhibition, since growth was only slightly inhibited by DCI and by triiodothyronine. Furthermore, although theophylline inhibited growth it increased the isocitrate lyase activity slightly, in harmony with the observed increase in glycogen content. Thus one effect of the adrenergically reactive drugs in *Tetrahymena* is to alter the activity of the first enzyme required for gluconeogenesis via the glyoxylate bypass.

Effects of adrenergic drugs on glycogen phosphorylase and glycogen synthetase. At the opposite end of the gluconeogenic pathway from isocitrate lyase are the enzymes involved in the metabolism of glycogen. The presence of phosphorylase in *Tetrahymena* was first established by Ryley (1952), but since then no further work has been done on this enzyme. Cook, et al. (1968) and, independently, Kahn and Blum (see Blum, 1970) found that *Tetrahymena* had glycogen synthetase activity and that UDPG was much more effective than ADPG as a glycosyl donor. In view of the effects of adrenergic compounds on the regulation of glycogen metabolism in metazoan tissues and on growth, glycogen content and isocitrate lyase activity in *Tetrahymena*, an investigation of the effects of several adrenergic drugs on the glycogen metabolising enzymes of *Tetrahymena* was undertaken (Blum, 1970).

It was found that theophylline altered the glycogen

TABLE 1

Effect of adrenergic drugs on peroxisomal enzymes of *Tetrahymena*

Drug	Concentration M × 10⁻⁴	Protein content % of control	Isocitrate lyase % of control	D-Amino acid oxidase % of control	Catalase % of control
Reserpine	0.34	101	77*	91	85*
Theophylline	14.0	160*	112*	93	99
DCI	1.4	98	85*	112	121
T₃	0.18	93	69*	95	102
DCI + T₃	1.4 + 0.18	91	62*	105	104
Propanolol	1.5	112	56*	117	106

Cells were grown in proteose-peptone medium (cf. legend to Fig. 1) liter Erlenmeyer flasks at 25° without shaking. Culture volumes were between 42 and 44 ml, including drugs at the concentrations shown. After exposure to the drugs for about 17 hr cells were collected, treated with ultrasound, and assayed for the enzyme activities shown. Data is taken from Blum (1968) and from unpublished experiments.

*Indicates values appreciably different from the controls, which were taken at 100%.

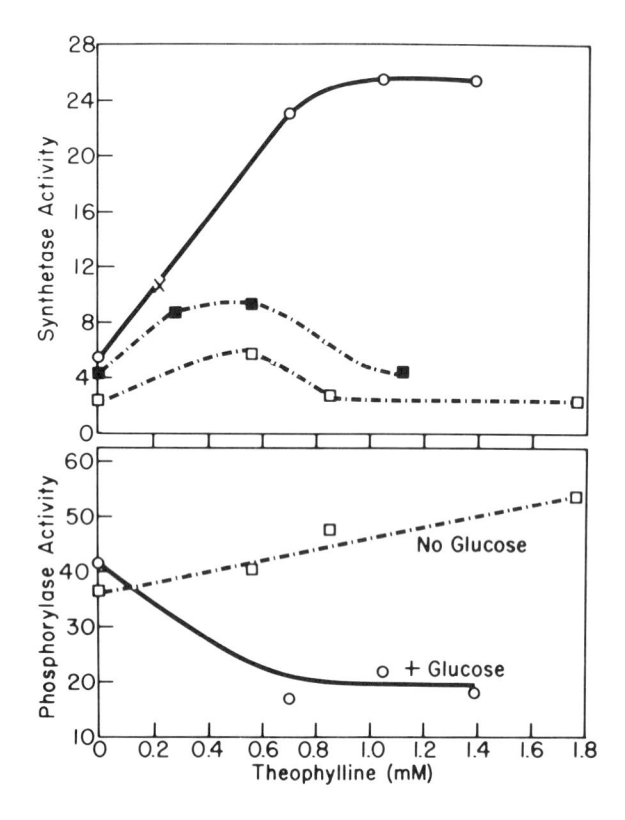

Figure 5. Effects of theophylline on glycogen synthetase and glycogen phosphorylase activities. Cells were exposed to theophylline and glucose for 17 hr under aerobic conditions, washed, resuspended in ice cold buffer (0.25 M imidazole, mM glutathione, pH 6.8) and treated with ultrasound for 60 sec. Phosphorylase activity was assayed in the direction of glycogen synthesis by measuring phosphate released from glucose-1-phosphate. Glycogen synthetase activity was assayed by measuring the rate of incorporation of ^{14}C-glucose from uridine-5'-diphosphoglucose-^{14}C (U) into glycogen. Activities are expressed as µM/min·mg protein. Three separate experiments are shown: Experiment I, O———O, 11.6 mM glucose; Experiment II, ■-·-·-■, no glucose; Experiment III, ▲-·-·-▲, no glucose.

synthetase and phosphorylase activities of *Tetrahymena*, and that the effect depended on culture conditions. When theophylline and glucose were added to aerobically grown cultures and the activities assayed 17 hr later, glycogen synthetase activity increased markedly (Fig. 5). About half of the increase was attained at 0.4 mM theophylline, where growth inhibition was small. The large increase in glycogen synthetase activity was accompanied by a loss of up to half of the phosphorylase activity. In cultures to which glucose was not added, low concentrations of theophylline also increased the glycogen synthetase activity, but the increase was prevented when higher concentrations of theophylline (>0.6 mM) were used. In these cultures, phosphorylase activity increased with increasing theophylline concentration (Fig. 5).

Glycogen synthetase activity was increased by growth in the presence of glucose even without theophylline addition (Fig. 5). In growing cultures, glucose addition usually increased the glycogen synthetase activity from 2- to 3-fold, but in aerobic stationary phase cultures, where glycogen synthetase activity drops to low levels, glucose addition frequently increased the activity more than 5-fold. No consistent effect of glucose on the level of phosphorylase activity was observed.

At the concentrations shown in Table 2 neither DCI nor reserpine, singly or together, significantly changed the glycogen synthetase activity of cells grown in the presence of glucose, but each drug increased the phosphorylase activity, and the effect of both drugs together was approximately additive. At higher concentrations, reserpine also decreased the glycogen synthetase activity. Triiodothyronine had no effect on the phosphorylase activity, but decreased the glycogen synthetase activity, even in cultures supplemented with glucose (Table 3). As might be expected, triiodothyronine was more potent in decreasing the glycogen synthetase activity of cells growing without glucose.

The effects of reserpine, DCI, triiodothyronine and theophylline on glycogen content and on the activities of isocitrate lyase, glycogen synthetase, and glycogen phosphorylase are summarized in Table 4. It can be

TABLE 2

Effect of reserpine and of DCI on glycogen synthetase and phosphorylase activities of *Tetrahymena*

Drugs	(M X 10^{-4})	Synthetase activity $\frac{m\mu M}{mg\ prot.\times min}$	Phosphorylase activity $\frac{m\mu M}{mg\ prot.\times min}$
None		6.7	29
DCI	1.1	6.4	37
Reserpine	0.14	6.3	45
Reserpine + DCI	0.14 + 1.1	6.5	52

Glycogen synthetase and glycogen phosphorylase activities were measured in separate experiments. Cells were grown as described in the legend to Table 1. In each experiment 11.6 mM glucose was added together with the drugs shown about 17 hr before the cells were collected for assay.

TABLE 3

Effect of triiodothyronine on glycogen synthetase and phosphorylase activities of *Tetrahymena*

Triiodo-thyronine M X 10^{-5}	Glucose mM	Glycogen synthetase $\frac{m\mu M}{mg\ prot.\times min}$	Glycogen phosphorylase $\frac{m\mu M}{mg\ prot.\times min}$
0	0	3.0	38
1.1	0	0.9	39
0	16	5.8	41
1.1	16	3.0	41

Cells were grown as described in the legend to Table 2, except for the concentration of glucose, which is as shown.

TABLE 4

Effects of drugs on glycogen content and on isocitrate lyase, glycogen synthetase and glycogen phosphorylase

Drug	Glycogen content	Isocitrate lyase	Glycogen phosphorylase	Glycogen synthetase
Reserpine	↓	↓	↑	0
DCI	↓	↓	↑	0
T_3	↓	↓	0	↓
Theophylline (in presence of glucose)	↑	↑	↓	↑

seen that the changes in enzyme activities caused by these drugs, though different for each drug, are in each case consistent with the observed effect on glycogen content. Thus reserpine and DCI, which reduced cell glycogen content, reduced the isocitrate lyase activity and raised the phosphorylase activity. Triiodothyronine, which also reduced cell glycogen content, decreased isocitrate lyase activity and glycogen synthetase activity. Theophylline, which increased glycogen content, increased isocitrate lyase activity and glycogen synthetase activity, and (for cells growing with glucose present) decreased phosphorylase activity. It should not, however, be assumed that other adrenergic drugs will also produce a coordinated metabolic response. When cells just entering stationary phase under aerobic conditions were exposed to desmethylimipramine for 17 hr, glycogen content decreased but glycogen synthetase activity increased (Blum, to be published).

Properties of the glycogen phosphorylase of *Tetrahymena*. The changes in the activities of glycogen synthetase and glycogen phosphorylase activities in response to the above mentioned drugs were consistent with the view that *Tetrahymena* might have a primitive adrenergic metabolic control system (Blum, 1967). These observations naturally raised questions about the phylogenetic origins of the intricate phosphorylase b to a and synthetase I to D conversions. It can safely be assumed that the complex series of events by which epinephrine increases adenyl cyclase activity in muscle, thereby increasing cyclic AMP concentration and activating phosphorylase kinase, which in turn catalyzes the conversion of phosphorylase b to the physiologically more active a form is the result of a long evolutionary process, and a similar comment applies to the regulation of glycogen synthetase activity. An obvious question is whether any part of this system arose at the single cell stage, or whether it arose only in response to the needs of a more complex metazoan organism. We therefore decided to study these enzymes, and have begun with the phosphorylase (Kahn and Blum, to be published).

Fractionation of acetone powders of *Tetrahymena* by ammonium sulfate, followed by ion exchange chromatography on DEAE-cellulose resulted in a 20-fold purification of the phosphorylase activity. The most interesting properties of the enzyme are summarized in Fig. 6. It can be seen that AMP had no effect on the activity of the partially purified phosphorylase. Indeed, no effect of AMP on activity was observed at any stage of purification. Incubation of the enzyme with either EDTA or ATP resulted in destruction of activity. Of other phosphorylases studied, only that from the algae *Oscillatoria princeps* is known to be inhibited by EDTA, and in this enzyme Mn^{++} reversed the inhibition (Fredrick, 1960). We have not, however, been able to reverse the EDTA or ATP caused loss of activity with metal ions or other procedures. Although the loss of activity appears to be irreversible, if AMP was added to the enzyme before either EDTA or AMP were added, protection was observed.

Since AMP did not affect the activity of the phosphorylase, another approach was needed to check on the possible presence of two forms of this enzyme in *Tetrahymena*. A clue was provided by the observation that when acetone powders prepared from cells grown in proteose-peptone medium were purified by chromatography

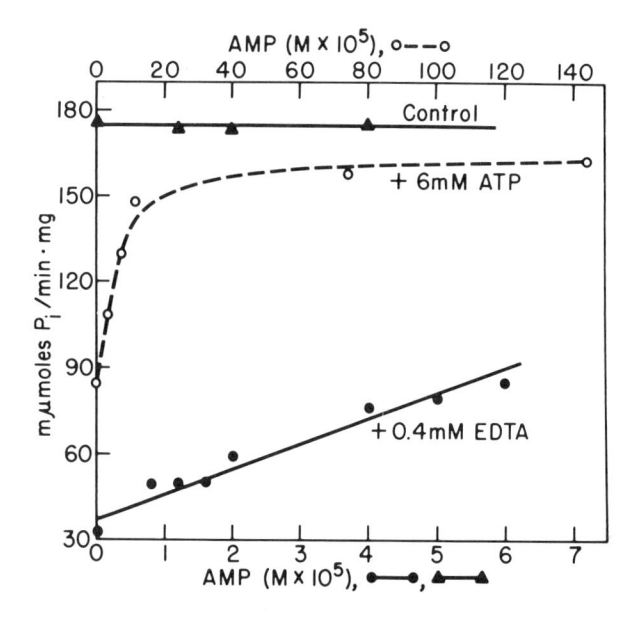

Figure 6. Protection by AMP against inhibitory effects of EDTA and of ATP on *Tetrahymena* phosphorylase. Reactions were initiated by adding a partially purified phosphorylase preparation to reaction mixtures containing EDTA or ATP as indicated and AMP as shown on the abscissae.

on DEAE, the phosphorylase activity frequently did not survive passage through the column, whereas no loss of activity resulted from this chromatographic step applied to acetone powders of cells grown in media supplemented with glucose. We therefore prepared acetone powders from cells grown with and without glucose supplementation and extracted the powders and performed ammonium sulfate fractionation and chromatography on paired DEAE cellulose columns as soon as possible after extraction. The fractions with peak activity were tested for their response to ATP and EDTA. It was found that 0.02 mM EDTA or 3.5 mM ATP completely inactivated the phosphorylase prepared from cells grown in the absence of glucose, but over 25% of the activity of the enzyme obtained from cells grown with glucose resisted inactivation by EDTA and by ATP (Fig. 7).

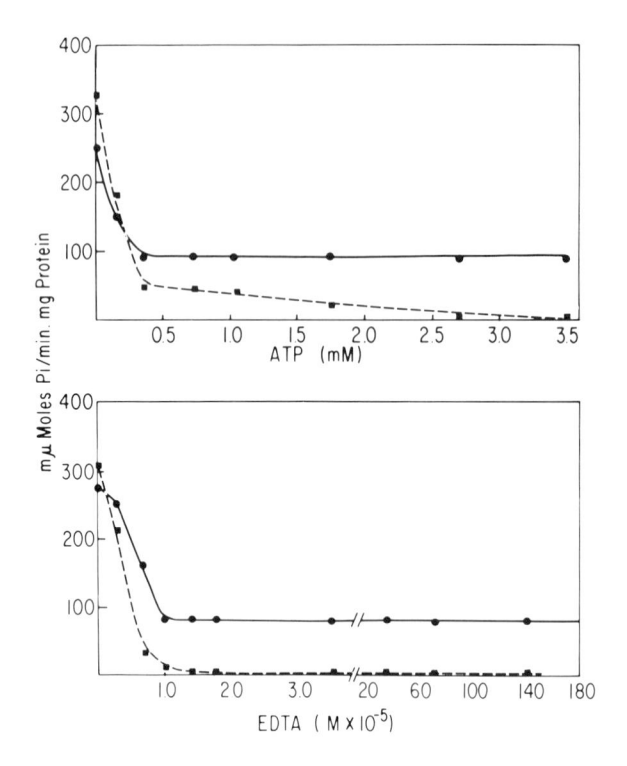

Figure 7. Effects of ATP and EDTA on phosphorylase from cells grown with and without glucose. Acetone powders were prepared from cultures grown in the absence (■----■) or presence (●——●) of glucose in the growth medium. After ammonium sulfate fractionation the enzymes were purified by DEAE-cellulose chromatography and the peak fractions were dialyzed overnight against 0.01 mM Tris-HCl, mM β-mercaptoethanol, pH 7.4. Assays were performed at 24° and initiated by adding aliquots of dialyzed enzyme to reaction mixtures containing ATP or EDTA at the concentrations shown.

Another indication of the presence of different forms of phosphorylase was obtained from experiments in which the partially purified enzymes from cells grown with and without glucose supplementation were subjected to heat denaturation in the presence and absence of AMP. It was found that the phosphorylase prepared from cells grown with glucose was more resistant to heat denaturation than the enzyme prepared from cells grown in the absence of glucose (Fig. 8). For both enzymes, however, prior addition of AMP prevented the loss of activity due to heating, thus providing further evidence for specific interaction of AMP with phosphorylase of *Tetrahymena*.

While the differential sensitivity of phosphorylase prepared from cells grown with and without glucose to ATP, EDTA, and heating could be an artifact of our preparative techniques (e.g. more proteolysis in extracts of cells grown without glucose) it is clear that *Tetrahymena* may have two kinds of phosphorylase activity. If so, much remains to be done to establish the differences between the two forms and to ascertain whether epinephrine, serotonin, or any of the drugs used in this study alter their amounts in a physiologically meaningful way.

General Considerations

In procaryotic cells such as *E. coli* the major mechanisms by which metabolism is adjusted to the changing environment are induction, repression, and the allosteric control of enzyme activities. In the higher metazoa, these control mechanisms are modulated and adjusted to the needs of the whole organism by intercellular regulators, both neural and hormonal. Except for the work of Mansour (reviewed in this volume), who found that serotonin regulated glycolysis in *Fasciola hepatica*, little is known about the steps leading to the formation of the enzyme systems involved in the fine control of glycogen metabolism in muscle. For reasons already mentioned, it was reasonable to suppose that *Tetrahymena* might possess elements of an adrenergic or serontoninergic metabolic control system, and the findings summarized above are consistent with this view. According to this hypothesis the metabolic effects of the adrenergic drugs on *Tetrahymena* result from an intracellular imbalance in the amount or location of the biogenic amines and consequent changes in

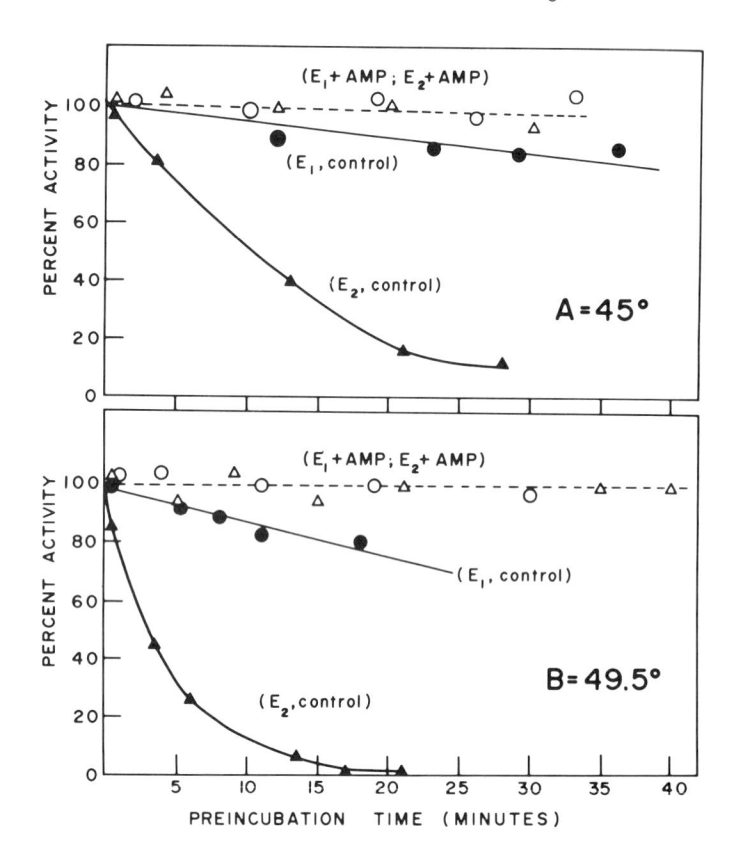

Figure 8. Heat stability of phosphorylase prepared from cells grown in the presence or absence of glucose. Phosphorylase was prepared as described in the legend to Fig. 7 from cells grown in the presence (E_1) or absence (E_2) of glucose. The purified phosphorylase (E_1, 1.8 mg protein; E_2, 1.35 mg protein) in 2.4 ml of 5 mM Tris-HCl, 0.5 mM β-mercaptoethanol, pH 7.4, with 25 mM AMP (o----o, △----△) or without AMP (●----●, ▲----▲) were incubated at 45.0°, as shown. At the times indicated on the abscissa aliquots were removed and phosphorylase activity was assayed at 24° in standard assay mixtures.

the states of enzymes such as isocitrate lyase, glycogen synthetase, or phosphorylase. Hogg and Kornberg (1963) observed that conversion of fat into glycogen by *Tetrahymena* required not only the presence of isocitrate lyase but also its incorporation into particles (now known to be peroxisomes). Evidence for a change in the state of enzymes involved in gluconeogenesis in *Tetrahymena* is also suggested by Levy's (1967) observation that neither puromycin nor actinomycin D prevented the rapid increase in gluconeogenesis which occurs when shaken cultures are transferred to static conditions.

Alternatively, one could assume that enzyme levels in *Tetrahymena* are determined primarily by changes in the level of protein synthesis, and that the biogenic amines control the transcription or translation of genetic information. Direct evidence for such a mechanism has been found by Miyamoto, *et al.* (1969), who showed that cyclic AMP increased the rate of histone phosphorylation catalyzed by an enzyme from rat brain, and by Pastan and Perlman (1969) who reported that cyclic AMP stimulated the rate of tryptophanase synthesis in *E. coli*, probably by an action at the level of the polysome. There is some evidence supporting a transcription-translation mechanism in *Tetrahymena*. Cells exposed to theophylline plus glucose require at least two hours before there is an appreciable rise in glycogen synthetase, a time long enough for protein synthesis to have occurred. In work to be reported elsewhere we have found that desmethylimipramine causes a large increase in the glycogen synthetase activity of early stationary phase *Tetrahymena* cells, and that this increase can be prevented by actinomycin D.

In view of the multiplicity of action of many drugs it could be argued that the effects of adrenergic drugs on *Tetrahymena* result from mechanisms not related to amine metabolism. Balzer, *et al.* (1968), for example, have shown that several of the drugs used in our studies bind to the membranes of the sarcoplasmic reticulum and inhibit the rate of calcium uptake by the vesicles and the calcium-induced increase in ATPase activity. While our failure so far to find reproducible effects of added serotonin or epinephrine on the metabolism or growth of *Tetrahymena* can easily be rationalized as being due to inadequate penetration into the cell, it is clear that such effects must be found if we are to gain a more detailed understanding of metabolic control processes in *Tetrahymena*.

Finally, it is worth stressing that *Tetrahymena* is not the only protozoan that contains serotonin and the catecholamines. In their original papers Janakidevi, et al. (1966a, 1966b) reported that the trypanosomatid *Crithidia fasciculata* also contained these amines. It was already known that serotonin and epinephrine played a role in the growth of *Crithidia* (Kidder and Dewey, 1963) and we therefore investigated the effects of adrenergic drugs on the growth of *Crithidia*. We found (Blum, 1969) that growth of *Crithidia* was inhibited by several drugs, but the pattern of sensitivity was different than for *Tetrahymena*. Thus growth of *Tetrahymena* was completely inhibited at 0.33 mM dibenzyline, but growth of *Crithidia* was only partially inhibited. The situation was reversed with respect to DCI, which was lethal for *Crithidia* at 0.2 mM but scarcely inhibited growth of *Tetrahymena*. Similarly, *Crithidia* was insensitive to reserpine whereas prenylamine inhibited growth of both organisms. These data further suggest that endogenous serotonin and epinephrine are involved in the control of growth (and, presumably, of metabolism) in this cell. If so, functional adrenergic and/or serotoninergic systems originated early in evolution, either in the ancestors of the ciliates and trypanosomatids or independently in each of these lines at a later stage of evolution. In either case many of the pharmacologic sensitivities found in mammalian cells are already present at the single cell stage of development. Since these two protozoa differ markedly in their pattern of drug sensitivities, they provide convenient tools for the study of intracellular metabolic control systems and, perhaps, for the study of certain drug receptor sites. There is no reason to believe that these are the only two protozoa with these properties, and it seems probable that further studies of the protozoa may reveal new insights into the origins of metabolic control systems.

REFERENCES

Balzer, H., M. Makinose and W. Hasselbach. 1968. *Naunyn-Schmiedebergs Arch. Pharmak. u. exp. Path.* 260: 444.
Barber, A.A., W.W. Harris and G.M. Padilla. 1965. *J. Cell Biol.* 27: 281.
Blum, J.J. 1967. *Proc. Natl. Acad. Sci. U.S.* 58: 81.
Blum, J.J. 1968. *Molec. Pharmacol.* 4: 247.
Blum, J.J. 1969. *J. Protozool.* 16: 317.
Blum, J.J. 1970. *Arch. Biochem. Biophys.* In press.
Blum, J.J., N. Kirshner and J. Utley. 1966. *Molec. Pharmacol.* 2: 606.
Brana, H. and F. Chytil. 1966. *Folia Microbiol.* 11: 43.
Butcher, R.W. and E.W. Sutherland. 1962. *J. Biol. Chem.* 237: 1244.
Cheung, W.Y. and L. Salganicoff. 1967. *Nature* 214: 90.
Cook, D.E., N.I. Rangaraj, N. Best and D.R. Wilken. 1968. *Arch. Biochem. Biophys.* 127: 27.
de Duve, C. 1969. *Proc. Roy. Soc.* Series B. 173: 71.
de Duve, C. and P.M. Baudhuin. 1966. *Physiol. Rev.* 46: 323.
Fredrick, J.F. 1960. *Ann. N.Y. Acad. Sci.* 88: 385.
Genghof, D.S. 1949. *Arch. Biochem. Biophys.* 23: 85.
Hadzi, J. 1963. *The Evolution of the Metazoa.* (N.Y.: The Macmillan Co.).
Haugaard, N. and M.E. Hess. 1966. *Pharmacol. Rev.* 18: 197.
Helmreich, E. and C.F. Cori. 1966. *Pharmacol. Rev.* 18: 189.
Hogg, J.F. and A.M. Elliott. 1951. *J. Biol. Chem.* 192: 131.
Hogg, J.F. and H.L. Kornberg. 1963. *Biochem. J.* 86: 462.
Holz, C.G. 1966. *J. Protozool.* 13: 2.
Iwata, H., K. Kariya and S. Fujimoto. 1967. *Jap. J. Pharmacol.* 17: 328.
Janakidevi, K., V.C. Dewey and G.W. Kidder. 1966a. *Arch. Biochem. Biophys.* 113: 758.
Janakidevi, K., V.C. Dewey and G.W. Kidder. 1966b. *J. Biol. Chem.* 241: 2516.
Kidder, G.W. 1967. *Chemical Zoology*, Vol. I, G.W. Kidder, ed. (New York: Academic Press), pp. 93.
Kidder, G.W. and V.C. Dewey. 1947. *Proc. Natl. Acad.*

Sci. U.S. 33: 95.
Kidder, G.W. and V.C. Dewey. 1963. *Biochem. Biophys. Res. Commun.* 12: 280.
Levey, G.S. and S.E. Epstein. 1968. *Biochem. Biophys. Res. Commun.* 33: 990.
Levy, M.R. 1967. *J. Cell. Physiol.* 69: 247.
Levy, M.R. and A.E. Hunt. 1967. *J. Cell Biol.* 34: 911.
Manners, D.J. and J.F. Ryley. 1952. *Biochem. J.* 52: 480.
Miyamoto, E., J.F. Kuo and P. Greengard. 1969. *Science* 165: 65.
Muller, M., J.F. Hogg and C. de Duve. 1968. *J. Biol. Chem.* 243: 5385.
Murad, F., Y.M. Chi, T.W. Rall and E.W. Sutherland. 1962. *J. Biol. Chem.* 237: 1233.
Pastan, I. and R.L. Perlman. 1969. *J. Biol. Chem.* 244: 2226.
Ryley, J.F. 1952. *Biochem. J.* 52: 483.
Ryley, J.F. 1967. *Chemical Zoology*, Vol. I, G.W. Kidder, ed. (New York: Academic Press), pp. 84
Sagan, L. 1967. *J. Theoret. Biol.* 14: 225.
Shrago, E., W. Brech and K. Templeton. 1967. *J. Biol. Chem.* 242: 4060.
Weber, A. 1968. *J. Gen. Physiol.* 52: 760.
Whittaker, R.H. 1969. *Science* 163: 150.
Wingo, W.J. and L.E. Cameron. 1952. *Texas Rept. Biol. Med.* 10: 1075.
Wu, C. and J.F. Hogg. 1956. *Arch. Biochem. Biophys.* 62: 70.

Biogenic Amines as Metabolic Regulators in Invertebrates

Tag E. Mansour

Dept. of Pharmacology
Stanford University
School of Medicine
Stanford, California

Control of different metabolic processes in living organisms occurs through two main mechanisms. The first is regulation of the activity of rate-limiting enzymes. The findings of Monod and his colleagues (1965) have clearly shown that there are complex regulatory systems which control the activity of these enzymes. Therefore, in addition to their function as catalysts of certain biochemical reactions they act as regulators of different metabolic pathways. A metabolite, which in many cases is neither the immediate substrate nor the direct product of the enzyme, can activate or inhibit its catalytic activity through changes

---

*Part of the author's research reviewed here was supported by U.S.P.H.S. Research Grant AI04214, National Institute of Allergy and Infectious Diseases, and PHS Research Career Development Award GM 3848, Division of General Medical Sciences.

in its kinetic properties. The mechanism by which these enzymes are regulated has been given the name "allosteric regulation." A classical example of such regulation can be seen in the glycogen phosphorylase system from mammalian muscle. In their extensive investigations of this enzyme during the past 30 years, Cori and his associates have shown that an inactive form of phosphorylase can be converted to an active form by AMP. Such activation does not involve a change in the degree of aggregation of the enzyme but merely a change in its kinetics. The dissociation constant of the enzyme is markedly reduced, which allows the enzyme to be active at physiological levels of its substrate. In this particular example, AMP, which is not involved in the reaction at all, is considered to be an allosteric effector. This type of regulation has been shown to occur in lower forms of life, such as bacteria, as well as in higher organisms such as mammalian cells. A second mechanism by which the organism can regulate its metabolic activity to meet its vital demands for survival is through the action of different hormones. In many cases the effect of these agents has been traced to action through a complex enzyme system. A good example of such a mechanism can also be seen in the glycogen phosphorylase system of skeletal muscle. Sutherland and co-workers (Sutherland and Rall, 1960; Robison, *et al.*, 1968) have shown that epinephrine increases the synthesis of cyclic AMP by a particulate fraction. The cyclic adenylic nucleotide in turn activates the phosphorylase kinase through a complex enzyme system, with the outcome being the conversion of the inactive form to the active form of the enzyme. Activation of the enzyme here involves phosphorylation and aggregation of the enzyme to the active form. Control of metabolism by hormones appears to occur only in higher forms of life. While much progress has been made in understanding the hormonal systems concerned with reproduction, metamorphosis and growth in invertebrates, very little is known about hormonal systems regulating metabolism in these organisms. The scanty reports in the literature indicate that mammalian hormones are either inactive or have little effect on the metabolism of many invertebrates.

This argues for the possibility that invertebrate hormones are different from those of mammals.

Starting from this point of view, I shall review some of the experiments which were carried out in our laboratory to investigate the role of serotonin (5-hydroxytryptamine) in regulating the carbohydrate metabolism in the trematode, the liver fluke, *Fasciola hepatica*.

Effect of Serontonin on Metabolism of the Liver Fluke

Occurrence of synthesis of serotonin. The first evidence for the possible presence of serotonin in the liver fluke came from the finding that serotonin, lysergic acid diethylamide (LSD) and related indolalkylamines stimulated rhythmical movement of these organisms at low concentrations (Mansour, 1957). None of the catecholamines tested had any effect. Subsequently, the presence of serotonin was demonstrated in extracts from flukes (Mansour, *et al.*, 1957; and Mansour and Stone, 1969). Further evidence for the presence of an active system for the synthesis of serotonin by intact flukes was obtained by incubating intact flukes with 5-hydroxytryptophan (5-HTP) for several periods of time. While control flukes contained 231 ± 38 ng/g wet weight, in flukes incubated with 10^{-3} M of 5-HTP for 4 hr the average level was increased as much as eight times (Mansour and Stone, 1969). Such an increase in the level of serotonin was found to be dependent on the concentration of the precursor, 5-HTP, and on time of incubation (Fig. 1).

Effect of amines on the carbohydrate metabolism of the liver fluke. Early investigation (Mansour, 1959b) on the carbohydrate metabolism of the liver fluke showed that the organisms are predominantly anaerobes. The organisms metabolize either glucose or glycogen at a high rate. When glucose was not included in the culture medium the parasites metabolized their own glycogen. Production of volatile fatty acids and CO_2 could account for all the carbohydrate metabolized. Only 4 to 9% of the metabolized carbohydrate could be accounted

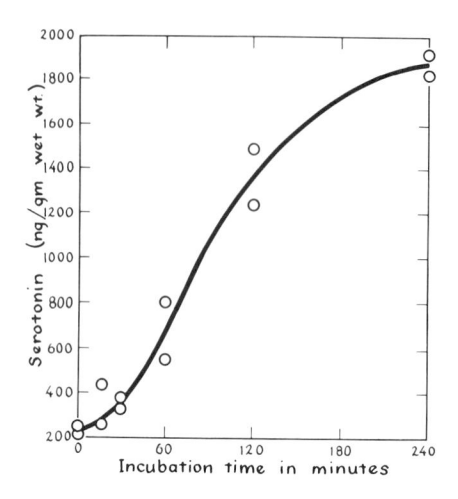

Figure 1. Levels of serotonin in the liver flukes after incubation with 10^{-3} M 5-hydroxytryptophan for different times. (From Mansour and Stone, 1969).

for as lactic acid.

The effect of several amines and other related compounds on the carbohydrate metabolism of these organisms under anaerobic conditions was tested (Table 1) (Mansour, 1959a; Mansour and Stone, 1969). In the absence of glucose, glycogen was utilized and serotonin caused an increase in its utilization. When glucose was present in the culture medium serotonin caused an increase in glucose uptake. The increase in carbohydrate utilization was always accompanied by a marked increase in lactic acid production (6-10 fold) while the production of volatile fatty acids was either not changed significantly or was increased slightly. The precursor of serotonin, 5-hydroxytryptophan, as well as lysergic acid diethylamide had a similar effect on the metabolism of the liver fluke. In contrast to serotonin, epinephrine did not cause a major change in the carbohydrate metabolism of these organisms when tested under the same conditions (Table 1).

The fact that serotonin increases the muscular activity of the liver fluke, in addition to its marked effect on the metabolism, made it imperative to find

TABLE 1

Effect of amines and other related compounds on the carbohydrate metabolism of *Fasciola hepatica*

Medium	Additions	Glucose uptake	Glycogen utilized	Lactic acid	Volatile fatty acids
With glucose	None	60	0	16	119
With glucose	5 X 10^{-4} M 5-HT	101	0	99	153
Without glucose	None		48	3	168
Without glucose	5 X 10^{-4} M 5-HT		77	30	220
With glucose	None	73		27	128
With glucose	10^{-3} M l-epinephrine	80		29	147
Without glucose	None		45	3	100
Without glucose	10^{-3} M l-epinephrine		55	3	110
With glucose	None	92		48	
With glucose	10^{-3} M 5-HTP	135		129	
With glucose	None	92		32	
With glucose	10^{-6} M LSD	124		63	

Liver flukes were incubated for 3 hr in a saline medium as described before. Medium contained 0.01 M glucose where indicated. All results are expressed as μmole/g wet weight. The following abbreviations are used: 5-HT, serotonin; 5-HTP, 5-hydroxytryptophan; LSD, lysergic acid diethylamide (from data of Mansour, 1959; Mansour and Stone, 1969).

out whether the action of the amine on the metabolism is primary or secondary to its effect on rhythmical movement. Subsequent experiments from our laboratory (Mansour, 1962) showed that stimulation of glycolysis by serotonin observed in intact organisms can also be demonstrated in cell-free homogenates. Thus the increase in glycolysis caused by serotonin is not merely due to an increase in rhythmical movement, but is due to an effect on one or more of the enzymes involved in glycolysis.

Effect of serotonin on glycolytic enzymes. The question then arose: What is the enzyme (or enzymes) which is responsible for such an increase in glycolysis? Glycogen phosphorylase was the first enzyme to be tested because of its involvement in the epinephrine effect on glycogenolysis in mammals. Assays of this enzyme in homogenates from control flukes as well as from flukes pre-incubated with serotonin showed that the enzyme was significantly increased after incubation with the indolamine (Mansour, *et al*., 1960). On the other hand, epinephrine did not have an effect on the fluke enzyme.

An increase in the activity of an enzyme (or enzymes) other than phosphorylase in the glycolytic scheme was predicted from the fact that lactic acid production was markedly increased by serotonin even in flukes which were metabolizing glucose instead of glycogen. Furthermore, in the cell-free extract system (Mansour, 1962) from serotonin-treated flukes lactic acid production was increased when any of the following substrates were used: glucose, glucose-6-P, or fructose-6-P (Table 2). On the other hand, when fructose-1,6-di-P was used as the substrate, lactic acid production was high in control experiments. Serotonin did not show its stimulatory effect on glycolysis in the presence of the hexose diphosphate (Table 2). These experiments indicated that the reaction, fructose-6-P to fructose-1,6-di-P, is rate limiting in these organisms. This was suggested by Cori (1942) in the case of the frog muscle. Serotonin then appeared to increase the activity of the enzyme catalyzing this reaction, phosphofructokinase. This was further con-

TABLE 2

Glycolysis under anaerobic conditions in homogenates from control and serotonin (5-HT) treated flukes in the presence of different substrates.

Substrate	No. of experiments	Lactic acid produced Control	5-HT treated
Glucose	4	11.3 ± 4.8*	45.3 ± 6.0*
Glucose-6-P	1	7.5	29.2
Glucose-6-P	1	4.7	30.0
Fructose-6-P	3	0	36.6 ± 3.5
Fructose-1,6-di-P	4	48.5 ± 3.9	56.3 ± 6.3

Taken from the data of Mansour (1962).
*All results are given as (μmole/g wet wt.) with standard errors.

firmed when it was found after assays of phosphofructokinase activity that the enzyme was practically inactive in control flukes and that in serotonin treated flukes the activity was very high (Mansour, 1962). Serotonin, added directly to cell-free extracts also activates the enzyme (Table 3). That an increase in glycolysis in intact organisms by serotonin could be explained by an increase in phosphofructokinase activity was further confirmed by the fact that serotonin caused a decrease in the levels of glucose-6-P and of fructose-6-P while the levels of fructose-1,6-di-P were markedly increased (Mansour, 1962). A conclusion can be drawn from these experiments that serotonin increases the activity of at least two glycolytic enzymes, namely phosphorylase and phosphofructokinase. These effects are not induced by epinephrine.

Activation of adenyl 3',5'-cyclase by serotonin. It was originally discovered by Sutherland and his group that activation of phosphorylase by epinephrine in cell-free preparations is brought about by an increase in the formation of adenosine 3',5'-phosphate (cyclic 3',5'-AMP) (Sutherland and Rall, 1958). Since in the fluke, serotonin and not epinephrine activates this en-

TABLE 3

Effect of serotonin and of cyclic-3',5'-AMP on phosphofructokinase activity in liver fluke homogenates

Additions	Phosphofructokinase activity mole fructose-1,6-di-P/g wet wt./30 min
None	4.8
Serotonin 5 X 10^{-5} M	27.8
Cyclic-3',5'-AMP (fluke)* 2 X 10^{-5} M	47.3
Cyclic-3',5'-AMP (synthetic) 2 X 10^{-5} M	46.0

Taken from the data of Mansour (1962).
*Cyclic-3',5'-AMP was synthesized by fluke particles according to the procedure of Mansour, et al. (1960).

zyme the hypothesis that serotonin is a specific activator of adenyl cyclase in the fluke was investigated. Experiments carried out in collaboration with Dr. Sutherland (Mansour, et al., 1960) demonstrated that serotonin caused a rapid and specific increase in the formation of adenosine 3',5'-phosphate by a particulate fraction from the liver fluke. Requirements for the system were similar to those of the mammalian system. The presence of a phosphodiesterase for the inactivation of cyclic AMP was also shown. None of the catecholamines had an effect on this system. Thus serotonin at that time became the fourth known chemical agent which increases the formation of the cyclic nucleotide. The others were the catecholamines, glucagon and adenocorticotropic hormone, which are active only when tested with certain mammalian tissues. Since these findings were reported the number of agents which have been found to affect the cyclase has increased to at least 14 (Robison, et al., 1968). All these are mammalian hormones and none had a reported effect on any invertebrate systems. The effect of serotonin was tested on cyclic AMP production by particles from *Schistosoma mansoni* and *Ascaris lumbricoides* under conditions which were similar to those used for the liver fluke. No increase in the formation of cyclic AMP the presence of either serotonin or epinephrine was observed (Mansour, *et al.*, 1960). These negative results might be explained by the possibility that these species accumulate cyclic AMP to an extent lower than could be detected by the assay method used. Furthermore, the conditions employed may not have been optimal for the accumulation of the nucleotide in preparations from these species.

The effect of lysergic acid diethylamide on the production of cyclic AMP from particulate fractions showed only very small, although detectable, increments in the cyclic nucleotide formation.

Effect of cyclic 3',5'-AMP on anaerobic glycolysis and on phosphofructokinase of the liver fluke. The finding that serotonin stimulates the production of cyclic AMP and that the amine increases glycolysis and phosphofructokinase activity raised the question whether the cyclic nucleotide activates both glycolysis and phosphofructokinase. Experiments on cell-free extracts from the flukes showed that indeed cyclic AMP, like serotonin, increases both glucose utilization and lactic acid production (Mansour, 1962). Furthermore, the cyclic adenylic nucleotide added directly to extracts from the flukes caused a marked increase in phosphofructokinase activity (Table 3). Thus a hypothesis similar to that explaining the action of epinephrine on glycogenolysis in the mammalian liver (Robison, *et al.*, 1968) can be put forward to explain the mechanism of action of serotonin on fluke glycolysis. According to this, serotonin stimulates glycolysis via the production of cyclic AMP. The cyclic adenylic nucleotide stimulates the activity of phosphofructokinase, a rate limiting enzyme, with the ultimate result being an increase in glycolysis. Thus, according to Sutherland's concept of the action of hormones, cyclic AMP is a second messenger, mediating the effect of serotonin in this invertebrate. The main difference between the liver fluke and mammalian systems lies in the specificity of the liver fluke adenylic cyclase to serotonin and its sensitivity to the catecholamines.

Control mechanisms involved in the regulation of phosphofructokinase. Activation of phosphofructokinase by cyclic AMP in the liver fluke appears to occur through two different mechanisms (Mansour and Mansour, 1962). First, the cyclic nucleotide activates an inactive form of phosphofructokinase. Second, cyclic AMP can activate phosphofructokinase which was inhibited by ATP through changes in the kinetics of the enzyme. The nature of cyclic AMP activation of the fluke enzyme by the two mechanisms has been the subject of recent investigations by Stone and Mansour (1967a, 1967b). A cellular fraction from the fluke containing inactive phosphofructokinase was separated by a procedure utilizing differential centrifugation. Activation of phosphofructokinase in this fraction required ATP, Mg^{++}, a polyvalent anion, and cyclic AMP. An absolute requirement for the cyclic nucleotide was shown since no other nucleotide could replace it. Serotonin activated the enzyme in this fraction only in the presence of a particulate fraction which catalyzes the synthesis of cyclic AMP. Thus, the effect of serotonin on fluke phosphofructokinase is mediated through cyclic AMP and in that respect is similar to the action of epinephrine on glycogen phosphorylase in liver and muscle from mammals. The similarity between the two systems is supported by the finding that activation of phosphofructokinase, like that of glycogen phosphorylase from skeletal muscle, involves monomer-polymer system conversion (Stone and Mansour, 1967a). The sedimentation coefficient of the inactive form of the enzyme was 5.5 S while that after full activation was 12.8 S.

The results summarized above indicate that phosphofructokinase from the liver fluke, like many other regulatory enzymes is reversibly converted from active to inactive enzyme. In addition to this form of regulation the enzyme once activated, is also endowed with another mechanism based on its allosteric kinetics. Different ligands were reported to affect the fully active enzyme through an apparent change in its conformation. ATP inhibits the activity of the enzyme while cyclic AMP, and fructose-6-P activate the ATP-inhibited enzyme. The kinetics of inhibition and activation here is of interest in understanding the mechanism of regulation of this enzyme (Stone and Mansour, 1967b). An example of the kinetics of inhibition by ATP is shown in Fig. 2. It can be seen that an in-

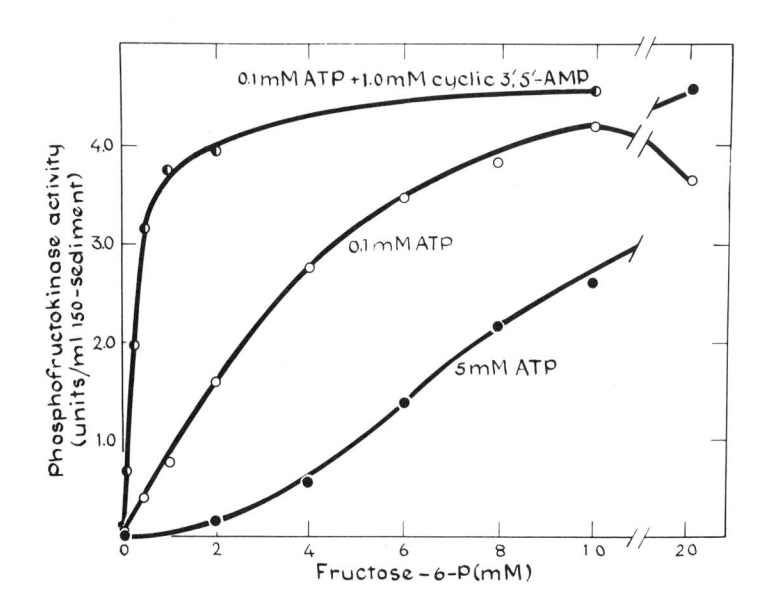

Figure 2. Effect of cyclic 3',5'-AMP and ATP on the saturation function for fructose-6-P. (From Stone and Mansour, 1967b).

crease in the concentration of fructose-6-P caused a significant increase in the concentration of ATP necessary to cause 50% inhibition. Similar results were obtained with other enzyme activators such as 5'-AMP Saturation curves for fructose-6-P were not hyperbolic in shape, indicating that the kinetics are not of the Michaelis-Menton type (Fig. 3). In the presence of cyclic AMP (an activator) the curve for fructose-6-P became closer to a hyperbolic shape with a marked decrease in K_m. In the presence of ATP (an inhibitor) the K_m for fructose-6-P increased and the curve became more sigmoidal. A summary of the effect of ATP and of

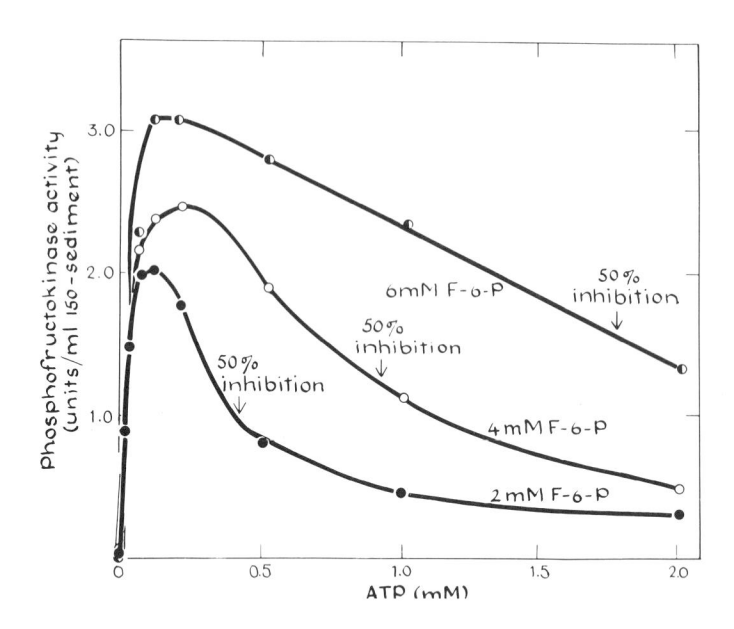

Figure 3. Effect of fructose-6-P concentration on the saturation function for ATP (From Stone and Mansour, 1967b).

cyclic AMP on the apparent dissociation constant for fructose-6-P is shown in Table 4. The substrate func-

TABLE 4

Effect of cyclic 3',5'-AMP and of ATP on fructose-6-P concentration required for half saturation for activated liver fluke phosphofructokinase

Addition to the assay mixture	ATP concentration	Concentration required for half-saturation
	(mM)	(mM)
None	0.1	2.9
1 mM cyclic 3',5'-AMP	0.1	0.3
None	5.0	8.5

Enzyme activation and assay were carried out as described before (from Stone and Mansour, 1967b).

tion of cyclic AMP was also shown to be influenced by both enzyme substrates. Fructose-6-P at high concentrations caused a decrease in the half-saturation point for cyclic AMP while ATP had the opposite effect. These kinetics have been analyzed in terms of the Monod-Changeux-Wyman model (Monod, *et al.*, 1965) for allosteric proteins. Saturation curves for fructose-6-P indicate cooperative homotropic interaction between multiple binding sites for this substrate. ATP at high concentrations increased the homotropic interactions while cyclic AMP had the opposite effect.

At present one can only speculate on the relationship between the above described kinetics for phosphofructokinase and the action of serotonin on glycolysis. Cyclic AMP can alter the kinetics of the enzyme when it is inhibited by ATP in such a way that it will catalyze the reaction in the presence of low substrate concentrations. Such an effect could be used to modulate the activity of the fully active form of the en-enzyme.

Effect of Serotonin on the Metabolism of Other Invertebrates

Much work has been carried out on the effect of serotonin on the gills of mussels. Activity of the cilia is controlled through innervation via a branch of the branchial nerve. Serotonin at an extremely low concentration (5 X 10^{-10} M) stimulates the activity of these cilia (Aiello, 1957; Gosselin, *et al.*, 1962). Serotonin or a compound very similar to it has been found in the gills and was postulated to be the natural local transmitter for the cilia. Intensity of ciliary motion approximately parallels the serotonin content in the gills. The effect of serotonin on the ciliary movement is reversible and there is a more prolonged effect by 5-hydroxytryptophan, the serotonin precursor. These observations suggest that there is a system for the synthesis and inactivation of the indolalkylamine.

The effect of exogenous serotonin and related agents was investigated by Moore and Gosselin (1962). They showed that when gills of *Modiolus demissus* were incubated in sea water containing serotonin (10^{-5} M) the rate of respiration, anaerobic glycogen utilization, and lactic acid production were stimulated. Other agents which accelerated ciliary movement such as K^+ (Gray, 1928), veratrum (Gray, 1928; Aiello, 1960), and lysergic acid diethylamide (Moore, et al., 1961) also caused an increase in gill respiration. 5-Hydroxytryptophan, following a latent period, had an effect similar to that of serotonin. Both epinephrine and norepinephrine stimulated lactic acid production by the gills, but only at concentrations 10-fold higher than that of serotonin. The increase in glycogen breakdown by serotonin, according to Gosselin, does not appear to be due to an increase in active phosphorylase or total phosphorylase activity. Although one cannot generalize here about a role for serotonin as a metabolic regulator in *Modiolus*, a point can be made about the high sensitivity of the metabolism of these organisms to serotonin when compared to the catecholamines. The obvious question is whether the effect is secondary to ciliary stimulation or whether it is a primary effect of the amine itself. The finding that serotonin stimulated glycolysis when the gills were incubated anaerobically—a condition under which ciliary movement is arrested—supports the latter view.

The effect of serotonin was tested on smooth muscle from *Mytilus*. The effect of the amine on the state of prolonged contraction of the muscle and its resistance to stretch (known as catch) was tested. Serotonin was shown to cause relaxation of the muscle, increased the spikes and stimulated its metabolism (Twarog, 1969). Nauss and Davis (1966) showed that when catch is relaxed by serotonin about 0.2 mole of phosphorylarginine per g of muscle is broken down. Prolonged exposure of the relaxed muscle to serotonin results in the release of inorganic phosphate and arginine indicating that serotonin stimulates the metbolism of the resting muscle. A gradual increase in oxygen consumption by *Mytilus* muscle in the presence of 10^{-6} M serotonin was also observed by Baguet and Gillis (1964).

The experiments of S.-Rózsa on the hearts of molluscs implicate serotonin and cyclic AMP in the excitatory effects on contractility and metabolism (S.-Rózsa and Zs.-Nagy, 1967; S.-Rózsa, 1969). Both agents were shown to have an excitatory effect at low concentration on the isolated *Lymnaea stagnalis* heart. Serotonin is regarded as a transmitter, not at the nerve endings, but in the myocardial cells. After studying the interaction between serotonin, cyclic AMP and other biologically active amines on the activity of the heart, it was concluded that the excitatory effects of serotonin are mediated through the adenine nucleotide. Evidence of the effect of the amine on the metabolism comes from a recent demonstration that serotonin, at a concentration of 10^{-6} M or higher, caused a complete conversion of inactive phosphorylase to the fully active form in the heart of the snail. A demonstration of the effect of serotonin on adenyl cyclase from the hearts of these molluscs is required before a definite conclusion can be drawn on the postulate that the effects of serotonin are mediated through cyclic AMP.

Serotonin has been implicated in the control of early embryogenesis (Buznikov, *et al.*, 1964). Concentrations of serotonin in developing sea urchin embryos were shown to follow a rhythmic pattern with peaks one hour after fertilization and then at or near the time of the first four cleavages. Whether serotonin here is involved in controlling mitotic activity or other developmental mechanisms is still not yet determined. The possibility also exists that serotonin stimulates the metabolism during these stages of development. This hypothesis is supported by the finding that the process of fertilization in the sea urchin eggs is followed by an increase in respiration (Aketa, *et al.*, 1964), and an increase in glycogen utilization (Ostrom and Lindberg, 1940). The levels of glucose-6-P were increased following fertilization.

The investigation of Kay (1963) on bioluminescence of a planktonic crustacean, *Meganyctophanes*, suggests that serotonin or a related compound may function as a natural control mechanism for this process. The organism in the presence of serotonin becomes much more responsive to topical stimulation of luminescence than do control crustaceans. The dependence of bioluminescence on ATP is now well established (McElroy and Seliger, 1963). This again raises the question of whether or not this effect of serotonin is a primary action on the metabolism.

Effect of Serotonin on Carbohydrate Metabolism of Mammals

A direct comparison between the effect of epinephrine and the effect of serotonin on mammalian carbohydrate metabolism is always handicapped by the finding that serotonin was reported to enhance the release of epinephrine from the adrenals (Hagen, 1959; Reid, 1952). Variable results were reported using relatively high amounts of serotonin. Correll, *et al.* (1952) showed an increase in plasma glucose level from an average of 157 to 269 mg percent. The presence of the adrenal medulla was not essential for the hyperglycemic effect. Columbo, *et al.* (1960) and Kobayashi, *et al.* (1960), however, attributed the hyperglycemic effect of serotonin to the release of epinephrine from the adrenal medulla. Others have found that serotonin either has no effect on blood glucose or produces hypoglycemia (Scaltrini, 1956; Weitzel, *et al.*, 1956; Mirsky, *et al.*, 1957; Sirek, 1957). A thorough study of the effect of serotonin on both perfused rat liver and on intact animals was reported by Levine, *et al.* (1964). Perfusion of the liver with a solution containing 4×10^{-4} M serotonin caused progressive glycogenolysis, stimulation of hepatic glycogen phosphorylase activity and concomitant hyperglycemia. There was, however, considerable variation in the rate of these changes, and the effect on glycogen metabolism was not related to the amount of serotonin present in the perfusate. The actions of serotonin were independent of observed vasometer effects of the amine on the liver. Furthermore, the glycogenolytic actions were blocked by the infusion of a potent serotonin antagonist, 1-methyl-(methylergonovine). These observations suggest that the effect of serotonin is a direct one and not mediated through the release of epinephrine from the adrenal medulla.

In vivo experiments in rats by the same group showed a consistent hyperglycemic effect when serotonin was given subcutaneously or intraperitoneally, but not after intravenous injection. Levine, *et al.* (1964) also clearly showed that in the liver perfusion experiments, epinephrine was much more potent than serotonin.

Control mechanisms in the regulation of mammalian phosphofructokinase. The role of phosphofructokinase as a rate-limiting enzyme in glycolysis was indicated first by Cori (1942) in the frog muscle and was subsequently confirmed in different mammalian cells. A clear understanding of the regulation of phosphofructokinase was achieved in our laboratory upon the purification and crystallization of the enzyme from sheep heart (Mansour, 1963; Mansour, et al., 1966; Mansour and Ahlfors, 1968). Studies on the purified enzyme from the sheep as well as from other sources (cf. Stadtman, 1967) demonstrated striking similarities between the allosteric properties of the mammalian and the liver fluke enzyme. ATP inhibits, while AMP, cyclic AMP and ADP are activators. The saturation curve for fructose-6-P was sigmoidal, indicating cooperative kinetics. ATP increases the sigmoidal kinetics while the activators have the opposite effect. One of the main distinguishing characteristics of these enzymes is that they have a site, different from the catalytic site, where enzyme modifiers act. The existence of such an allosteric site has recently been confirmed in our laboratory by desensitizing the enzyme to ATP inhibition without destroying the active site (Ahlfors and Mansour, 1969). This was achieved through photo-oxidation of the enzyme in the presence of methylene blue. The enzyme, after photo-oxidation, besides being insensitive to ATP, did not show cooperative kinetics. Maximal enzyme activity was only slightly affected by photo-oxidation.

Studies on the physical properties of the heart enzyme have shown that the enzyme can reversibly dissociate from the aggregated active enzyme to a dissociated inactive form. Dissociation of the enzyme is accelerated at an alkaline pH and upon the addition of ATP, ADP, cyclic AMP, fructose-6-P or fructose-1,6-di-P. The results obtained so far point to close similarities between the properties of the enzyme from many lower organisms (cf. Stadtman, 1967), including the liver fluke, and those of the mammalian enzyme.

Summary and Conclusions

Much of the work discussed above points to the possible importance of serotonin or a related indolalkylamine as a metabolic regulator in the carbohydrate metabolism of the liver fluke, *Fasciola hepatica*. The amine in the fluke appears to have the same functions that epinephrine has in mammals. These include: the formation of cyclic AMP, phosphorylase activation, and stimulation of glycogenolysis and of glycolysis. The foregoing discussion of the literature indicates that serotonin also is a metabolic regulator in other invertebrates. Furthermore, evidence has been accumulating that serotonin performs other physiological functions in invertebrates similar to those performed by epinephrine in higher animals. For example, serotonin stimulates the hearts of molluscs and crustaceans (Erspamer, 1966). In some of these species, the stimulation of the heart can be achieved at much lower concentrations than with catecholamines (Welsh, 1953). These findings strengthen the case for a hormonal control by serotonin in these lower organisms. At a much higher phylogenetic level, serotonin effects on the carbohydrate metabolism are either much inferior to those of the catecholamines or may not have been found.

On a molecular level, serotonin appears to regulate the carbohydrate metabolism through cyclic AMP. Thus, both serotonin and epinephrine have the same second messenger which is responsible for their regulatory effects. Allosteric control by cyclic AMP and other nucleotides of phosphofructokinase from a wide variety of organisms including the liver fluke and mammals reveals striking similarities. Thus, while the hormonal control of glycolysis in the liver fluke could be different from that in mammals, the allosteric control inherent in the enzymes appears to be basically the same at both phylogenetic levels.

REFERENCES

Ahlfors, C.E. and T.E. Mansour. 1969. *J. Biol. Chem.* 244: 1247.

Aiello, E.L. 1957. *Biol. Bull. Woods Hole* 113: 325.
Aiello, E.L. 1960. *Physiol. Zool.* 33: 120.
Aketa, K., R. Bianchetti, E. Marre and A. Monroy. 1964. *Biochim. Biophys. Acta* 86: 211.
Baguet, F. and J.M. Gillis. 1964. *Arch. Intern. Physiol. Biochim.* 72: 351.
Buznikov, G.A., I.V. Chudakova and N.D. Zvezdina. 1964. *J. Embryol. Exptl. Morphol.* 12: 563.
Columbo, J.P., G. Weber, G. Guidotti, D. Kanameishi and P.P. Foa. 1960. *Endocrinology* 67: 693.
Cori, C.F. 1942. In *A Symposium on Respiratory Enzymes*, (Madison, Wisconsin: University of Wisconsin Press), p. 175,
Correll, J.T., L.F. Lyth, S. Long and J.C. Vanderpoel. 1952. *Amer. J. Physiol.* 169: 537.
Erspamer, V. 1966. *Handbüch der Experimentellen Pharmacologie*. O. Eichler and A. Farah, eds. Vol. XIX, p. 132.
Gosselin, R.E., K.G. Moore and A. Milton. 1962. *J. Gen. Physiol.* 46: 277.
Gray, J. 1928. In *Ciliary Movement*. (London, New York: Cambridge University Press).
Hagen, P. 1959. *Pharmacol. Rev.* 11: 361.
Kay, R.H. 1963. *J. Physiol.* (London) 165: 63P.
Kobayashi, B., M. Ui and Y. Warashima. 1960. *Endocr. Jap.* 7: 225.
Levine, R.A., L.S. Pesch, G. Klatskin and N.J. Giarman. 1964. *J. Clin. Invest.* 43: 797.
Mansour, T.E. 1957. *Brit. J. Pharmacol.* 12: 406.
Mansour, T.E. 1959a. *J. Pharmacol. Exptl. Therap.* 126: 212.
Mansour, T.E. 1959b. *Biochim. Biophys. Acta* 34: 456.
Mansour, T.E. 1962. *J. Pharmacol. Exptl. Therap.* 135: 94.
Mansour, T.E. 1963. *J. Biol. Chem.* 238: 2285.
Mansour, T.E., A.D. Lago and J.L. Hawkins. 1957. *Federation Proc.* 16: 319.
Mansour, T.E., E.W. Sutherland, T.W. Rall and E. Bueding. 1960. *J. Biol. Chem.* 235: 466.
Mansour, T.E. and J.M. Mansour. 1962. *J. Biol. Chem.* 237: 629.
Mansour, T.E., N. Wakid and H.M. Sprouse. 1966. *J. Biol. Chem.* 241: 1512.

Mansour, T.E. and C.E. Ahlfors. 1968. *J. Biol. Chem.* 243: 2523.
Mansour, T.E. and D.B. Stone. 1969. *Biochem. Pharmacol.* In Press.
McElroy, W.D. and H.H. Seliger. 1963. *Advan. Enzymol.* 25: 119.
Mirsky, I.A., G. Perisutti and R. Jinks. 1957. *Endocrinology* 60: 318.
Monod, J., J. Wyman and J. Changeux. 1965. *Mol. Biol.* 12: 88.
Moore, K.E., A.S. Milton and R.E. Gosselin. 1961. *Brit. J. Pharmacol.* 17: 278.
Moore, K.E. and R.E. Gosselin. 1962. *J. Pharmacol. Exptl. Therap.* 138: 145.
Nauss, K.M. and R.E. Davies. 1966. *Biochem. Z.* 345: 173.
Örström, A. and O. Lindberg. 1940. *Enzymologia* 8: 367.
Reid, G. 1952. *J. Physiol.* 118: 435.
Robison, G.A., R.W. Butcher and E.W. Sutherland. 1969. *Ann. Rev. Biochem.* 37: 149.
Scaltrini, G.C. 1956. *Haematologica* 41: 681.
Sirek, A. 1957. *Nature* (London) 179: 376.
S.-Rózsa, K. and I. Zs.-Nagy. 1967. *Comp. Biochem. Physiol.* 23: 351.
S.-Rózsa, K. 1969. *Life Sciences* 8: 229.
Stadtman, E.R. 1967. *Advan. Enzymol.* 28: 41.
Stone, D.B. and T.E. Mansour. 1967a. *Mol. Pharmacol.* 3: 161.
Stone, D.B. and T.E. Mansour. 1967b. *Mol. Pharmacol.* 3: 177.
Sutherland, E.W. and T.W. Rall. 1958. *J. Biol. Chem.* 232: 1065.
Sutherland, E.W. and T.W. Rall. 1960. *Pharmacol. Rev.* 12: 265.
Twarog, B.M. 1969. *Advances in Pharmacol.* 6B: 5.
Weitzel, G., U. Roester, E. Buddecke and F.J. Strecker. 1956. *Hoppe-Seylers Z. Physiol. Chem.* 303: 161.
Welsh, J.H. 1953. *Anat. Record* 117: 637.

Regulation of Cardiac and Skeletal Muscle Glycogen Metabolism by Biogenic Amines

Steven E. Mayer

School of Medicine
University of California, San Diego
La Jolla, California

The purpose of this paper is to discuss certain metabolic effects of biogenic amines in various types of striated muscle, and to relate these effects to the physiological functions of the muscle types. The discussion will be limited to the actions of the biogenic amines on carbohydrate metabolism, specifically the initiation of glycogenolysis by drugs and hormones. Other sites of action of these substances such as cardiac lipid metabolism have been described but will not be discussed here.

Adrenergic amines are the most thoroughly explored biogenic amines in terms of physiological, pharmacological and biochemical effects on muscle (Fig. 1).

*The major portion of the author's research was carried out at Emory University and was supported by a grant from the United States Public Health Service, HE-04626.

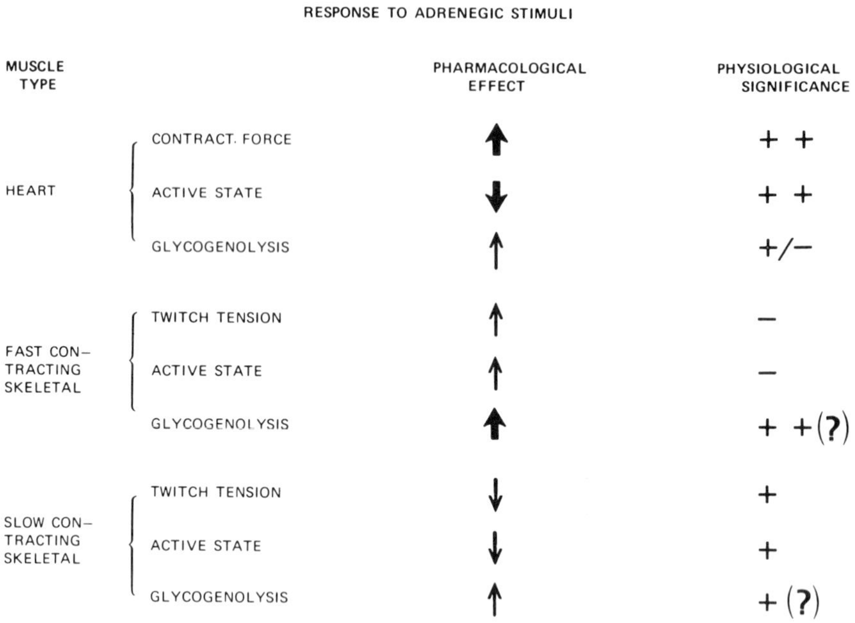

Regulation of Glycogen Metabolism

RESPONSE TO ADRENEGIC STIMULI

MUSCLE TYPE		PHARMACOLOGICAL EFFECT	PHYSIOLOGICAL SIGNIFICANCE
HEART	CONTRACT. FORCE	↑	+ +
	ACTIVE STATE	↓	+ +
	GLYCOGENOLYSIS	↑	+/−
FAST CONTRACTING SKELETAL	TWITCH TENSION	↑	−
	ACTIVE STATE	↑	−
	GLYCOGENOLYSIS	↑	+ + (?)
SLOW CONTRACTING SKELETAL	TWITCH TENSION	↓	+
	ACTIVE STATE	↓	+
	GLYCOGENOLYSIS	↑	+ (?)

Figure 1. Effects of adrenergic stimuli on muscle contraction and glycogenolysis. See text for details.

The physiological significance and pharmacological utility of adrenergic stimuli on the force of contraction of cardiac muscle is well known. However, while the ability of these substances to activate the biochemical system responsible for the breakdown of cardiac glycogen has been well documented its physiological significance has been controversial. Variability in results can be attributed to differences in the basal physiological state of the cardiac muscle preparations which have been used. Intact heart developing tension within physiologically normal limits and well supplied with oxygen relies almost entirely on the oxidation of glucose and of fatty acids taken up from the circulation. Under these conditions, net breakdown of glycogen is usually not demonstrable (see reviews by Mayer, 1967, 1969 for references). Thus the infusion of increasing doses of epinephrine had no effect on dog heart glycogen (Table 1). The largest dose used had toxic effects on cardiac contraction but in-

TABLE 1

Effect of epinephrine on glycogenolysis in dog heart *in situ*

Dose of epinephrine µg/kg·min	R. ventricular contractile force (relative)	Phosphorylase a % of total	Glucose-6-P in intracellular H$_2$O mM	Glycogen as glucose mM
Control	100	6.6	0.13 ± .2	35 ± 3
0.3	120	6.2	0.13 ± .2	36 ± 4
1.0	176	19	0.14 ± .2	35 ± 3
3.0	227	43	0.14 ± .3	36 ± 3
10	210	93	0.48 ± .7	36 ± 4

Epinephrine was infused intravenously at varying rates for 5 min, the time required for peak contractile force and phosphorylase activity changes to occur. Contractile force was measured with a strain gauge sewn to the right ventricle. Enzyme activity and metabolite concentrations were determined on biopsies (20-40 mg) obtained from the right ventricle and immediately frozen in dichlorodifluoromethane at -150°. Data are presented as ± S.E. Those underlined are significantly different from control with $P < .05$.

creased the concentration of myocardial glucose-6-phosphate, a more sensitive index of glycogenolysis than is the direct measurement of concentration of the polysaccharide. Glycogen of *in situ* rat hearts was more susceptible to degradation in the presence of epinephrine than that of the dog, but even here large doses were required in excess of those needed to cause considerable diminution in skeletal muscle glycogen (Williams and Mayer, 1966). On the other hand, glycogenolysis is readily demonstrable in various *in vitro* preparations (see reviews by Ellis, 1956, 1959). Epinephrine caused a greater than 50% reduction in cardiac glycogen stores in perfused isolated rat hearts (Williamson, 1964). This effect was transient and was followed by a more prolonged increase in oxidative metabolism, probably as a consequence of increased mechanical activity. Furthermore, the breakdown of glycogen was accompanied by considerable alteration in adenine nucleotide concentrations (Williamson, 1966). This suggests that the supply of oxygen was not adequate to maintain the increased force of contraction of isolated perfused hearts subjected to epinephrine and that the glycogenolytic response to epinephrine was secondary to hypoxia. These and other results which will be discussed below have led to the development of the hypothesis that metabolic control mechanisms may be superimposed upon the biochemical system, the activation of which is initiated by an adrenergic amine.

The effects of these amines on the physiological state and on glycogenolysis in skeletal muscles can be contrasted to those on the heart. The former have recently been reviewed by Bowman and Nott (1969). The direct action of epinephrine on fast contracting mammalian muscle is to increase twitch tension and the amplitude of tetanic contraction associated with an increase in the duration of the active state. The physiological significance of these effects is questionable. Bowman and Nott have pointed out that the doses of epinephrine required to increase the tension and duration of twitches of fast contracting muscle are much larger than those which can be anticipated to be released by the adrenal medulla during stress. On the other hand, this type of muscle is very rich in the enzymes asso-

ciated with the breakdown of glycogen (Stubbs and
Blanchaer, 1965). It is generally accepted that this
system is stimulated by epinephrine released from the
medulla under stress. However, careful studies relating dose of the amine to the degree of stimulation
of the entire biochemical sequence and the actual
breakdown of glycogen have not yet been performed. The
response of slow contracting skeletal muscle to catecholamines, such as the soleus of the cat, is the opposite of that of fast contracting muscle, i.e. a decrease of twitch tension and in the duration of the -
active state. The enzymes of the phosphorylase activating pathway in these red muscles are about 50% lower
than that found in white muscles (Stubbs and Blanchaer,
1965; unpublished observations of the author). In
contrast, the activity of the rate limiting step in
glycogen synthesis is higher in red muscles and glycogen synthesis proceeds more rapidly after glucose administration (Bar and Blanchaer, 1965).

Epinephrine exerts an initial facilitatory action
on neuromuscular transmission in all types of skeletal
muscle (Bowman and Nott, 1969). While this effect appears to involve a different type of adrenergic receptor, that is, *alpha*, when compared to the direct effect of catecholamine on skeletal muscle (*beta* type of
receptor), Bowman and Nott have speculated that both
effects may be mediated by cyclic adenosine 3',5'-phosphate (cyclic AMP). The role of this cyclic nucleotide
will be discussed in more detail below. The facilitatory effect of epinephrine on neuromuscular transmission has been shown to require inhibition of acetylcholinesterase, to be facilitated by theophylline, and
abolished by curare (Breckenridge, *et al.*, 1967).
These investigators proposed that epinephrine caused
an increase in formation of cyclic AMP at the neuromuscular junction and that the nucleotide in turn was
involved in the release of acetylcholine.

Among the other biogenic amines, the effects of
acetylcholine are potentially the most interesting
(Fig. 2). This amine antagonized the glycogenolytic
effect of epinephrine in perfused hearts (Vincent and
Ellis, 1963). However, it has not yet been possible
to establish whether or not this action is related to

Regulation of Glycogen Metabolism

BIOGENIC AMINES ON HEART AND SKELETAL MUSCLE

		ADRENERGIC	CHOLINERGIC	SEROTONIN	GLUCAGON
HEART	CONTRACTION	↑	↓	↑	↑
	GLYCOGENOLYSIS	(↑)	(↓)	O	↑
SKELETAL MUSCLE	CONTRACTION	↓^{a}, ↑^{a}, ↑^{b}	↑^{b}	O	↑^{a}
	GLYCOGENOLYSIS	↑	?, ↑^{b}	O	O^a

a direct
b neuromuscular

Figure 2. Effects of biogenic amines on contraction and glycogenolysis in cardiac and skeletal muscles. See text for details.

the negative inotropic effect of acetylcholine on cardiac muscle. The role of acetylcholine in neuromuscular transmission must be an important factor in the glycogenolytic response of skeletal muscle to tetanic stimulation initiated in a motor nerve. I am not aware of any direct effects of acetylcholine on skeletal muscle carbohydrate metabolism independent of its action at the end plate.

In contrast to epinephrine and acetylcholine, serotonin (5-hydroxytryptamine) does not have a clearly established biochemical effect on striated muscles. This amine was found to be about 1/200 as potent as epinephrine in eliciting a positive inotropic effect in the intact dog heart, and this was not accompanied by any activation of cardiac phosphorylase (Mayer and Moran, 1960). No effects of serotonin on skeletal muscle are known to me.

Glucagon does not fall strictly within the category

of biogenic amines although four of its 29 amino acids are amides. The major significance of glucagon in relation to the topic of this paper is that its biochemical actions are similar to epinephrine in some ways but not in others. Glucagon is approximately as potent as epinephrine in increasing the contractile amplitude of cardiac muscle (Mayer, *et al.*, 1969). It also stimulates cardiac glycogenolysis but glucagon probably acts at a different receptor site from that of epinephrine. The effect of epinephrine can be blocked by a *beta* adrenergic blocking agent while the effect of glucagon cannot. Glucagon, as well as epinephrine, also restores twitches of rat diaphragm depressed by excess potassium ion (Bowman and Raper, 1964). This facilitatory effect on skeletal muscle contraction, however, is not paralleled by a stimulation of glycogenolysis (Sutherland and Rall, 1960) nor activation of adenyl cyclase from skeletal muscle *in vitro* (unpublished observations of the author).

The ability of adrenergic stimuli to directly increase the force of contraction of cardiac muscle and to facilitate electrically induced contraction of skeletal muscle may be related to their action on glycogen metabolism in several ways.

The pros and cons of these alternative hypotheses have been reviewed recently (Mayer, 1967; Mayer, 1969; Sutherland, *et al.*, 1968).

1. It is conceivable that the physiological response to epinephrine is due to the availability of ATP synthesized as a result of increased utilization of the glucosyl units liberated from glycogen.
2. It has been proposed by Ellis (1956) that hexosmonophosphate, particularly glucose-6-phosphate, produced during glycogenolysis is an important intermediate in the effects of catecholamines on muscle contraction.
3. Cyclic AMP, the mediator of epinephrine action on the phosphorylase activating pathway, may also mediate an action on excitation-contraction coupling or on the contractile machinery itself.
4. The catecholamine may affect the physiological parameters in a manner that is entirely indepen-

dent of its action on the formation of cyclic AMP.

The details of the sequence of events beginning with the formation of cyclic AMP and ending with the phosphorylase catalyzed breakdown of glycogen has been worked out primarily by Edwin Krebs and his colleagues. These are presented schematically in Fig. 3. The mech-

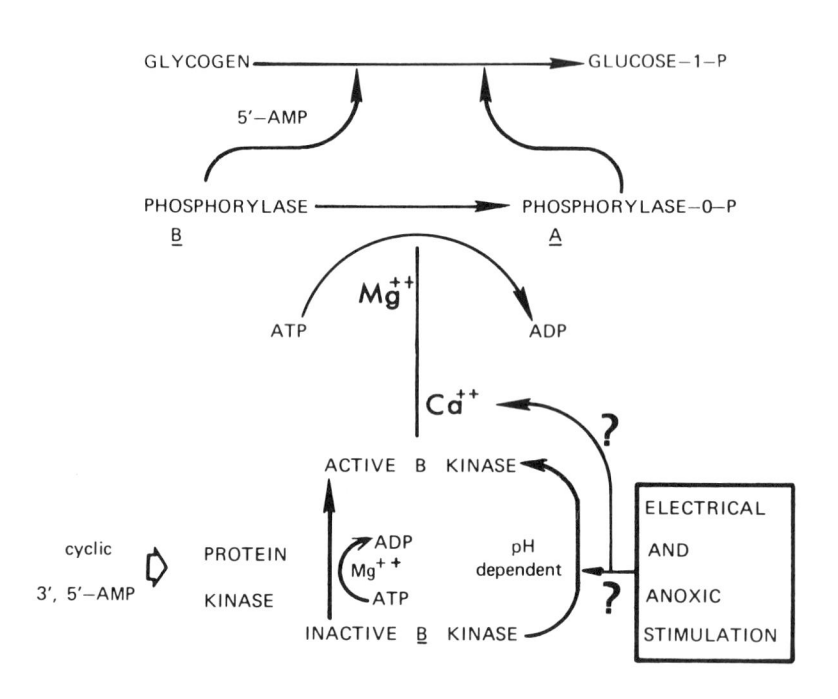

```
GLYCOGEN ─────────────────────→ GLUCOSE-1-P
              5'-AMP
       PHOSPHORYLASE ──────→ PHOSPHORYLASE-O-P
            B                         A
                    Mg++
            ATP            ADP
                    Ca++
            ACTIVE B KINASE
                                    ELECTRICAL
  cyclic    PROTEIN   ADP     pH    AND
  3',5'-AMP KINASE    Mg++   dependent
                      ATP            ANOXIC
            INACTIVE B KINASE        STIMULATION
```

Figure 3. The phosphorylase activating system in skeletal and possibly cardiac muscle. Modified from Mayer, 1969.

anism of action of cyclic AMP appears to be the activation of a protein kinase. This enzyme in turn catalyzes the activation of phosphorylase kinase in the presence of ATP. The phosphorylated kinase then catalyzes another protein phosphorylation, namely the conversion of phosphorylase b to a.

An alternative route to part of this pathway can be demonstrated *in vitro* and the whole scheme can be modified at several places. Activation of phosphorylase kinase can occur without phosphorylation of the enzyme by an alkaline shift in pH (Krebs, et al., 1964). Whatever the mechanism of activation, the catalytic activity of phosphorylase kinase depends upon the presence of magnesium and also on calcium ion. Since the effective concentration of the latter ion is about 10^{-7} M, the activity of phosphorylase kinase may be dependent on the availability of free calcium within the muscle fiber (Krebs, et al., 1969). Additional regulation may take place by allosteric effects on phosphorylase. Allosteric transitions can be induced by changes in the energy charge of the adenylate pool (Atkinson, 1968) and interactions between the subunits of the enzyme (Helmreich, et al., 1967). The activities of either the b or a form of the enzyme are affected by the relative concentrations of the different adenine nucleotides and of inorganic phosphate (Morgan and Parmeggiani, 1964). A decrease in ATP concentration and increase in AMP and P_i should result in a shift from allosteric inhibition to allosteric activation of phosphorylase b and provide substrate for the phosphorolytic cleavage of glycogen. In the presence of hormones such as epinephrine, glucagon, and insulin, that increase intercellular glucose-6-phosphate concentration without necessarily altering the relative concentration of adenyl nucleotides, the a form of the enzyme would be the more important. Glucose-6-phosphate is a potent inhibitor of phosphorylase b. The binding of glycogen may affect the interaction of the subunits of phosphorylase and thereby modify the activity of either phosphorylase b or phosphorylase a. Thus the intercellular distribution of the enzymes, substrates, regulatory metabolites and metal ion cofactors may play an important role in the regulation of the phosphorylase activating pathway. Changes in these parameters may occur as a function of the physiological state of the muscle fiber. The process of excitation and of contraction may alter the availability of cations and of the relative concentrations of the components of the adenylate pool. Therefore, feedback, both on a chemical and on a physiological level may considerably modify the response of the systems, apart from any initial stimulation by a biogenic amine of the formation of cyclic AMP.

Regulation of Cardiac Glycogen Metabolism

Much more information about the regulation of glycogenolysis and its relation to muscle contraction is available for the heart than for skeletal muscle, mostly due to the extensive work of Morgan and colleagues and of Williamson. At the beginning of this article it was stated, that provided the heart is adequately oxygenated, cardiac glycogenolysis occurs only after large doses of catecholamines. Evidence to support this conclusion is presented in Table 1 and has been published elsewhere (Williams and Mayer, 1966). Ischemia or anoxia are potent and very rapid stimulants of the breakdown of cardiac glycogen. Either or both of these pathological conditions may play a role whenever a stimulant biogenic amine is added to an inadequately oxygenated preparation of cardiac muscle. The stimulation by ischemia of phosphorylase activity and of the entire glycolytic pathway is largely but not completely dependent upon the release of norepinephrine from sympathetic nerve endings in the heart (Wollenberger and Krause, 1968). In contrast, when anoxia was produced in animals with an intact circulatory system, most of the stimulation of glycogenolysis could not be attributed to sympathetic stimulation or the release of epinephrine from the adrenal medulla (Mayer, *et al.*, 1967; Mayer, 1967). It seems likely that in this situation, activation of the pathway leading to the transformation of phosphorylase b to a does not involve the formation of cyclic AMP nor an activation of phosphorylase kinase dependent on phosphorylation of the enzyme. This hypothesis is supported by the observation of Robison, *et al.* (1967) that cyclic AMP concentration did not increase during anoxia of the isolated perfused rat heart. The mechanism of activation of phosphorylase kinase under these conditions is not known although two factors, a possible change in intracellular pH and in free calcium ion, may be important. The limitations on the activity of phosphorylase b and a imposed by availability of substrate and inhibition by adenylates is removed in anoxia (Cornblath, *et al.*, 1963).

Whatever the mechanism of activation of the phosphorylase pathway in the ischemic or anoxic heart, the mammalian myocardium is not capable of maintaining contractions with glycogen as the sole substrate for the synthesis of ATP at normal body temperatures. In contrast, the poikilothermic heart of the turtle contracts rhythmically for a long time, solely through the anaerobic utilization of glycogen, provided that the tension developed does not exceed a critical value (Reeves, 1963).

What then is the significance of the phosphorylase activating pathway in cardiac muscle? The activation of phosphorylase by catecholamines has been dissociated both in terms of the dose-response relationship and of the time-response relationship from the augmentation of contractile force produced by this amine (see Mayer, 1969 for references). On the other hand, the rate of production of cyclic AMP and of the activation of phosphorylase kinase is such as to suggest that these two events may play roles in the effect of epinephrine on cardiac contraction. This was first demonstrated by Drummond, *et al.* (1966) with the perfused isolated rat heart and by Robison, *et al.* (1965). The effect of graded doses of epinephrine on the activities of phosphorylase kinase and phosphorylase relative to the isometric force developed by the intact dog heart is shown in Fig. 4. Any dose of epinephrine that produced a measurable increase in contractile force of the right ventricle also caused a significant increase in the activity of phosphorylase kinase. Activation of phosphorylase did not occur until a dose of epinephrine one order of magnitude greater than that required to augment the other two parameters was administered. Measurements of changes in cyclic AMP are lacking in these experiments. This is due to the necessity of freezing cardiac muscle *in situ* to detect such changes whereas changes in enzyme activities were detected in biopsy samples frozen after excision. Furthermore, small changes in the activities of phosphorylase kinase and phosphorylase were measured with good precision whereas the concentration of cyclic AMP had to increase by an apparent value of at least 15-20% before statistically significant differences

INTACT DOG HEART

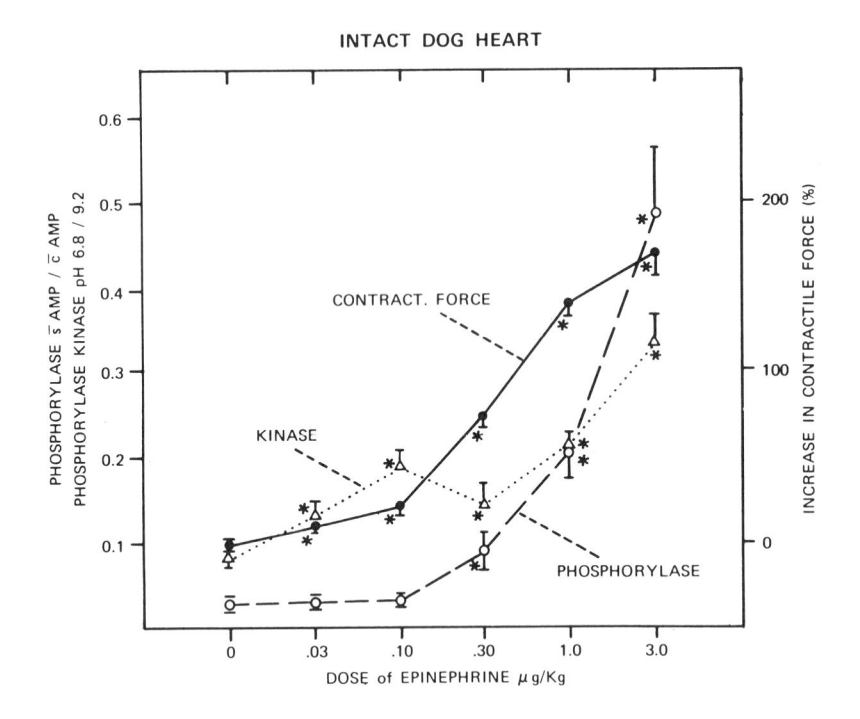

Figure 4. Effect of graded doses of epinephrine on cardiac contractile force, phosphorylase and phosphorylase kinase activities in intact dogs. Epinephrine was administered in single rapid injections intravenously. Biopsy samples for kinase activity measurements were taken 10 sec later, for phosphorylase after 25 sec. These times correspond to peak responses of each enzyme. (Data from unpublished experiments of J.P. Hickenbottom and S.E. Mayer.)

from control values were observed (Mayer, et al., 1969) It is thus not possible at the present time to state with certainty that the formation of cyclic AMP is essential for the augmentation of cardiac contraction by catecholamines.

On the other hand it is evident that the response of the heart to a catecholamine can be markedly modified by procedures such as anoxia, which was discussed above, and by alterations in the ionic environment (Namm, et al., 1968). The removal of calcium from the medium perfusing isolated rat hearts resulted in a loss of response of phosphorylase activation to epinephrine. This occurred at a rate consistent with the time required to wash out the extracellular space. The inability of epinephrine to promote the transformation of phosphorylase b to a under these conditions was associated with some modification of the time course and degree of activation of phosphorylase kinase. However, epinephrine was at least as effective in increasing the formation of cyclic AMP in these hearts when compared to those perfused with a medium of normal composition.

In contrast to the effect of deprivation of calcium on the action of epinephrine, an increase in concentration of this ion to three times its normal value of 3.2 mEq/liter produced of itself an activation of phosphorylase, a suggestive but statistically insignificant increase in phosphorylase kinase activation and a significant decrease in the concentration of cyclic AMP in the perfused rat hearts. These data are consistent with observations on enzyme preparations *in vitro*. The activity of phosphorylase kinase in catalyzing the transformation of phosphorylase b to a requires free calcium ion. The concentration of this ion appears to be limiting in the nonstimulated heart so that introduction of a higher concentration of calcium augments phosphorylase b to a conversion. The observations are also consistent with the hypothesis that calcium inhibits adenyl cyclase. The activity of this enzyme in a particulate preparation from rat heart was inhibited upon the addition of calcium to the medium but augmented by scavengering of contaminating calcium by EGTA (Mayer, 1969).

Evidence that the response of the phosphorylase activating pathway in heart to a catecholamine may be regulated at a step beyond the amine receptor site is also suggested by results on *in situ* dog hearts Fig. 4. Phosphorylase kinase activation was demonstrated without conversion of phosphorylase b to a after small doses of epinephrine. This again suggests that the catalytic activity of kinase is limited by other factors than the degree of activation mediated by cyclic AMP.

Potassium ion appears to play a role in the effect of epinephrine on the activation of cardiac adenyl cyclase. Elevation of potassium ion from 5.6 to 56 mEq/liter produced depolarization. Epinephrine was then no longer effective in increasing cyclic AMP concentration. Its effect on phosphorylase kinase activity and on the activation of phosphorylase were markedly reduced (Namm, *et al.*, 1968). However, elevation of potassium ion in the medium containing a rat heart adenyl cyclase preparation had only a stimulatory effect on the activity of the enzyme. Depolarization of the cardiac muscle fiber membrane may be the critical component of the altered response to epinephrine in intact heart. Loss of contractility *per se* is not likely to be important since the removal of calcium accomplished this but did not inhibit the formation of cyclic AMP in response to epinephrine.

Regulation of Skeletal Muscle Metabolism

The breakdown of muscle glycogen in response to epinephrine administration is a well-documented phenomenon. Evidence obtained from both intact muscle and from enzymes purified from skeletal muscle indicates that the mechanism of the effect of epinephrine involves the production of cyclic AMP, the activation of a protein kinase, and the subsequent transformation of phosphorylase kinase from an inactive to an active form (Posner, *et al.*, 1965 DeLange, *et al.*, 1968; Walsh, *et al.*, 1968). It is likely that in skeletal muscle as well as in cardiac muscle, this system of enzymes is controlled at other than the initial site of action of the hormone or drug, the activation of adenyl cyclase. The catalytic activity of skeletal muscle phosphorylase *b* kinase *in vitro* requires low concentrations of calcium ion (Ozawa, *et al.*, 1967). This effect of calcium is quite distinct from the one which has been described by Huston and Krebs (1968). Calcium ion is a cofactor for a proteolytic enzyme in skeletal muscle which produces irreversible activation of phosphorylase kinase. It is unlikely that this mechanism constitutes a physiologically significant regulatory device because of its irreversibility and the relatively high concentration of Ca ion required (mM). In contrast, the action of calcium ion on kinase activity is likely to be important. However, it is not yet clear to what extent calcium or the regulation of phosphorylase activity by substrate availability and by allosteric transitions actually influences the activation by epinephrine of the entire phosphorylase activating pathway in intact skeletal muscle. Drummond and his collaborators (1969) have recently verified that epinephrine induced conversion of phosphorylase *b* to *a* was accompanied by increased cyclic AMP and activation of phosphorylase kinase. In contrast, during electrically induced muscle contraction, conversion of phosphorylase *b* to *a* took place with no increase in the concentration of cyclic AMP nor detectable activation of phosphorylase *b* kinase. Helmreich and collaborators (1966) have concluded that the conversion of phosphorylase *b* to *a* in response to electrical stimulation is best explained by an effect of calcium ion on the activity of phosphorylase *b* kinase. The activities of the rate limiting steps of glycolysis, phosphorylase and phosphofructokinase, increased synchronously with contraction of frog sartorius muscle under anaerobic conditions. Thus there appear to be two pathways of phosphorylase activation, one dependent on and the other independent of cyclic AMP in skeletal muscle. The extent of integration of these two pathways remains to be determined.

Another biochemical effect of epinephrine on carbohydrate metabolism in skeletal muscle is inhibition of the pathway resulting in the synthesis of glycogen. This was first demonstrated by Craig and Larner (1964). Epinephrine promoted the conversion of the I form (physiologically relatively active) of glycogen synthetase to the form (D) active only in the presence of glucose-6-phosphate. This reciprocal relationship between glycogen synthesis and glycogenolysis was confirmed by observations of the effect of epinephrine in intact rats (Williams and Mayer, 1966). It has also been observed subsequent to electrical stimulation of rat muscle (Piras and Staneloni, 1969). Direct electrical stimulation of leg muscle resulted in an increase in phosphorylase a, whereas during the recovery period following cessation of stimulation phosphorylase a returned to control value and the activity of glycogen synthetase I increased gradually to about twice resting value.

The reciprocal relation between phosphorylase and glycogen synthetase activities in response to epinephrine in skeletal muscle can be explained on the basis of a single common control point: the production of cyclic AMP. The *in vitro* conversion of rat muscle glycogen synthetase from the I to the less active, D form, requires ATP and magnesium ion and is stimulated by cyclic AMP (Appleman, *et al.*, 1966). However, the hypothesis that epinephrine turns glycogenolysis on and glycogen synthesis off solely through the formation of cyclic AMP is probably an over-simplification. Calcium ion has been shown to affect the I to D conversion (Appleman, *et al.*, 1966). Regulation of glycogen synthetase activity during and following muscular contraction appears to depend not only on the interconversion of the two forms of the enzyme, but also upon differential inhibition by adenylates, glucose-6-phosphate and other phosphate compounds (Piras and Staneloni, 1969). Glycogen has also been implicated in exerting a negative feedback control over its own synthesis through an effect on the activity of glycogen synthetase (Danforth, 1965). These regulatory effects superimposed on the epinephrine induced formation of cyclic AMP may be important in cardiac as well as in skeletal muscle. The failure of epinephrine to produce net breakdown of glycogen in the rat heart *in situ* was associated with an absence of the expected reciprocal relation between phosphorylase activity and glycogen synthetase activity (Williams and Mayer, 1966).

Two inheritable alterations in the activities of enzymes of the skeletal muscle phosphorylase system have raised some interesting questions about the role of the system. The human disease of phosphorylase deficient myopathy (McArdle's disease) is characterized by the development of contracture during exercise. Muscle glycogenolysis does not occur in response to the administration of epinephrine. Caffeine or calcium pro-

duced contracture in preparations of cell membrane-
free muscle removed from a patient with this disease.
The effect was potentiated by electrical stimulation.
It was concluded that clacium reaccumulation by sarcoplasmic reticulum was defective (Gruener, *et al.*, 1968).
These observations may be explained by the absence of
the enzymes catalyzing the most critical step in glycogenolysis, phosphorylase b and a, or may reflect
another genetic deficiency.

In contrast, it is conceivable that phosphorylase
kinase is not essential for skeletal muscle glycogenolysis. This enzyme is not detectable in skeletal
muscle of the I/FnLn strain of mice. Conversion of
phosphorylase b to a did not occur in response to either epinephrine or electrical stimulation (Lyon and
Porter, 1963; Danforth and Lyon, 1964). Yet, after
administration of epinephrine to these animals, the
rate of formation of cyclic AMP in gastrocnemius was
more rapid and as intensive as that observed in a control strain of mice, the $C_{57}BL/FnLn$ (Lyon and Mayer,
1969). However, epinephrine exerted some effect on
glycogenolysis in the I strain mice (Lyon and Porter,
1963). Therefore, the possibilities must be enter-
tained that (1) cyclic AMP plays a role in glycogenolysis independent of its effect on phosphorylase kinase activation or (2) that the enzymatic degradation
of cyclic AMP produces a sufficiently high localized
concentration of 5'-AMP to activate phosphorylase b.
Thus, the mechanism of action of catecholamines on
muscle glycogenolysis may not be as well defined as
it appears.

Suggestive, but not conclusive evidence links the
formation of cyclic AMP in cardiac muscle to the ability of catecholamines to increase the force of contraction. The question then arises: to what extent
are the physiological effects of these amines on skeletal muscle due to a role of the cyclic nucleotide in
excitation or contraction. Some distinctions must be
made between the effects of the two naturally occuring
catecholamines, norepinephrine and epinephrine. Norepinephrine, either as a circulating hormone or as a
neurohumor released from skeletal muscle sympathetic
nerve ending, is probably less significant. This agent

is much less potent than epinephrine in increasing plasma lactate concentration probably because it limits its own access to skeletal muscle fibers as a result of its powerful vasoconstrictor effect. Furthermore, release of norepinephrine from sympathetic nerve endings in muscle appears to be limited to the smooth muscle of blood vessels. Finally, norepinephrine appears to be inherently a weaker agonist than is epinephrine on skeletal muscle glycogenolysis.

It seems likely, although not yet incontrovertably proven, that epinephrine released from the adrenal medulla in response to stress, stimulated glycogenolysis, especially in fast contracting muscle while inhibiting the tension developed by slow contracting muscles. The facilitatory effect of epinephrine on tension developed by fast contracting muscle is not likely to be physiologically significant (Bowman and Nott, 1969). Epinephrine also facilitates neuromuscular transmission. Cyclic AMP has been implicated in this process (Breckenridge, *et al.*, 1967). More information is needed to assess the biochemical and physiological significance of this concept.

The ability of adrenergic stimuli to affect skeletal muscle contraction can be examined in terms of the hypothesis stated earlier in this paper. (1) It is possible that the effects of epinephrine on fast contracting muscle and on neuromuscular transmission are simply due to increased ATP synthesized during anaerobic glycogenolysis. This remains to be proven. (2) While hexosemonophosphates, particularly glucose-6-phosphate, accumulate when glycogenolysis is stimulated in skeletal muscle, this proves only that phosphofructokinase remains a rate limiting step (Wilson, *et al.*, 1967). No action of these compounds on excitation or contraction has been demonstrated. (3) While cyclic AMP has been implicated in the facilitatory effect of epinephrine on neuromuscular transmission, it is difficult to assign a role in both the facilitation of twitch tension in fast contracting muscle and in inhibition of the same in slow contracting muscle. Evidence has recently been presented that the reduction of the calcium accumulating capacity of rabbit muscle sarcoplasmic reticulum by caffeine cannot be attributed to cyclic

AMP. That is, the action of caffeine on skeletal muscle contraction does not appear to be due to an inhibition of the enzymatic degradation of cyclic AMP by phosphodiesterase (Weber and Herz, 1968; Weber, 1968).

Conclusion

The best understood effects of biogenic amines on carbohydrate metabolism of striated muscles are those of epinephrine upon the pathway that results in the breakdown of glycogen. This involves the sequential activation of four enzymes: adenyl cyclase, a protein kinase, phosphorylase b kinase and the transformation of phosphorylase b to a. However, this system probably is subject to considerable regulation in addition to the interaction of the amine with a receptor site by cofactors such as calcium, metabolites such as the components of the adenylate pool, and intracellular pH. Alterations in these superimposed controlling factors may be the result of biogenic amine action at some other site than adenyl cyclase or may reflect feedback control by the biochemical changes associated with or consequent to the process of muscular contraction.

REFERENCES

Appleman, M.M., L. Birnbaumer and H.N. Torres. 1966. *Arch. Biochem. Biophys.* 116: 390.
Atkinson, D.E. 1968. *Biochemistry* 7: 4030.
Bar, U. and M.C. Blanchaer. 1965. *Am. J. Physiol.* 209: 905.
Bowman, W.C. and M.W. Nott. 1969. *Pharmacol. Rev.* 21: 27.
Bowman, W.C. and C. Raper. 1964. *Brit. J. Pharmacol. Chemother.* 23: 184.
Breckenridge, B. McL., J.H. Burn and F.M. Matschinsky. 1967. *Proc. Natl. Acad. Sci. U.S.* 57: 1893.
Cornblath, M., P.J. Randle, A. Parmeggiani and H.E. Morgan. 1963. *J. Biol. Chem.* 238: 1592.
Craig, J.W. and J. Larner. 1964. *Nature* (London) 202: 971.
Danforth, W.H. 1965. *J. Biol. Chem.* 240: 588.

Danforth, W.H. and J.B. Lyon, Jr. 1964. *J. Biol. Chem.* 239: 4047.
DeLange, R.J., R.G. Kemp, W.B. Riley, R.A. Cooper and E.G. Krebs. 1968. *J. Biol. Chem.* 243: 2200.
Drummond, G.I., L. Duncan and E. Hertzman. 1966. *J. Biol. Chem.* 241: 5899.
Drummond, G.I., J.P. Harwood and C.A. Powell. 1969. *J. Biol. Chem.* In press.
Ellis, S. 1956. *Pharmacol. Rev.* 8: 485.
Ellis, S. 1959. *Pharmacol. Rev.* 11: 469.
Gruener, R., B. McArdle, B.E. Ryman and R.O. Weller. 1968. *J. Neurol. Neurosurg. Psychiat.* 31: 268.
Helmreich, E., W.H. Danforth, S. Karpatkin and C.F. Cori. 1966. *Control of Energy Metabolism, Colloquium.* p. 299.
Helmreich, E., M.C. Michaelides and C.F. Cori. 1967. *Biochemistry* 6: 3695.
Huston, R.B. and E.G. Krebs. 1968. *Biochemistry* 7: 2116.
Krebs, E.G., D.S. Love, G.E. Bratvold, K.A. Trayser, W.L. Meyer and E.H. Fischer. 1964. *Biochemistry* 3: 1020.
Krebs, E.G., R.B. Huston and F.L. Hunkeler. 1969. *Adv. in Enzymol.* In press.
Lyon, J.B., Jr. and S.E. Mayer. 1969. *Biochem. Biophys. Res. Commun.* 34: 459.
Lyon, J.B., Jr. and J. Porter. 1963. *J. Biol. Chem.* 238: 1.
Mayer, S.E. 1967. In *Factors Influencing Myocardial Contractility*, R.D. Tanz, F. Kavaler and J. Roberts, eds. (New York: Academic Press), p. 443.
Mayer, S.E. 1969. *Federation Proc.* In press.
Mayer, S.E. and N.C. Moran. 1960. *J. Pharmacol. Exptl. Therap.* 129: 271.
Mayer, S.E., D.H. Namm and L. Rice. 1969. *Circulation Res.* Submitted for publication.
Mayer, S.E., B.J. Williams and J.M. Smith. 1967. *Ann. N.Y. Acad. Sci.* 139: 686.
Morgan, H.E. and A. Parmeggiani. 1964. *J. Biol. Chem.* 239: 2435.
Namm, D.H., S.E. Mayer and M. Maltbie. 1968. *Mol. Ph rmacol.* 4: 522.
Ozawa, E., K. Hosoi and S. Ebashi. 1967. *J. Biochem.*

(Tokyo) 61: 531.
Piras, R. and R. Staneloni. 1969. *Biochemistry* 8: 2153.
Posner, J.B., R. Stern and E.G. Krebs. 1965. *J. Biol. Chem.* 240: 982.
Reeves, R.B. 1963. *Am. J. Physiol.* 205: 23.
Robison, G.A., R.W. Butcher, I. Øye, H.E. Morgan and E.W. Sutherland. 1965. *Mol. Pharmacol.* 1: 168.
Robison, G.A., R.W. Butcher and E.W. Sutherland. 1967. *Ann. N.Y. Acad. Sci.* 139: 703.
Sutherland, E.W. and T.W. Rall. 1960. *Pharmacol. Rev.* 12: 287.
Sutherland, E.W., G.A. Robison and R.W. Butcher. 1968. *Circulation* 37: 279.
Stubbs, S. St.G. and M.C. Blanchaer. 1965. *Canad. J. Biochem.* 43: 463.
Vincent, N. and S. Ellis. 1963. *J. Pharmacol. Exptl. Therap.* 139: 60.
Walsh, D.A., J.P. Perkins and E.G. Krebs. *J. Biol. Chem.* 243: 3763.
Weber, A. 1968. *J. Gen. Physiol.* 52: 760.
Weber, A. and R. Herz. 1968. *J. Gen. Physiol.* 52: 750.
Williams, B.J. and S.E. Mayer. 1966. *Mol. Pharmacol.* 2: 454.
Williamson, J.R. 1964. *J. Biol. Chem.* 239: 2721.
Williamson, J.R. 1966. *Mol. Pharmacol.* 2: 206.
Wilson, J.E., B. Sacktor and C.G. Tiekert. 1967. *Arch. Biochem. Biophys.* 120: 542.
Wollenberger, A. and E.-G. Krause. 1968. *Am. J. Cardiol.* 22: 349.

Hormonal Control of Glycogen Synthetase Interconversions

G. Villar-Palasi, N. D. Goldberg, J. S. Bishop
F. Q. Nuttall, K. K. Schlender, and J. Larner

Departments of Biochemistry and Pharmacology
College of Medical Sciences
University of Minnesota
Minneapolis

The first effect of a hormone on glycogen synthetase (UDP glucose α-1,4-glycogen α-4-glucosyltransferase, E.C.2.4.1.11) was uncovered by Villar-Palasi and Larner in 1960. Addition of insulin to the incubation media of rat hemidiaphragms promoted an increase in the activity in glycogen synthetase in the diaphragm extracts when assayed in the absence of glucose-6-phosphate. No difference in synthetase activity between control and insulin treated diaparagms was found, however, when glucose-6-phosphate was added in millimolar concentrations to the assay mixture. Subsequent experiments showed that this increase in synthetase activity was a stable change induced in the tissue and not the result of procedures involved in the preparation of the extracts.

From these results, Villar-Palasi and Larner (1960) postulated the existence of two interconvertible forms of glycogen synthetase in muscle, one requiring glucose-6-phosphate for activity (glucose-6-phosphate *dependent*, *D*), and another active in the absence of this activator (glucose-6-phosphate *independent*, *I*). Friedman and Larner (1962, 1963) were able to show that, in muscle, synthetase *I* was converted into synthetase *D* by a phosphorylation requiring ATP and Mg^{++}. Larner and Sanger (1965) found that the γ phosphate of ATP was incorporated into a serine group of the synthetase enzyme protein. This reaction was catalyzed by a protein kinase (synthetase *I* kinase) present in muscle extracts. Addition of 3',5'-cyclic adenylate (cyclic AMP) was found (Rosell-Perez and Larner, 1964; Appleman, *et al.*, 1964) to stimulate the phosphorylation by ATP of synthetase *I*, apparently by increasing the affinity of synthetase *I* kinase for Mg^{++} (Huijing and Larner, 1966a, 1966b). The reverse conversion, which involves the dephosphorylation of synthetase *D* to yield synthetase *I* was shown to be catalyzed by a phosphatase also present in muscle extracts. The interconversion mechanism of muscle glucogen synthetase is summarized in Fig. 1. Similar enzymic reactions appear to mediate

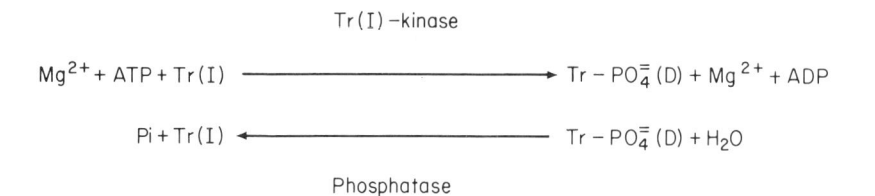

$$Mg^{2+} + ATP + Tr(I) \xrightarrow{Tr(I)-kinase} Tr - PO_4^= (D) + Mg^{2+} + ADP$$

$$Pi + Tr(I) \xleftarrow{Phosphatase} Tr - PO_4^= (D) + H_2O$$

Figure 1

the interconversions of synthetase in heart, liver and brain. Muscle synthetase *I* kinase has been found to catalyze the conversion of liver synthetase *I* into *D* in the presence of ATP-Mg^{++}; *vice versa*, liver synthetase *I* kinase catalyzes the phosphorylation of muscle synthetase *I* (Yip and Larner, 1969).

Effects of Individual Hormones on Glycogen Synthetase Activity

Following the initial finding of an insulin mediated effect on glycogen synthetase in isolated cut rat diaphragm, similar or identical effects of insulin were observed in liver *in vivo* (Steiner, *et al.*, 1961; Bishop and Larner, 1964), in open chest rat heart (Williams and Mayer, 1966), in the *in vitro* fat pad (Jungas, 1966), and in *in vivo* rat skeletal muscle (Goldberg, *et al.*, 1967). Infusion of glucose (De Wulf and Hers, 1967a) or administration of glucocorticoids to the intact animal has also been found to induce an increase in glucose-6-phosphate independent synthetase activity in liver (De Wulf and Hers, 1967b; Kreutner and Goldberg, 1967).

Belocopitow (1961) first showed, using rat hemidiaphragms, that epinephrine decreased total synthetase activity. Craig and Larner (1964), using intact rat diaphragms, were able to confirm this finding and, in addition, demonstrated that synthetase *I* activity was greatly decreased. Injection of epinephrine also reduced the levels of synthetase *I* activity in mouse skeletal muscle (Danforth, 1965) and liver (De Wulfe and Hers, 1968). Glucagon administration to normal dogs was shown by Bishop and Larner (1964) to cancel the activation of glycogen synthetase induced by insulin. There appears to be general agreement that these effects of epinephrine and glucagon are due to an increased synthetase *I* kinase activity mediated by an elevation of the intracellular concentration of its activator, 3',5'-cyclic adenylate, promoted by the known stimulation of adenyl cyclase by epinephrine and glucagon.

Using a rather highly purified protein kinase obtained from rabbit muscle which had an almost absolute cyclic AMP requirement for activity, the cyclic nucleotide specificity of the reaction has been studied (Schlender, *et al.*, 1969). As shown in Table 1, all of the 3',5' cyclic nucleotides were active and gave V_{max} values that varied only about 20%. The K_a values differed widely with the lowest constant, 6.7 X 10^{-8} M, that for cyclic AMP. The activation constant for cyc-

TABLE 1

Nucleotide specificity of synthetase I kinase

Compound	K_a	V_{max}
	M	$-\Delta$ milliunits transferase I/5 min
Cyclic AMP	6.7×10^{-8}	0.962
Cyclic GMP	9.9×10^{-6}	0.869
Cyclic CMP	8.9×10^{-6}	0.807
Cyclic UMP	6.5×10^{-6}	0.792
Cyclic deoxy-AMP	1.1×10^{-4}	0.843
Cyclic TMP	1.3×10^{-3}	0.854
3' AMP	—	—
5' AMP	—	—
2',3' Cyclic AMP	—	—

K_a is the dissociation constant for the binding of the indicated nucleotide to the purified synthetase I kinase, and V_{max} is the maximum rate of kinase activity at high nucleotide concentrations. For further details, see Schlender, et al., 1969.

lic GMP was over 100-fold higher, as were the constants for cyclic CMP and cyclic UMP. The constants for cyclic deoxy-AMP and cyclic TMP were even higher than the previous group, indicating that the structure of the base and of the sugar have a very great influence on the binding of the cyclic nucleotide to the enzyme.

Some questions have been raised in regard to the cause of the increase in synthetase I in liver observed after insulin injection. The work here presented is oriented to clarify three immediate questions, namely: a) what is the effector—hormone or metabolite—responsible for the observed increase in synthetase I; b) is a decrease in the tissue concentration of 3',5'-cyclic adenylate involved in the insulin mediated effects on synthetase?; and c) through which one of the two enzymes which control synthetase interconversions is the insulin effect expressed?

There is no doubt that insulin promotes an increase in synthetase I in muscle. This effect has been found by a large number of investigators in the isolated intact or cut rat diaphragm incubated with or without glucose, and also *in vivo* in skeletal muscle or normal or alloxan-diabetic rats (Fig. 2). In this regard, it should be pointed out that when normal fasted rats are used there appears to be no difference in the final levels of synthetase I or in the approximate speed of conversions of D to I by administering either a large glucose load or glucose supplemented with insulin. However, when alloxan-diabetic rats are used only insulin can induce the elevation of synthetase I levels. It would appear that the endogenous release of insulin in normal rats induced as a result of an elevation of blood glucose levels is enough to promote maximal effects on the synthetase system in muscle.

This effect of insulin is not mediated by a decrease in 3,5'-cyclic adenylate levels in muscle. Intraperitoneal injection of insulin to normal anesthetized rats promoted an increase in synthetase I activity in skeletal muscle within 5 minutes which was sustained during the 20 min period tested. This increase in synthetase occurred in the presence of *increased* levels of 3',5'-cyclic adenylate (30 to 60% over the controls). In this same series of experiments, epinephrine injec-

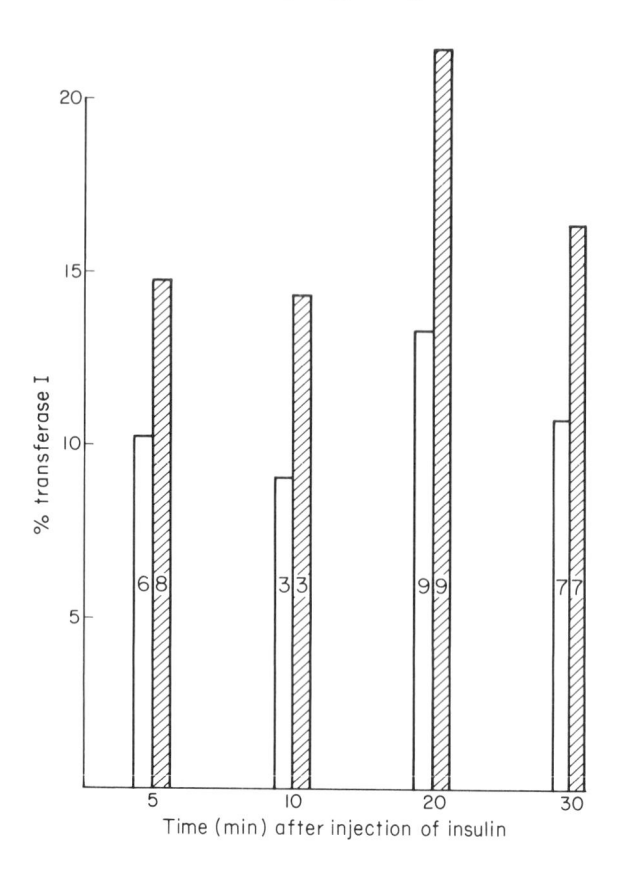

Figure 2. Percent glycogen synthetase in rat skeletal muscle extracts after intraperitoneal injection of saline (unshaded) or insulin (shaded) bars. Numbers in bars represent the number of animals studied. The insulin concentration was 2-4 units/kg.

tion produced a 3-fold increase in skeletal muscle concentration of cyclic AMP (Fig. 3). Most probably, the increased muscle concentrations of cyclic AMP found after insulin injection were due to a release of small amounts of epinephrine in response to the hypoglycemia which developed.

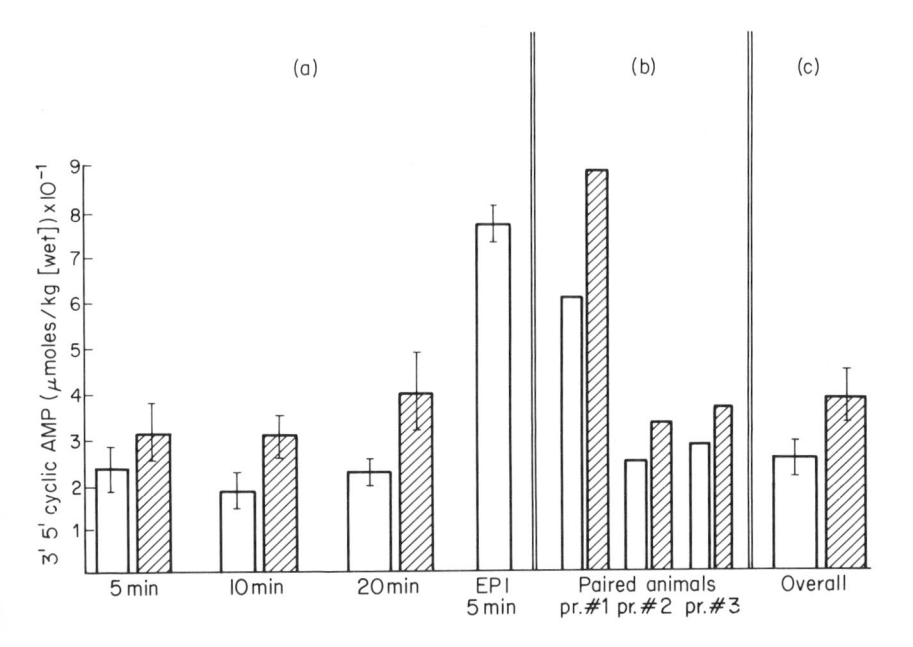

Figure 3. Concentrations of 3',5'-cyclic AMP in rat skeletal muscle after injection of saline (unshaded), insulin (shaded), or epinephrine (EPI). *A.* Each bar represents values based on averages for at least 3 animals. *B.* Each bar represents values based on 1 animal in a pair studied at 20 min. *C.* Bars from all animals studied. The insulin concentration was 2-4 units/kg.

"Two Hormone" Effects on Glycogen Synthetase Activity

When isolated rat diaphragms were incubated with insulin, there was no change in intracellular levels of cyclic adenylate; furthermore, insulin did not decrease the elevated concentrations of cyclic AMP induced by epinephrine (Fig. 4). In contrast to these usual results, under certain other specified conditions, which I will call for sake of discussion, the "two hormone conditions," effects of insulin on cyclic AMP levels in tissues have been observed. For example, if diaphragms are first preexposed for 30 min to insu-

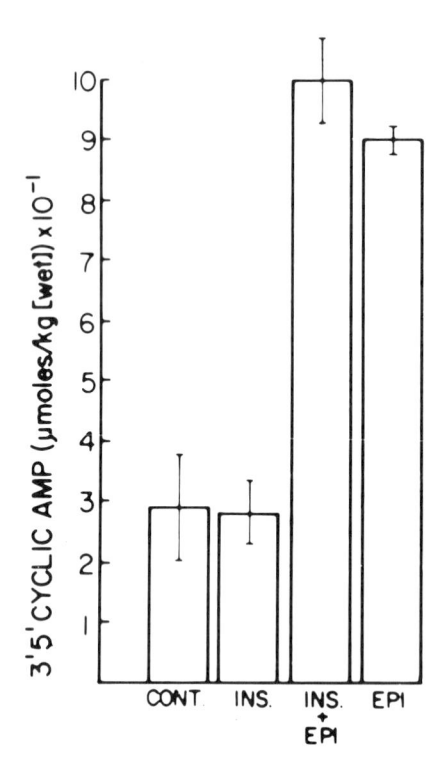

[Figure: Bar chart showing 3',5' CYCLIC AMP (μmoles/kg [wet]) × 10⁻¹ on y-axis (0 to 10) with four bars: CONT (~3), INS (~3), INS+EPI (~10), EPI (~9)]

Figure 4. Concentrations of 3',5'-cyclic AMP in rat diaphragms after incubation in buffer (CONT), insulin (INS), epinephrine (EPI), or a combination of the hormones. Incubations were for 30 min following a 30 min wash period *without hormones*. The concentrations of insulin and epinephrine were 0.1 units/ml and 11 μg/ml, respectively.

lin, and then transferred to fresh media and exposed to epinephrine, as seen in Fig. 5, the diaphragms pre-exposed to insulin show a blunted epinephrine induced rise of cyclic AMP levels. It is to be emphasized that the effect is a rapid one, seen only within 1 to 2 min following epinephrine treatment, and disappearing within 5 min (Craig, *et al.*, 1969). I would also like to include in this category of "two hormone conditions," the experiments in which the levels of cyclic AMP in

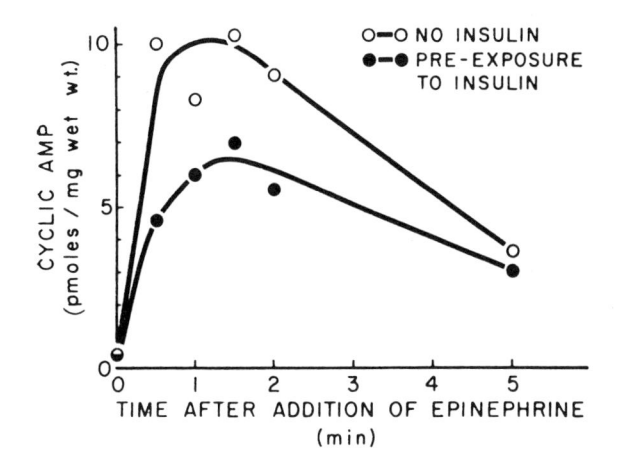

Figure 5. Concentrations of 3',5'-cyclic AMP in rat diaphragm after short intervals of incubation with epinephrine (60 µM) with and without preexposure to insulin (1 unit/ml - 30 min).

adipose tissue were first elevated by epinephrine or some other lipolytic hormones which stimulate lipolysis and the insulin added at the height of the cyclic AMP concentration curve. Under these special "two hormone conditions" levels of cyclic AMP were seen to decrease under the influence of insulin.

Changes in Synthetase I Kinase

The question regarding the specific synthetase interconverting enzyme through which the effect of insulin is expressed has also been investigated. Villar-Palasi and Wenger (1967) found that in muscle extracts from normal, fed, anesthetized rats, in which there was an increase in synthetase I phosphatase activity (90%), ten minutes after insulin injection, there was no change in synthetase D phosphatase activity. Synthetase I kinase activity, however, was found to be decreased in muscle extracts from insulin treated rats when the kinase was tested in the absence of added 3',5'-cyclic adenylate. When the kinase was maximally stimulated by the addition of cyclic AMP (5×10^{-5} M), no differences were observed in the kinase activity from control or insulin treated rat muscle extracts. In other studies using rabbit muscle, partially purified synthetase *I* kinase preparations were obtained that were either totally 3',5'-cyclic adenylate dependent or independent. It was suggested that the effect of insulin might be mediated through an alteration of synthetase *I* kinase sensitivity to 3',5'-cyclic adenylate (Villar-Palasi and Wenger, 1967); the hormone would induce a kinase activity more cyclic AMP dependent.

Isolated perfused rat hearts show an increased uptake of glucose when insulin is present in the perfusion media; however, insulin does not change the levels of glycogen synthetase *I* in these isolated heart preparations. Rat hearts *in vivo*, on the other hand, respond to insulin with an increase in the percentage of synthetase in the *I* form. The reason for this difference in sensitivity to insulin of the heart synthetase *in vivo* and *in vitro* can perhaps be explained on the basis of a difference in the synthetase *I* kinase originally present in these two preparations. Total synthetase *I* kinase activity, tested in the presence of 10^{-6} M cyclic AMP, was essentially the same in the perfused and nonperfused hearts (0.381 *vs*. 0.358 μMole/min/g wet weight change in synthetase *D* activity in the nonperfused and perfused, respectively). However, 91% of the total kinase activity in the nonperfused heart was independent of added 3',5'-cyclic adenylate, whereas less than 50% of the total synthetase *I* kinase activity was detectable in perfused rat heart extracts in the absence of added cyclic nucleotide. Thus, during *in vitro* perfusion the heart kinase appears to revert to a more cyclic AMP dependent state. The reason for this change is not yet understood. Preliminary results indicate that an effect of perfusion on endogenous cyclic AMP levels cannot explain the difference. As in the case of skeletal muscle extracts, a brief incubation of crude perfused rat heart extracts renders the synthetase *I* kinase activity less 3',5'-cyclic adenylate dependent. Thus, there are now 2 instances in which an altered sensitivity of synthetase *I* kinase to cyclic AMP has been observed, insulin treatment, and "*in vitro*" heart perfusion. This altered sensitivity to the kinase appears to be a stable enzymic marker of the perturbations, and suggests that the kinase itself is under control by a mechanism different from and in addition to the level of cyclic AMP in the tissue.

Insulin administered *in vivo* to normal animals appears to raise the levels of synthetase I in liver with little or no change in total synthetase activity (Bishop and Larner, 1964). Synthetase I kinase (Bishop and Larner, 1969) and synthetase D phosphatase (Bishop, *et al.*, 1969) activities have been found in liver and purified from liver extracts. In contradistinction to the case in muscle, however, the insulin effect on liver synthetase and its mechanism appears to be more controversial. Glucose injection in non-diabetic rats has been found to increase synthetase I levels in liver (De Wulf and Hers, 1967b), and it has been suggested that this effect is mediated by a lowering in the blood levels of glucagon (De Wulf and Hers, 1968; Hers and De Wulf, 1969), rather than through a more direct effect of insulin itself on the glycogen synthetase mechanism. Increased intracellular concentrations of 3',5'-cyclic adenylate have been found in liver as a result of alloxan diabetes or anti-insulin serum injection, and a 20% decrease in cyclic AMP levels has been observed in the normal isolated perfused liver after insulin administration (Jefferson, *et al.*, 1968). Questions have also arisen regarding the role of glucocorticoids on the activation of hepatic glycogen synthetase and whether the mechanism of the steroid effect is independent or in part dependent upon insulin (De Wulf and Hers, 1968; Kreutner and Goldberg, 1967; Mersmann and Segal, 1968; Hornbrook, *et al.*, 1966). The following data will be presented with the hope of clarifying some of the controversial aspects in this area.

It has been shown previously (Kreutner and Goldberg, 1967) that an early phase of the apparent steroid stimulation of glycogen synthetase D to I conversion is dependent upon the presence of insulin. The changes in some of the key elements in the glycogenic mechanism were measured in adrenalectomized-alloxan diabetic rats (deprived of food for only 4 hr), receiving either saline (control), glucose (after 2 hr), glucose + 2 units of insulin/kg (after 75 min), or glucose + 10 mg hydrocortisone/kg (after 2 hr). It can be seen (Fig. 6) that under these conditions only the animals receiving insulin showed a stimulation of the glycogenic system, which was reflected in: a) a new deposition of glycogen, b) an increase in the percentage of synthetase I from 10 to 80% with no change in total synthetase, and c) a characteristic drop in hepatic UDP-glucose concentration. Glucose-6-phosphate levels were also diminished significantly after insulin injection.

With glucose alone or glucose and hydrocortisone, there was no change in synthetase I activity of net deposition of glycogen, in spite of the fact that hepatic glucose and glucose-6-phosphate levels were elevated with each of these treatments. Therefore, glucose alone, or with hydrocortisone, fails to activate liver synthetase under these conditions.

It is on interest to mention here the recent experiments reported by Blatt, Scamahorn and Kim which were carried out in tadpoles (1969). The initial observation was that puromycin injection into tadpoles brought about glycogen synthetase activation in liver. Tracing this down, it was found that the effect depended on the presence of the pancreas. If tadpoles were pancreatectomized or treated with alloxan, this effect of puromycin was not present. It had been known for some time that puromycin injection in rats brings about a liver glycogenolysis together with hyperglycemia. These workers demonstrated the same phenomenon in tadpoles and showed in addition that the hyperglycemia activated pancreatic insulin release, and that the insulin specifically activated liver glycogen synthetase. Controls were done with glucose injection alone. In the normal tadpoles, liver glycogen synthetase was activated. In the diabetic tadpoles, there was no activation, thus showing again that insulin was required for the enzyme activation, and that glucose alone was not sufficient.

That the synthetase activation by hydrocortisone is dependent upon the presence of insulin can perhaps be better demonstrated by the progressive disappearance of the steroid effect with the onset of the diabetic state after alloxan administration. The response of

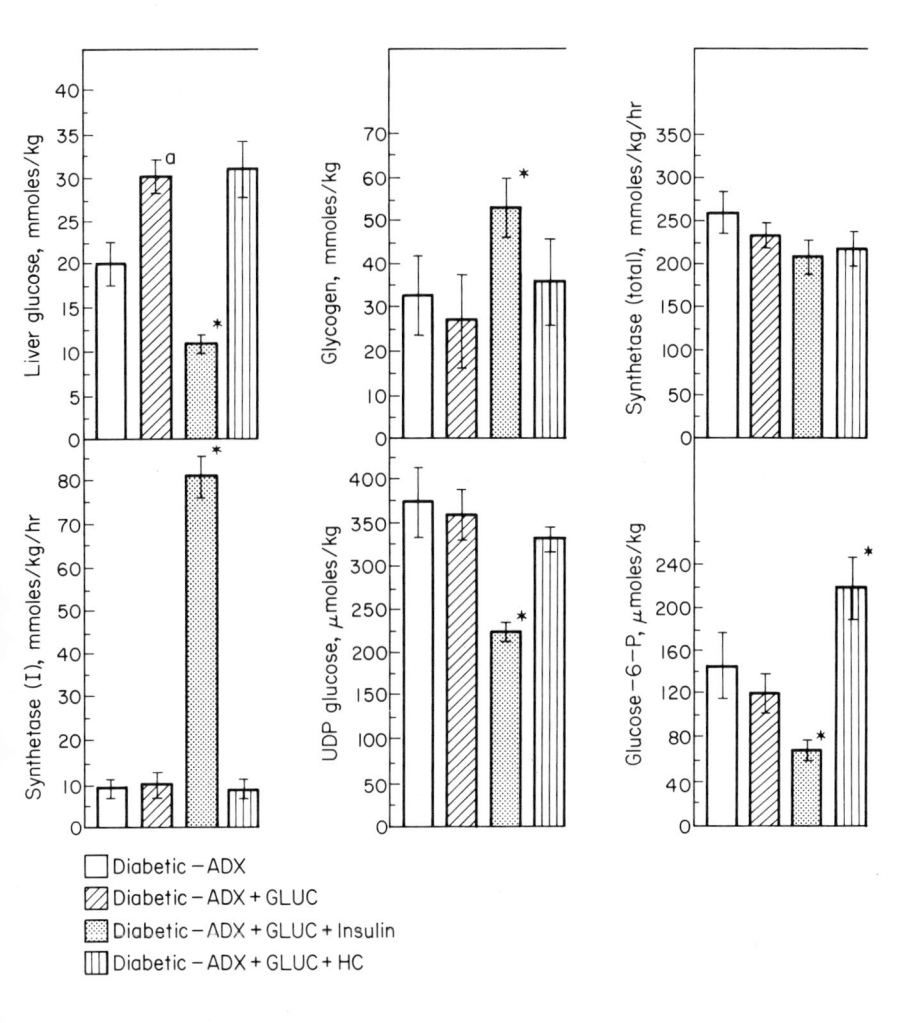

☐ Diabetic – ADX
▨ Diabetic – ADX + GLUC
▦ Diabetic – ADX + GLUC + Insulin
▥ Diabetic – ADX + GLUC + HC

a $P < 0.05$ as compared with diabetic – ADX control
* $P < 0.05$ as compared with diabetic – ADX + glucose control

Figure 6. Effect of insulin or hydrocortisone and/or glucose administration to diabetic adrenalectomized rats. Insulin, 6 units/kg, was administered intraperitoneally 75 min prior to sacrifice; hydrocortisone, 10 mg/kg subcutaneously 2 hr prior to sacrifice; glucose, 5 mg/kg orally 2 hr prior to sacrifice.

174 Control of Glycogen Synthetase

hepatic synthetase activity to hydrocortisone was tested 4, 8, 12, 24 and 48 hr after alloxan administration (40 mg/kg, intravenously) to 16 hr fasted rats. In each case, the response was measured two hr after the injection of 10 mg of hydrocortisone/kg, subcutaneously, or after saline, at the designated times after alloxan injection (Fig. 7). The response in the saline

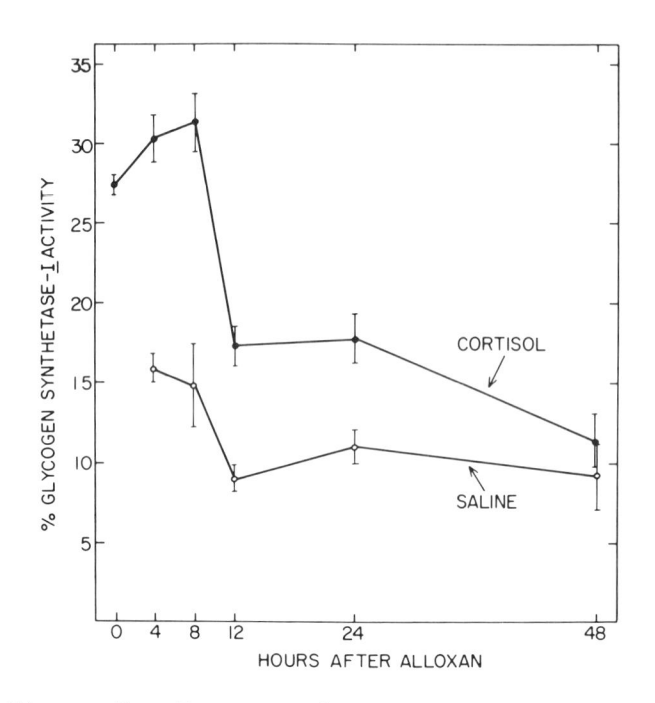

Figure 7. Percent glycogen synthetase in rat liver after alloxan treatment, following saline or cortisol (10 mg/kg subcutaneously, 2 hr prior to sacrifice) injection.

treated animals during the early time periods undoubtedly reflects the effect on the hepatic synthetase of the alloxan induced insulin release. It can be seen that the response to steroid diminished after 12 hr and was essentially absent 48 hr after the injection of alloxan.

Because 3',5'-cyclic adenylate is a known activator of synthetase I kinase, it was conceivable that the action of insulin to promote an increase in the percentage of synthetase in the I form derived from a lowering of the hepatic levels of cyclic AMP by the hormone. This possibility seemed especially important to consider because of the known effect of insulin to antagonize the effects of glucagon in liver (Bishop and Larner, 1964) and the elevating influence of alloxan diabetes or anti-insulin serum injection on the levels of hepatic 3',5'-cyclic adenylate (Jefferson, et al., 1968). Indeed, it was found that the levels of hepatic 3',5'-cyclic adenylate rose approximately 3-fold when the diabetic state was produced in rats by alloxan treatment. Furthermore, 75 min after the administration of insulin, the levels of hepatic cyclic AMP in the diabetic animals returned to almost control levels. It had, however, been previously demonstrated that the response of hepatic glycogen synthetase to insulin administration occurs very rapidly—within minutes after injection of the hormone. The effect of insulin to normalize elevated 3',5'-cyclic adenylate levels was therefore measured considerably later than when the early effects of the hormone should be demonstrable. It was obviously important to correlate the earlier changes in synthetase I levels with the tissue concentrations of 3',5'-cyclic adenylate after insulin administration. As early as 2 min after intravenous injection of 1 unit of insulin/kg to alloxan diabetic, adrenalectomized rats, there was about a 2-fold rise in the percentage of synthetase in the I form and a progressive increase to about 40% by 10 min. During this early phase, there was little or no change in hepatic cyclic AMP levels (Fig. 8). At 2 and 4 min, the levels were unchanged and at 6 and 8 min, there was an apparent, transient fall of no more than 15%, which was not statistically significant even with a total of 24 animals in this experiment. At 10 min the levels of hepatic 3',5'-cyclic adenylate were not different than those of adrenalectomized diabetic controls not receiving insulin.

It should also be pointed out that the blood glucose levels of these animals are considerably (7- to 9-fold) greater than those usually found in non-diabetic animals, but this marked hyperglycemia does not

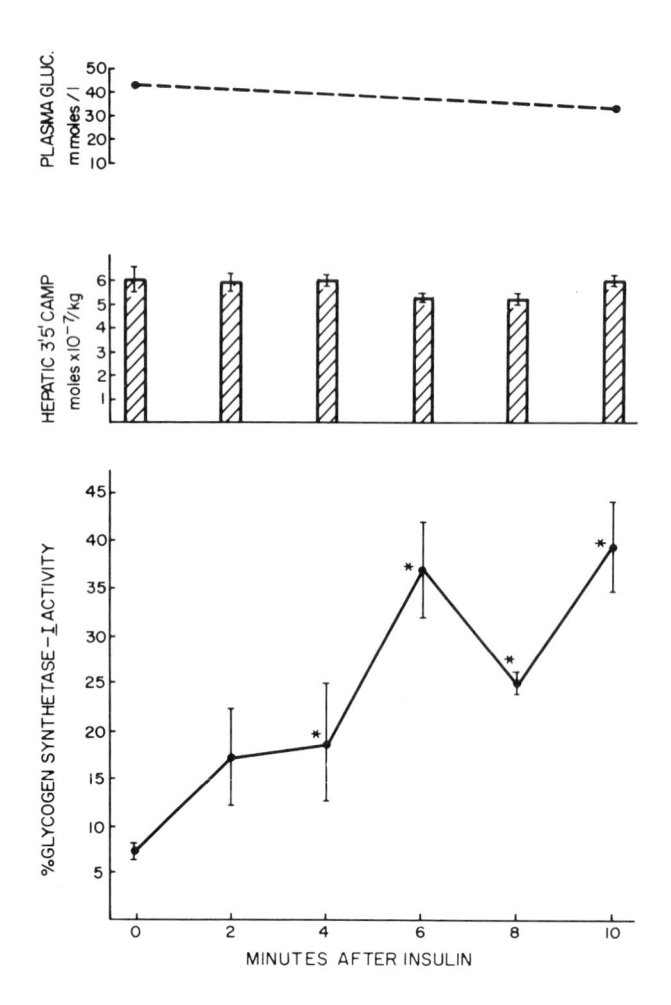

Figure 8. Time course of percent glycogen synthetase I change and 3',5'-AMP change in rat liver after insulin administration (1 unit/kg, intravenously).

influence by itself the state of the hepatic synthetase in the absence of insulin.

In the pancreatectomized dog maintained with insulin, the levels of hepatic 3',5'-cyclic adenylate are lower (about 50%) than those found in normal dogs. Infusion of glucose alone to these animals results in no change in hepatic glycogen synthetase I levels. However, infusion of insulin + glucose promotes an immediate and frequently large increase in the percentage of hepatic synthetase in the I form (Fig. 9). It

PANCREATECTOMIZED DOGS
(Diabetes Controlled with Insulin)

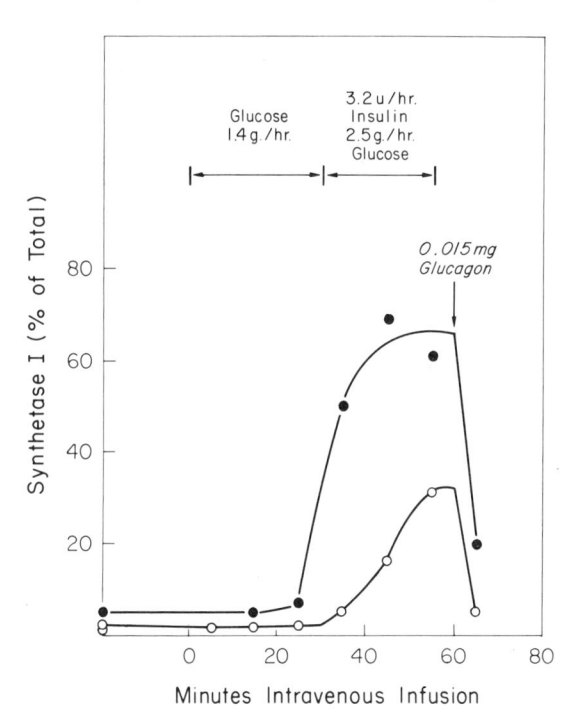

Figure 9. Graph recording responses to two different experiments. Time course of percent glycogen synthetase I change in pancreatectomized dog liver following glucose, glucose + insulin, and glucagon administration.

is evident that, under these conditions, neither glucose nor glucagon can be implicated or responsible for the increase in synthetase I activity in liver. Glucagon injection decreases, as expected, the elevated synthetase I levels induced by insulin.

The extent to which hepatic synthetase in the pancreatectomized dog responds to insulin infusion appears to be dependent on the status of the animals, especial-

178   Control of Glycogen Synthetase

ly if there is a period of insulin deprivation (72 to 96 hr) previous to the experiment. Preliminary data indicate that in these animals, the rate of increase in percentage of synthetase in the I form in the liver in response to insulin is considerably less than the increases found in normal or pancreatectomized insulin maintained dogs (Fig. 10). The activity of synthetase

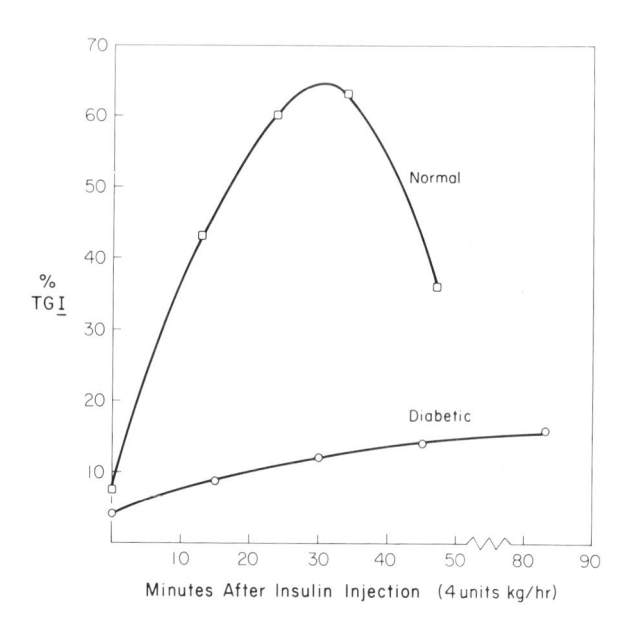

Figure 10. Time course of percent glycogen synthetase I change in normal and pancreatectomized dog liver following insulin administration.

I kinase from the liver of these diabetic dogs appeared to be normal, but the activity of synthetase D phosphatase was found to be less than 1/2 that occurring in extracts of normal liver. A decrease in synthetase D phosphatase could account at least for part of the decrease in insulin responsiveness of the glycogen synthesizing system in the pancreatectomized, insulin deprived dog livers.

The conclusion from these experiments is that insulin induces an increase in the percentage of glycogen synthetase I, not only in muscle, but also *in vivo* in heart and liver. In the absence of insulin, glucose alone does not appear to affect the levels of glycogen synthetase *I*. In the absence of glucagon (pancreatectomized dog) insulin promotes an increase in hepatic synthetase *I*. The early effects of insulin on synthetase are not mediated by a lowering of the intracellular levels of 3',5'-cyclic adenylate in muscle and liver. Finally, the early effects of corticosteroids appear to be dependent to a significant degree on the presence of insulin.

REFERENCES

Appleman, M.M., E. Belocopitow and H.M. Torres. 1964. *Biochem. Biophys. Res. Commun.* 14: 550.
Belocopitow, E. 1961. *Arch. Biochem. Biophys.* 93: 457.
Bishop, J.S., N.D. Goldberg, F. Grande and J. Larner. 1969. *Diabetes*. In press.
Bishop, J.S. and J. Larner. 1964. *J. Biol. Chem.* 242: 1935.
Bishop, J.S. and J. Larner. 1969. *Biochim. Biophys. Acta* 171: 374.
Blatt, L.M., J.O. Scamahorn and K.H. Kim. 1969. *Biochim. Biophys. Acta* 177: 553.
Craig, J.W. and J. Larner. 1964. *Nature* 202: 971.
Craig, J.W., T.W. Rall and J. Larner. 1969. *Biochim. Biophys. Acta* 177: 213.
Danforth, W.H. 1965. *J. Biol. Chem.* 240: 588.
De Wulf, H. and H.C. Hers. 1967a. *European J. Biochem.* 2: 50.
De Wulf, H. and H.C. Hers. 1967b. *European J. Biochem.* 2: 57.
De Wulf, H. and H.C. Hers. 1968. *European J. Biochem.* 6: 558.
Friedman, D.L. and J. Larner. 1962. *Biochim. Biophys. Acta* 64: 185.
Friedman, D.L. and J. Larner. 1963. *Biochemistry* 2: 669.
Goldberg, N.D., C. Villar-Palasi, H. Sasko and J. Larner. 1967. *Biochim. Biophys. Acta* 148: 665.
Hers, H.C. and H. De Wulf. 1969. *Control of Glycogen Metabolism*, W.J. Whelan, ed. (Oslo: Universitetsfor-

laget), p. 65.
Hornbrook, K.R., H.B. Burch and O.H. Lowry. 1966. *Mol. Pharmacol.* 2: 106
Huijing, F. and J. Larner. 1966a. *Biochem. Biophys. Res. Commun.* 23: 259.
Huijing, F. and J. Larner. 1966b. *Proc. Natl. Acad. Sci. U.S.A.* 56: 647.
Jefferson, L.S., J.H. Exton, R.W. Butcher, E.W. Sutherland and C.R. Park. 1968. *J. Biol. Chem.* 243: 1031.
Jungas, R.L. 1966. *Proc. Natl. Acad. Sci. U.S.A.* 56: 757.
Kreutner, W. and N.D. Goldberg. 1967. *Proc. Natl. Acad. Sci. U.S.A.* 58: 1515.
Larner, J. and F. Sanger. 1965. *J. Mol. Biol.* 11: 491.
Mersmann, H.J. and H.L. Segal. 1969. *J. Biol. Chem.* 244: 1701.
Rosell-Perez, M. and J. Larner. 1964. *Biochemistry* 3: 773.
Schlender, K.K., S.H. Wei and C. Villar-Palasi. 1969. *Biochim. Biophys. Acta.* In press.
Steiner, D.F., V. Rauda and R.W. Williams. 1961. *J. Biol. Chem.* 236: 299.
Villar-Palasi, C. and J. Larner. 1960. *Federation Proc.* 19: 1.
Villar-Palasi, C. and J. Larner. 1960. *Biochim. Biophys. Acta* 39: 171.
Villar-Palasi, C. and J. Larner. 1961. *Arch. Biochem. Biophys.* 94: 536.
Villar-Palasi, C. and J.I. Wenger. 1967. *Federation Proc.* 26: 63.
Williams, B.J. and S.E. Mayer. 1966. *Molec. Pharmacol.* 2: 454.
Yip, Agnes T. and J. Larner. 1969. *Phys. Chem. and Physics.* In press.

Effects of Biogenic Amines on Adipose Tissue Metabolism

Robert L. Jungas

Dept. of Biological Chemistry
Harvard Medical School
Boston

Adipose tissue is now widely recognized to be a site of great metabolic activity which is subject to the influences of a large variety of hormones and pharmacologic agents. The enormous importance of the metabolic processes occurring in this tissue to the overall energy metabolism of the animal was clearly recognized in the twenties by Schur and Löw (1928) when they stated "Das Fettgewebe...steht...im Zentrum der Stoffwechselvorgänge," although this fact did not become widely known until the review of Wertheimer and Shapiro appeared in 1948 (Wertheimer and Shapiro, 1948). Arndt (1926) had shown in 1926 that insulin increased the glycogen content of dog adipose tissue and Schur and Löw in 1928 urged investigators to examine adipose tissue in seeking out the action of insulin. Another 30 years were to pass however, before Winegrad and Renold clearly established the exquisite sensitivity of adipose tissue to insulin (Winegrad and Renold, 1958).

Almost simultaneously with this work reports appeared from three other laboratories indicating that the catecholamines and a variety of peptide hormones of pituitary origin caused the tissue to release free (unesterified) fatty acids (FFA) into the incubation medium (Shapiro, 1957; Reshef, *et al.*, 1958; Gordon and Cherkes, 1958; White and Engel, 1958; Engel and White, 1958). Following these reports literally dozens of laboratories leaped into the field with the result that only 7 years later a comprehensive account of the metabolism of adipose tissue and its regulation by hormones could be written (Renold and Cahill, 1965).

Fat Mobilization

The picture that has emerged from these studies of the action of the catecholamines on adipose tissue metabolism may be summarized as follows. First, and apparently of primary importance, these hormones belong to the rather large class of substances which have the property of accelerating the rate at which the tissue's endogenous stores of triglyceride are hydrolyzed to glycerol and FFA. The intricacies involved in the activation of tissue lipolytic activity by these agents are not yet well understood. An old drawing first prepared five years ago still seems to summarize adequately our current views of this process and is shown in Fig. 1. I should say immediately that several key features of the diagram have not yet been convincingly demonstrated. Shown schematically in the figure is a portion of a fat cell and the surrounding plasma and basement membranes which separate it from its blood supply. An active lipase able to attack the globules of depot fat is pictured as producing glycerol and FFA. Owing to a deficiency of glycerol kinase (Wieland and Suyter, 1957) the bulk of the glycerol so produced is not metabolized further by the fat cell (Shapiro, *et al.*, 1957; Lynn, *et al.*, 1960) and must exit. The FFA may either exit, provided albumin is present to accept them, or be reesterified to triglyceride provided a supply of glycerol phosphate is available. In order to account for the rapid changes in the activity of this depot fat lipase upon addition of any of the lip-

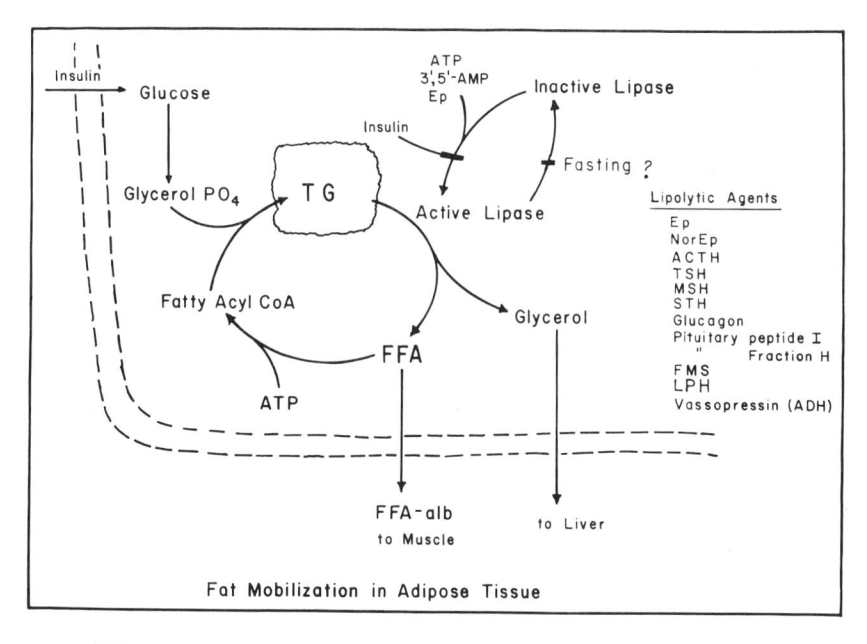

Figure 1. Schematic illustration of triglyceride metabolism. Abbreviations: TG, triglyceride; FMS, fat mobilizing substance from urine (Chalmers, et al., 1960); LPH, lipotropin (Li, 1964); Fraction H (Rudman, et al., 1961); Peptide I (Barret, et al., 1962).

olysis activators listed in the figure it seems necessary to represent the lipase as capable of existing in an inactive form. It is supposed that the lipolytic agents shift the equilibrium between these forms of the lipase toward the active state. Since all of the substances which activate lipolysis which have been tested also activate the glycogen phosphorylase of adipose tissue (Vaughan, 1960; Frerichs and Ball, 1962) it was assumed that these substances do not themselves interact directly with the lipase but that their effects are mediated by cyclic-3',5'-AMP (cyclic AMP). The measurements of cellular levels of cyclic AMP by Butcher and Sutherland (Butcher, et al., 1965) soon substantiated this supposition and taken together with the finding that methyl xanthines also accelerate lipolysis (Vaughan, 1961; Blecher, 1967) have led to general acceptance of the proposed role of cyclic AMP in this process. The catecholamines apparently increase the cellular levels of cyclic AMP by enhancing the adenyl cyclase activity of the tissue (Klainer, et al., 1962; Birnbaumer and Rodbell, 1969). Whether additional regulation is effected by alterations in the activity of the specific diesterase responsible for destroying cyclic AMP is not yet clear.

In contrast to the long list of agents able to evoke increases in the rates of lipolysis only one hormone normally found in plasma, namely insulin, is capable of inhibiting the process. This action of insulin is not well understood. Since insulin is able under some conditions to diminish the tissue adenyl cyclase activity (Jungas, 1966) and does lower cellular cyclic AMP concentrations (Butcher, et al., 1966) it seems reasonable to postulate that this hormone also acts by way of cyclic AMP. Because insulin also increases the glucose uptake of the tissue it may also increase the supply of glycerol phosphate and thus diminish FFA release by virtue of enhancing the reesterification process (Steinberg, et al., 1960; Jungas and Ball, 1963). Measurements of glycerol production however, both *in vitro* (Jungas and Ball, 1963; Mahler, et al., 1964) and *in vivo* (Carlson and Orö, 1963; Hagen, 1963) make it clear that the major effect of insulin is to suppress the activity of the depot fat lipase. In fact by promoting reesterification insulin lowers the tissue levels of FFA which in turn are inhibitors of the depot fat lipase. Some data illustrating this property of FFA are shown in Fig. 2. Thus were it not for its strong antilipolytic activity insulin would be a very inefficient agent for suppressing fat mobilization.

Figure 1 also includes a speculation as to the manner in which lipolysis might be accelerated during fasting. The basis for making this suggestion is the behavior of tissue from fasted animals when the tissue is deprived of oxygen. Illustrated in Fig. 3 is the basal glycerol production of normal, fasted, and fasted-refed rats. Note that anaerobiosis diminishes the lipase activity of normal tissue and even more strikingly suppresses the enhanced lipolysis seen in the

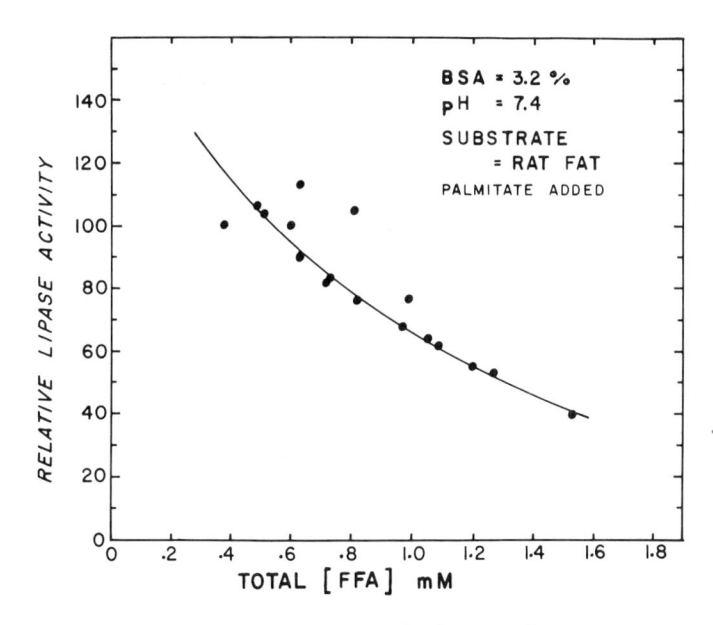

Figure 2. Inhibition of depot fat lipase by free fatty acids. The enzyme source was a crude homogenate of rat epididymal adipose tissue prepared in 0.05 M sodium phosphate, pH 7.4, containing 3.2% bovine serum albumin (BSA) and an excess of a fat emulsion. The emulsion was prepared by sonicating the fat pads of another rat in water and heating 5 min at 50°C to destroy endogenous lipases. The initial FFA content of the homogenate was adjusted to the values shown on the abscissa by addition of potassium palmitate. The content of FFA was assayed before and after a 15 min incubation at 37°C to determine the extent of lipolysis.

fasted-refed tissue. Thus energy in the form of ATP may be required to maintain the lipase in its active form in agreement with the studies of Rizack in cell-free systems (Rizack, 1964). However, when tissue from fasted rats is incubated under nitrogen there is no reduction in the rate of glycerol production. The lipase of fasted tissue seems to be "frozen" in the active form. This result could be readily accounted

186   Biogenic Amines and Adipose Tissue

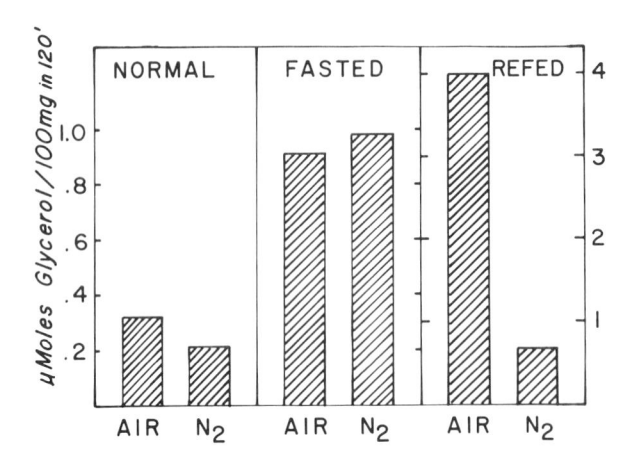

Figure 3. Effect of anaerobiosis on basal lipolysis. Data redrawn from Ball and Jungas (1963).

for if lipolysis were initiated during fasting by the gradual decay of the process (enzyme?) responsible for inactivating the lipase. The fact that neither the catecholamines nor the pituitary hormones appear to be essential for increased fat mobilization during fasting (Goodman and Knobil, 1959; 1961) would also be explained. Lipase activation by such a decay would likely be much slower in onset than that which follows exposure to the catecholamines as would be appropriate to the fasting situation. Moreover the resistance of the lipase from fasted animals to the antilipolytic action of insulin (Jungas and Ball, 1964) and of prostaglandin (Stock and Westermann, 1966) would be accounted for. Unfortunately, to date our attempts to demonstrate an enzymatic inactivation of the lipase even in extracts of tissue from fed animals have been unsuccessful (Schwartz, J. and Jungas, 1968, unpublished observations).

The hydrolysis of triglycerides is actually a more complicated affair than indicated in Fig. 1. In Fig. 4 this portion of the process is presented in more detail to make it clear that several enzymes are involved. The studies of Vaughan and Steinberg and their associates (Strand, et al., 1964; Vaughan, et al., 1964) have revealed that the depot fat lipase removes a fatty

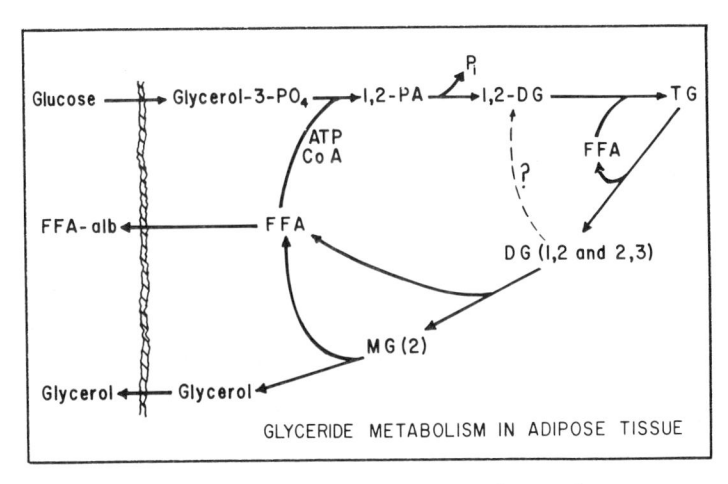

Figure 4. Reactions involved in fat mobilization. Abbreviations: MG, DG, TG, mono, di, and triglycerides.

acid from either the 1 or 3 positions of the triglyceride yielding a diglyceride. Another lipase more active on diglycerides than on triglycerides then removes the remaining alpha-fatty acid yielding a 2-monoglyceride. This is then subject to attack by a monoglyceride lipase to complete the process. Both the monoglyceride and diglyceride lipase activities appear to be present in excess and thus the breakdown of a triglyceride molecule, once initiated by the depot fat lipase tends to go immediately to completion. Partial glycerides do not accumulate in the tissue except under unusual conditions (Scow, et al., 1965). Thus the reports of Gorin and Shafrir (1967) and of Vaughan, et al. (1964) that the monoglyceride lipase may also be subject to activation by the catecholamines are of uncertain significance.

Other Metabolic Effects

Although the role of norepinephrine and epinephrine in the regulation of lipolysis is of such outstanding importance, these hormones do influence a variety of other processes in adipose tissue. Many of these additional effects appear to be consequences of the increased rate of triglyceride hydrolysis or of increased levels of cyclic AMP and may not reflect additional sites of influence of the hormones. For example, the ability of epinephrine to decrease the entry of α-aminoisobutyric acid into isolated fat cells (Touabi and Jeanrenaud, 1969) could not be observed in a preparation of fat cell "ghosts" which are largely depleted of their triglyceride content (Clausen and Rodbell, 1969). The effect has therefore been regarded as secondary to the rise in FFA in the fat cells since the adenyl cyclase of the ghosts remained responsive to the catecholamines (Birnbaumer and Rodbell, 1969). Alternatively, factors essential for the response may have been lost from the ghost preparation.

The inhibition of protein synthesis by epinephrine (Herrera and Renold, 1960) may be secondary to the rise in FFA but the possibility that cyclic AMP may be involved must still be considered.

The alterations in carbohydrate metabolism in response to the catecholamines are more complex and are summarized in Fig. 5, taken from Jeanrenaud's recent review (Jeanrenaud, 1968). First, the glucose uptake is increased but higher concentrations of the hormones are needed to demonstrate this effect than are necessary to give pronounced effects on lipolysis (Vaughan, 1961; Blecher, *et al.*, 1969). This result may simply reflect the technical difficulty of measuring small changes in glucose consumption. The esterification of FFA is increased in the presence of epinephrine (Vaughan and Steinberg, 1963) and a substantial fraction of the rise in glucose consumed must be diverted to supply the glycerol 3-phosphate needed for this process.

Notice also the ability of epinephrine to diminish the fraction of entering glucose which is metabolized via the pentose cycle (Katz, *et al.*, 1966). It now appears likely that this action results from the rise in cyclic AMP caused by the catecholamines. The phosphofructokinase of adipose tissue is unusually sensitive to the activating influence of this cyclic nucleotide (Söling, H.D., 1966, personal communication; Denton and Randle, 1966). Recent measurements by Halperin and Denton (1969) of the cellular levels of hexose phosphates in tissue exposed to epinephrine have now provided direct evidence for an activation of this en-

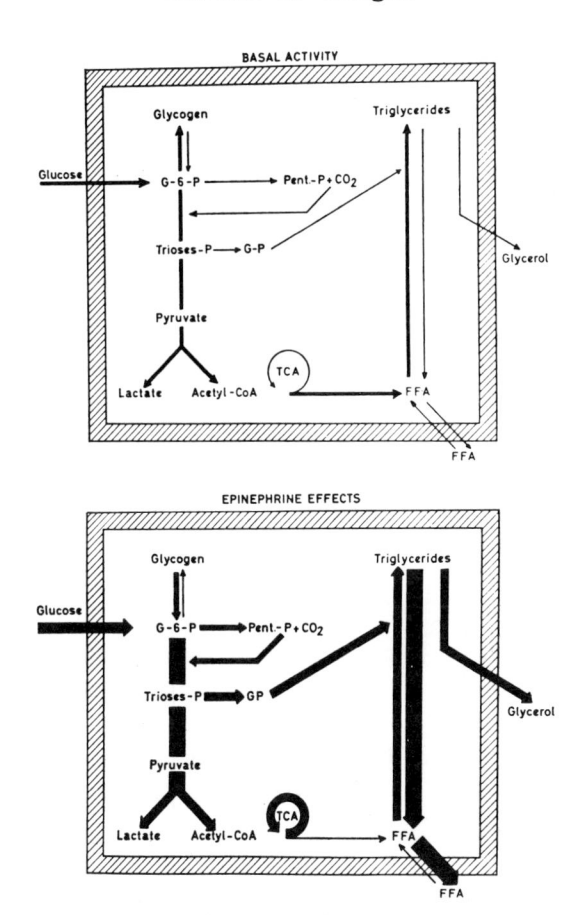

Figure 5. Diagrammatic representation of
glucose metabolism in adipose tissue.
From Jeanrenaud (1968).

zyme. Activation of phosphofructokinase lowers the
steady-state level of the hexose required for a given
throughput of the Embden-Meyerhof glycolytic pathway.
Consequently relatively less substrate is available
for entry into the pentose cycle. Insulin, it may be
noted, has just the opposite effect. The least satisfactory part of this explanation is the lack of evidence that entry into the pentose cycle is in fact controlled by the supply of glucose-6-phosphate. A more
important factor may be the decrease in the conversion
of glucose to fatty acids seen in the presence of epinephrine (Vaughan, 1961; Katz, et al., 1966). This

would diminish the supply of TPN$^+$ and thus tend to reduce the contribution of the pentose cycle. Currently there is wide disagreement as to why lipogenesis from glucose should be diminished by the catecholamines. Most investigators would agree however that the effect is secondary to the rise in cellular FFA.

The other actions of epinephrine shown in Fig. 5 are less difficult to understand. Glycogen metabolism is of course altered secondary to the changes in cyclic AMP. The increase in the citric acid cycle is needed to provide the ATP required to support the enhanced rate of FFA esterification. This requirement is reflected also in an increased rate of oxygen consumption in the presence of catecholamines (Hagen and Ball, 1961) whose magnitude can be quantitatively accounted for on this basis (Ball and Jungas, 1964). The reason for the enhancement in lactic acid production is obscure and presently under active investigation.

A more complete discussion of the metabolic effects of the catecholamines on adipose tissue metabolism may be found in several excellent recent reviews (Wenke, 1966; Himms-Hagen, 1967; Rudman and Di Girolamo, 1967; Jeanrenaud, 1968; Rodbell, et al., 1968).

Receptors

I would like to turn now from this general overview and consider in more detail several aspects of the action of the catecholamines on adipose tissue metabolism which have been of special interest the last few years. As with other metabolic actions of the catecholamines (Himms-Hagen, 1967) there has been some difficulty in classifying the receptor involved in adipose tissue into either the *alpha* or *beta* category of Ahlquist (1948). Several lines of evidence now make it apparent that these receptors more closely resemble the *beta* type. For example, butyl and isopropyl arterenols accelerate lipolysis at lower concentrations than epinephrine and norepinephrine, the latter two being approximately equivalent in their effects (White and Engel, 1958; De Caro, 1967). *Beta*-blocking agents in general are effective in much lower doses than are *alpha* blockers and show competitive behavior with the catecholamines except at very high concentrations (Wenke, 1966). *Alpha* sympatholytics on the other hand behave as noncompetitive and nonspecific antagonists. For example concentrations of the *alpha*-lytic agent phentolamine sufficient to inhibit catecholamine-induced lipolysis also inhibit the lipolytic action of theophylline whereas *beta*-lytic agents inhibit adrenergic lipid mobilization at concentrations without effect on lipolysis evoked by theophylline (Brodie, *et al.*, 1966).

Important quantitative distinctions may however be drawn between the *beta* receptors of adipose tissue and other more typical receptors of the same classification in other tissues. I have reproduced in Fig. 6 a drawing taken from the excellent review of this subject by Wenke (1966). You will note that the lipolytic response of adipose tissue to the series of substituted catecholamines falls between that of typical *alpha* and *beta* functions but more closely resembles the *beta* type. The explanation for these and other quantitative differences (Lincova, *et al.*, 1968) is not known. It seems likely that they may arise from interactions between adjacent receptor sites of the kind frequently encountered between binding sites of enzymes.

Innervation

The discovery of the manifold actions of the catecholamines on adipose tissue has given impetus to the investigations of the sympathetic innervation of the tissue (Havel, 1965). The importance of the nerve supply to the mobilization of fat and storage of glycogen has been recognized for many years (Hausberger and Neuenschwander-Lemmer, 1939). The demonstration by Correll (1961) that the release of FFA from rat or rabbit adipose tissue *in vitro* could be slightly increased by electrical stimulation of the nerve supply has been confirmed by Weiss and Maickel (1968) and extended to dog epigastric adipose tissue *in situ* (Orö, *et al.*, 1965). Uncertainty has persisted with respect to the extent to which the neural regulation *in vivo* is dependent upon changes in blood flow as opposed to

192   Biogenic Amines and Adipose Tissue

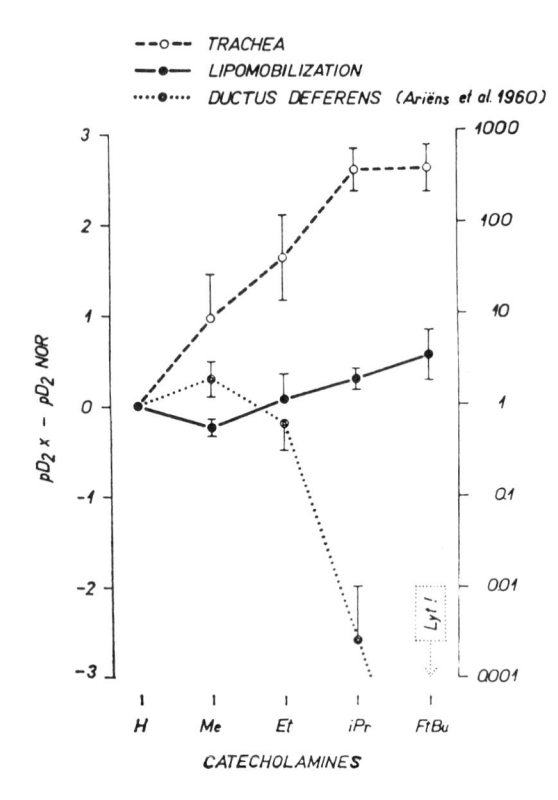

Figure 6. Relative parameters of affinity
(ΔpD_2) of various N-substituted catecholamines. Abbrevations: H, hydrogen; Me,
methyl; Et, ethyl; iPr, isopropyl; FtBu,
phenyl-tert-butyl. The abbrevation Lyt
signifies that FtBu is an *alpha* sympatho-
lytic agent for the ductus deferens.
(Means ± 95% fiducity intervals are given.)
From Wenke (1966).

direct neural influences upon fat cell metabolism
(Wirsén, 1965). Recently, studies with the electron
microscope have greatly expanded our knowledge of the
innervation of this tissue. The existence of nerve
fibers branching out from the vascular innervation to
contact individual fat cells has been clearly demonstrated in brown adipose tissue. Only the basement membrane surrounding the individual cells separates their

plasma membrane from these nerve endings (Bargmann, et al., 1968).
Using the technique of fluorescent microscopy of tissue specimens exposed to formaldehyde vapors, Derry, et al. (1969) have obtained evidence that the nerves supplying the fat cells may constitute a unique system of innervation distinct from the major vascular supply. Following either surgical or immunological sympathectomy the vascular innervation atrophied substantially whereas the fat cell innervation survived. These workers observed "intrinsic ganglia" in the fatty tissue and suggested that short post-synaptic fibers might emerge from them to innervate the fat cells. Only about 5% of the total catecholamine content of the tissue could be ascribed to this unusual system. These observations made on the interscapular brown fat of the rat need to be extended to the more common white adipose tissue. In view of the great differences in the metabolic functions of these two types of fatty tissue (Bizzi, et al., 1969) it is not unlikely that their neural innervation may be quite different. Convincing evidence for a parasympathetic innervation of either tissue is still lacking. The presence of the "intrinsic ganglia" in adipose tissue makes the significance of the presence of acetylcholine esterase even more uncertain (Salvador and Kuntzman, 1965).

Only via Cyclic AMP?

Another topic of current interest relates to the role of cyclic AMP in the response to the catecholamines. There are two questions in this regard which I would like to discuss. First, do the catecholamines influence adipose tissue metabolism solely by virtue of their ability to increase the tissue levels of cyclic AMP, or are there actions of these hormones which are independent of cyclic AMP? Several observations have been reported recently which lead me to suggest that not all of the effects of epinephrine on adipose tissue are mediated by cyclic AMP.
Suspicions in this regard were first aroused when it was found that, contrary to its effects on muscle (Walaas and Walaas, 1950), epinephrine increased the

glucose uptake of adipose tissue (LeBoeuf, *et al.*, 1959). Initially the fact that other hormones which increased lipolysis also increased glucose uptake, and that added FFA had a similar though smaller effect on glucose consumption led to the interpretation that the epinephrine effect on glucose utilization was merely a consequence of the increase it caused in the supply of intracellular FFA (Cahill, *et al.*, 1960; LeBoeuf and Cahill, 1961). When it was later found that other agents which are known to increase the tissue cyclic AMP concentration such as caffeine and theophylline in fact decreased the glucose uptake (Vaughan, 1961; Anderson, *et al.*, 1966; Blecher, 1967) as did added dibutyryl-cyclic AMP (Blecher, 1967; Bray, 1967; Blecher, 1968), it became apparent that a dual action of epinephrine had to be seriously considered.

Bray (Goodman and Bray, 1966; Bray, 1967; Bray and Goodman, 1968) and then Blecher (Blecher, *et al.*, 1969) took up the challenge of demonstrating the independence of the action of epinephrine on glucose utilization from that on lipolysis. Probably the strongest evidence was provided by Bray (1967) who was able to show that in the presence of propranolol high concentrations of epinephrine would increase glucose oxidation without causing a detectable rise in tissue or medium FFA. This observation showing a preferential blockage by propranolol of the effect elicited by the lower dose of the hormone (Vaughan, 1961; Blecher, *et al.*, 1969), i.e., lipolysis, implies that epinephrine accelerates glucose uptake via a different receptor.

A good system in which to study this problem would seem to be the fat cell ghost preparation of Rodbell since these membranous bags are essentially devoid of triglyceride but retain sufficient integrity to respond to the osmolarity of the medium and to insulin (Rodbell, *et al.*, 1968). An examination of the effects of epinephrine in this system failed to clarify the situation for it was found that epinephrine actually decreased the glucose utilization of such preparations (Rodbell, 1967).

Perhaps more significant was the report by Bray and Goodman (1968) that epinephrine increases the L-arabinose space of adipose tissue in a manner similar to insulin. The discovery that epinephrine also increases the penetration of non-utilizable sugars into skeletal muscle (Saha, et al., 1968) encourages the view that the transport of sugars across membranes may be commonly altered by epinephrine. Park, et al. (1969) however, could find no effect of this hormone on galactose or 3-0-methylglucose transport into adipose tissue.

Recently Girardier, et al. (1968) have provided stronger evidence for a membrane effect of epinephrine independent of cyclic AMP. These workers used a microelectrode to record the potential across the plasma membrane of fat cells of brown adipose tissue. They found that low concentrations of epinephrine would depolarize the membrane while increasing the flux of potassium ions through it. It was proposed that this action of the hormone might play an important role in its calorigenic effect (Himms-Hagen, 1967). Maintaining a gradient across the leaky membrane would in effect constitute an ATPase and hence result in the production of heat. They reported moreover that theophylline did not share this action of epinephrine nor did added cyclic AMP. Thus the depolarizing action of epinephrine does not appear to result solely from increased cellular levels of cyclic AMP or FFA.

A second action of epinephrine which is even more clearly independent of cyclic AMP may be of less physiological significance, but is nevertheless highly relevant to recent studies of the activation of lipolysis by epinephrine in cell-free systems. Fujii and his collaborators (Okuda, et al., 1966; Yanagi, et al., 1968) have shown that epinephrine somehow alters the physical state of certain fat emulsions so as to increase their susceptibility to hydrolysis by lipases. When an homogenate of adipose tissue is centrifuged a large cake of fat floats to the top and solidifies when cooled. When this fat layer, treated to eliminate its content of lipases, is subsequently rehomogenized in Tris buffer and incubated with large amounts of epinephrine for one hour, it becomes a better substrate when later tested with a variety of lipases as shown in Table 1. Note that this is not an effect on the lipases but on the substrate. Cyclic AMP does not share this effect. These observations have recently been

TABLE 1

Examination of the epinephrine
effect with various lipases

Lipase	Fatty acid released + Epi- nephrine	Fatty acid released - Epi- nephrine	Epinephrine effect (µEq/g) Δ
Pancreatic	44.6*	18.8	25.8
Microbial	16.3	7.4	8.9
Adipose tissue	18.8	3.6	15.2

*All values are microequivalents per g of adipose tissue.

The procedure was as described in the text, except that the indicated lipases were used. One hundred µg of pancreatic lipase or 1 mg of microbial lipase dissolved in 1 ml of 0.25 M sucrose was used as enzyme solution.

confirmed by Mosinger (1969) and indicate that other reports of an activation by epinephrine of lipolysis in cell-free systems must be evaluated critically (Rizack, 1961; Rubinstein, et al., 1964; Ho, et al., 1967). To compound the potentital confusion the *beta*-blocking agent dichloroisopropylarterenol exerts the opposite influence tending to clear unstable fat emulsions, thereby reducing their efficiency as substrates for lipases (Jungas, 1962; unpublished observations).

Role of calcium. The second question involving cyclic AMP is this. If, as we have just seen, epinephrine can activate lipolysis *in vitro* in a manner independent of cyclic AMP, can we be certain that cyclic AMP in fact plays a role in the regulation of lipolysis *in vivo*? First let us note that the cyclic AMP hypothesis is not yet universally accepted. Fujii and his collaborators (Yanagi, et al., 1967) have recently proposed that ACTH increases lipolysis not via cyclic AMP but as a result of its ability to increase the entry of calcium into adipocytes. Calcium is known to facilitate the action of pancreatic lipase (Schonheyder and Volqvartz, 1945) and lipoprotein lipase (Korn, 1955) as well as the lipase activity of adipose tissue homogenates (Rizack, 1964; Bjorntorp and Furman, 1962; Boyer, 1967; Yanagi, *et al.*, 1967). The Japanese workers were able to demonstrate the increased entry of calcium in response to ACTH in rat adipose tissue segments *in vitro* and Akgün and Rudman (1969) have just confirmed this finding in rabbits both *in vivo* and *in vitro*. Epinephrine also caused an increased entry of calcium into adipose tissue. Although added calcium does not appear to be required for the lipolytic action of epinephrine its omission does diminish the response (Mosinger and Vaughan, 1967a). Calcium also enhances the lipolysis evoked by raising the potassium concentration of the incubation medium (Bleicher, *et al.*, 1966). Thus the proposal of Fujii remains a plausible though currently less popular alternative to the cyclic AMP hypothesis.

Gaps in the evidence. Quite apart from the merits of this alternative proposal there are a number of considerations which merit our attention before the rather strong evidence for the cyclic AMP hypothesis (Butcher, *et al.*, 1968) can be regarded as conclusive. I have alluded earlier to the fact that no direct evidence for an inactive form of the depot fat lipase is yet available. Also no low molecular weight cofactors or regulators analogous to those known to be involved in the phosphorylase, phosphofructokinase and glycogen synthetase systems have been discovered. Thus the depot fat lipase cannot yet be said to fit into the pattern of other enzymes affected by cyclic AMP.

Moreover, the action of cyclic AMP on lipolysis in cell-free systems has been very difficult to demonstrate. The success of Rizack (1964) has been widely quoted but other investigators have had difficulty in observing this phenomenon. Among those unsuccessful in this regard I might list my own laboratory and several others whose failures I am not authorized to quote. Even Rizack did not obtain a reasonable concentration dependence for the effect of cyclic AMP. Concentrations of 13 and 20 micromolar stimulated the lipase in the presence of ATP but all other concentrations failed.

Measurements of cellular cyclic AMP. Some of the strongest evidence in support of the role of cyclic AMP in lipolysis has come from the measurements of levels of this nucleotide in adipose tissue segments (Butcher, et al., 1965) and isolated fat cells (Butcher, et al., 1968). These data also have not been without complications. First, the control levels of about 5 micromolar were rather high compared to those observed in other tissues and increases beyond about 50% were not closely correlated with the rates of lipolysis (Park, et al., 1969). It must be added however, that these observed intracellular concentrations of cyclic AMP fit remarkably well with those reported by Rizack (1964) to be effective in his cell-free system. Some recent measurements of Kuo and DeRenzo (1969) shown in Fig. 7 illustrate the difficulties involved. Notice that either norepinephrine or theophylline exerted its maximum effect on FFA release without altering the adipocyte levels of cyclic AMP detectably. Thus at the moment it seems necessary to resort to a model involving some type of cellular compartmentation of cyclic AMP to reconcile the data with the hypothesis (Park, et al., 1969). This in turn of course would spoil the fit between the cyclic AMP concentrations effective in the cell-free system and these measured in fat cells.

Serotonin

Vaughan (1960) reported that serotonin activated the phosphorylase of adipose tissue but did not increase the release of FFA, whereas low concentrations of epinephrine caused an increased release of FFA without activating phosphorylase. Her conclusion (Vaughan, 1961) was that cyclic AMP was unlikely to be involved in the regulation of lipolysis. More recent studies have revealed that serotonin does have a weak lipolytic action as revealed by measurements of glycerol production (Vaughan and Steinberg, 1965). This activity can be observed at extremely low concentrations of serotonin (0.3 micromolar) provided an in-

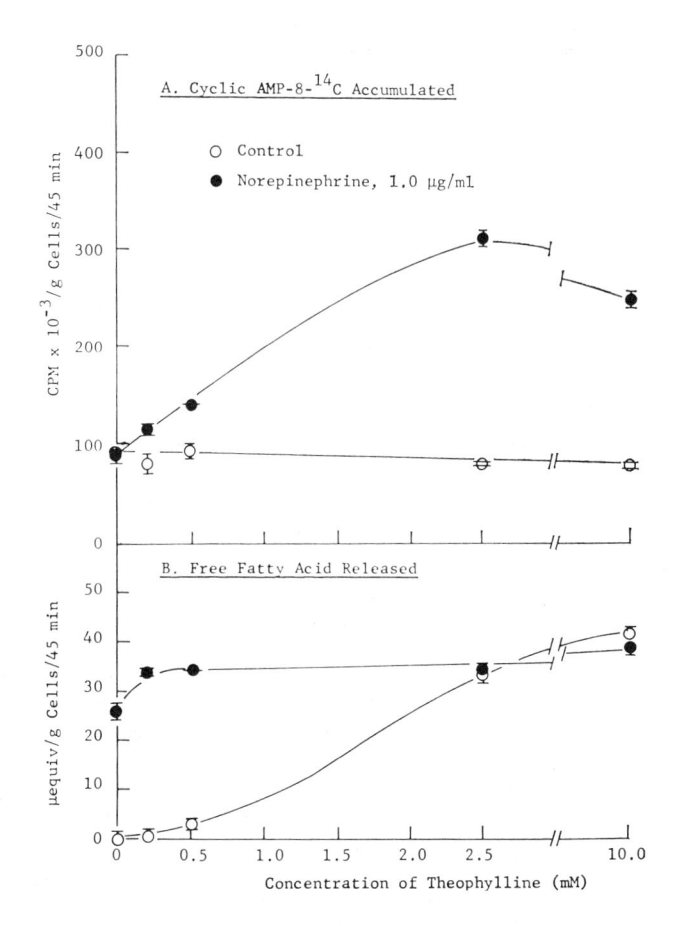

Figure 7. Effects of norepinephrine and theophylline on the release of FFA from isolated adipose cells and their content of cyclic AMP. From Kuo and DeRenzo (1969).

hibitor of monoamine oxidase and theophylline are also present (Bieck, et al., 1967). Melatonin does not share this effect of serotonin (Vaughan and Barchas, 1966). The cyclic AMP hypothesis can of course be rescued by proposing that the concentration of cyclic AMP needed to activate phosphorylase is below that which is effective on the lipolytic system. Another explanation would then be required to account for the action of low concentrations of epinephrine. The fact that it is less difficult to demonstrate the inhibitory influence of insulin on the lipase than on the phosphorylase activity of the tissue (Jungas, 1966) adds further credence to this proposal.

Action of Added Cyclic AMP

When the action of cyclic AMP itself on adipose tissue segments or isolated fat cells is examined the results further complicate the picture. Mosinger and Vaughan (1967b) have reported that the addition of 100 micromolar cyclic AMP to adipose tissue incubated in sodium phosphate buffer results in enhanced lipolysis. Either the removal of the phosphate, or addition of calcium, magnesium or potassium greatly reduces the effectiveness of the nucleotide (Fig. 8). These ionic manipulations exert just the *opposite* effect on the lipolytic action of epinephrine. Moreover, lipolysis evoked by epinephrine or theophyllin or accelerated by fasting could be *inhibited* by the addition of cyclic AMP. The dibutyryl analog of cyclic AMP behaved quite differently in that its lipolytic action was unaffected by the ionic environment (Mosinger and Vaughan, 1967a) or by insulin (Goodman, 1969). Mosinger and Vaughan (1967b) and Goodman (1969) agree that insulin blocks the lipolytic action of added cyclic AMP. High concentrations of imidazole (0.15 M) which are thought to activate the tissue diesterase activity (Butcher and Sutherland, 1962) blocked the action of both cyclic AMP and its dibutyryl analog (Goodman, 1969). These complex findings make it clear that the metabolic consequences of increased cellular concentrations of cyclic AMP cannot be reliably investigated simply by adding cyclic AMP to the bathing medium. Whether or not the dibutyryl analog can be so used is still an open question.

Conclusions

What finally, may be concluded with respect to the involvement of cyclic AMP? (1) The action of the catecholamines as well as that of the other "fast-acting" (Jeanrenaud, 1968) hormones on lipolysis is most likely

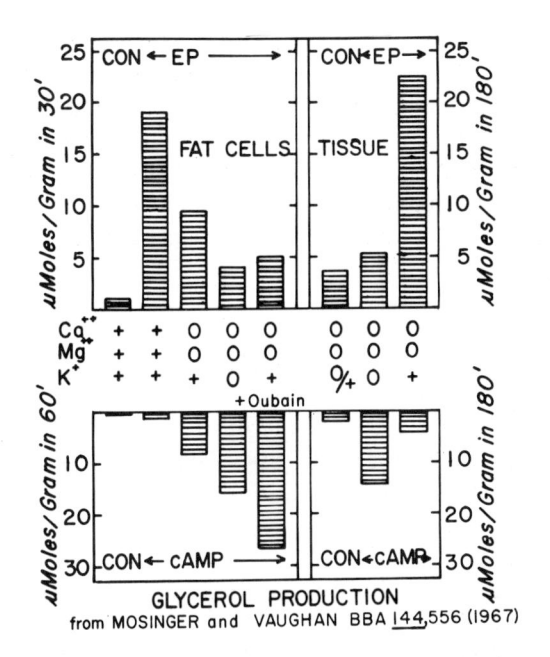

Figure 8. Comparison of the effects of the ionic composition of the incubation medium on lipolysis evoked by epinephrine or by cyclic AMP. Data are redrawn from Mosinger and Vaughan (1967a). Concentrations employed are: epinephrine, 0.05 µg/ml; cyclic AMP, 1 mM; oubain, 0.5 mM; ions, as in Krebs-Ringer phosphate.

mediated by cyclic AMP. The details of the overall process of lipase activation however are still sufficiently obscure to prevent the uncritical acceptance of this hypothesis at the present time. (2) Nearly all of the other metabolic changes observed in adipose tissue exposed to the catecholamines are probably secondary changes resulting either from the increased intracellular supply of cyclic AMP or of free fatty acids. The depolarizing effect of epinephrine on the membranes of brown fat cells appears to be an exception, and the facilitation of sugar entry may also be. (3) The manner in which the interaction of a catecholamine with its receptor on the plasma membrane of a fat cell produces the above changes is totally obscure. No alterations prior to the change in adenyl cyclase activity have yet been identified. The possibility that a conformational change in the plasma membrane could result from the interaction of hormone and receptor, and that this change might be sufficient to alter the activity of membrane-associated enzymes such as the cyclase stands as the outstanding investigative challenge in this field.

REFERENCES

Ahlquist, W.B. 1948. *Am. J. Physiol.* 153: 586.
Akgün, S. and D. Rudman. 1969. *Endocrinology* 84: 926.
Anderson, J., G. Hollifield and J.A. Owen, Jr. 1966. *Metabolism* 15: 30.
Arndt, H.J. 1926. *Verhandl. Deut. Pathol. Ges.* 21: 297.
Ball, E.G. and R.L. Jungas. 1963. *Biochemistry* 2: 586.
Ball, E.G. and R.L. Jungas. 1964. *Recent Progr. Hormone Res.* 20: 183.
Bargmann, W., G.v. Hehn and E. Lindner. 1968. *Z. Zellforsch.* 85: 601.
Barrett, R.J., H. Friesen and E.B. Astwood. 1962. *J. Biol. Chem.* 237: 432.
Bieck, P., K. Stock and E. Westermann. 1967. *Arch. Exp Exptl. Pathol. Pharmakol.* 256: 218.
Birnbaumer, L. and M. Rodbell. 1969. *J. Biol. Chem.* 244: 3477.
Bizzi, A., A.M. Codegoni, A. Lietti and S. Garattini. 1969. *Biochem. Pharmacol.* 17: 2407.
Blecher, M. 1967. *Biochem. Biophys. Res. Commun.* 27: 560.
Blecher, M. 1968. *Gunma Symp. Endocrinol.* 5: 145.
Blecher, M., N.S. Merlino, J.T. Ro'Ane and P.D. Flynn. 1969. *J. Biol. Chem.* 244: 3423.
Bleicher, S.J., L. Farber, A. Lewis and M.G. Goldner. 1966. *Metabolism* 15: 742.
Björntorp, P. and R.H. Furman. 1962. *Am. J. Physiol.* 203: 316.
Boyer, J. 1967. *Biochim. Biophys. Acta* 137: 59.
Bray, G.A. 1967. *Biochem. Biophys. Res. Commun.* 28: 621.
Bray, G.A. 1967. *J. Lipid Res.* 8: 300.

Bray, G.A. and H.M. Goodman. 1968. *J. Lipid Res.* 9: 714.
Brodie, B.B., J.I. Davies, S. Hynie, G. Krishna and B. Weiss. 1966. *Pharmacol. Rev.* 18: 273.
Butcher, R.W., C.E. Baird and E.W. Sutherland. 1968. *J. Biol. Chem.* 243: 1705.
Butcher, R.W., R.J. Ho, H.C. Meng and E.W. Sutherland. 1965. *J. Biol. Chem.* 240: 4515.
Butcher, R.W., J.G.T. Sneyd, C.R. Park and E.W. Sutherland, Jr. 1966. *J. Biol. Chem.* 241: 1651.
Butcher, R.W. and E.W. Sutherland. 1962. *J. Biol. Chem.* 237: 1244.
Cahill, G.F., Jr., B. LeBoeuf and R.B. Flinn. 1960. *J. Biol. Chem.* 235: 1246.
Carlson, L.A. and L. Orö. 1963. *Metabolism* 12: 132.
Chalmers, T.M., A. Kekwick, G.L.S. Pawan and I. Smith. 1958. *Lancet* 1: 866.
Clausen, T. and M. Rodbell. 1969. *J. Biol. Chem.* 244: 1258.
Correll, J.W. 1961. *Trans. Am. Neurol. Assoc.* 86: 178.
DeCaro, L.G. 1967. *Acta Neuroveget.* (Vienna) 30: 44.
Denton, R.M. and P.J. Randle. 1966. *Biochem. J.* 100: 420.
Derry, D.M., E. Schonbaum and G. Steiner. 1969. *Can. J. Pharmacol. Physiol.* 47: 57.
Engel, F.L. and J.E. White. 1958. *J. Clin. Invest.* 37: 1556.
Frerichs, H. and E.G. Ball. 1962. *Biochemistry* 1: 501.
Girardier, L., J. Seydoux and T. Clausen. 1968. *J. Gen Gen. Physiol.* 52: 925.
Goodman, H.M. 1969. *Proc. Soc. Exptl. Biol. Med.* 130: 97.
Goodman, H.M. and G.A. Bray. 1966. *Am. J. Physiol.* 210: 1053.
Goodman, H.M. and E. Knobil. 1959. *Proc. Soc. Exptl. Biol. Med.* 102: 493.
Goodman, H.M. and E. Knobil. 1961. *Am. J. Physiol.* 201: 1.
Gordon, R.S., Jr. and A. Cherkes. 1958. *Proc. Soc. Exptl. Biol. Med.* 97: 150.
Gorin, E. and E. Shafrir. 1967. *Biochim. Biophys. Acta* 137: 189.
Hagen, J.H. 1963. *J. Lipid Res.* 4: 46.

Hagen, J.H. and E.G. Ball. 1961. *Endocrinology* 69: 752.
Halperin, M.L. and R.M. Denton. 1969. *Biochem. J.* 113: 207.
Hausberger, F.X. and N. Neuenschwander-Lemmer. 1939. *Arch. Exptl. Pathol. Pharmakol.* 192: 530.
Havel, R.J. 1965. In *Handbook of Physiol.*, Sect. 5: *Adipose Tissue*, A.E. Renold and G.F. Cahill, Jr. eds. (Washington: Am. Physiol. Soc.) p. 575.
Herrera, M.G. and A.E. Renold. 1960. *Biochim. Biophys. Acta* 44: 165.
Himms-Hagen, J. 1967. *Pharmacol. Rev.* 19: 367.
Ho, S.J., R.J. Ho and H.C. Meng. 1967. *Am. J. Physiol.* 212: 284.
Jeanrenaud, B. 1968. *Ergeb. Physiol. Biol. Chem. Exptl. Pharmakol.* 60: 57.
Jungas, R.L. 1966. *Proc. Nat. Acad. Sci.* 56: 757.
Jungas, R.L. and E.G. Ball. 1963. *Biochemistry* 2: 383.
Jungas, R.L. and E.G. Ball. 1964. *Biochemistry* 3: 1696.
Katz, J., B.R. Landau and G.E. Bartsch. 1966. *J. Biol. Chem.* 241: 727.
Klainer, L.M., Y.M. Chi, S.L. Freidberg, T.W. Rall, and E.W. Sutherland. 1962. *J. Biol. Chem.* 237: 1239.
Korn, E.D. 1955. *J. Biol. Chem.* 215: 1.
Kuo, J.F. and E.C. De Renzo. 1969. *J. Biol. Chem.* 244: 2252.
LeBoeuf, B., R.B. Flinn and G.F. Cahill, Jr. 1959. *Proc. Soc. Exptl. Biol. Med.* 102: 527.
LeBoeuf, B.,and G.F. Cahill, Jr. 1961. *J. Biol. Chem.* 236: 41.
Li, C.H. 1964. *Nature* 201: 924.
Lincová, D., J. Čepelík, M. Černohorský and K. Elisová. 1968. *Biochem. Pharmacol.* 17: 2291.
Lynn, W.S., R.M. MacLeod and R.H. Brown. 1960. *J. Biol. Chem.* 235: 1904.
Mahler, R., W.S. Stafford, M.E. Tarrant and J. Ashmore. 1964. *Diabetes* 13: 297.
Mosinger, B. 1969. *Life Sci.* 8: 137.
Mosinger, B. and M. Vaughan. 1967a. *Biochim. Biophys. Acta* 144: 556.
Mosinger, B. and M. Vaughan. 1967b. *Biochim. Biophys. Acta* 144: 569.
Okuda, H., I. Yanagi and S. Fujii. 1966. *J. Biochem.* 59: 438.

Orö, L., L. Wallenberg and S. Rosell. 1965. *Nature* 205: 178.
Park, C.R., J.G.T. Sneyd, J.D. Corbin, L.S. Jefferson and J.H. Exton. 1969. In *Diabetes, Proc. Sixth Congr. Intern. Diabetes Federation, Stockholm, 1967*, J. Ostman, ed. (Amsterdam: Excerpta Medica), p. 5.
Reshef, L., E. Shafrir and B. Shapiro. 1958. *Metabolism* 7: 723.
Rizack, M.A. 1961. *J. Biol. Chem.* 236: 657.
Rizack, M.A. 1964. *J. Biol. Chem.* 239: 392.
Rodbell, M. 1967. *J. Biol. Chem.* 242: 5721.
Rodbell, M., A.B. Jones, G.E. Chiappe de Cingolani and L. Birnbaumer. 1968. *Recent Progr. Hormone Res.* 24: 215.
Rubinstein, D., S. Chie, J. Naylor and J.C. Beck. 1964. *Am. J. Physiol.* 206: 149.
Rudman, D., and M. Di Girolamo. 1967. *Advan. Lipid Res.* 5: 36.
Rudman, D., M.B. Reid, F. Seidman, M. Di Girolamo, A.R. Wertheim and S. Bern. 1961. *Endocrinology* 68: 273.
Saha, J., R. Lopez-Mondragon and H.T. Narahara. 1968. *J. Biol. Chem.* 243: 521.
Salvador, R.A. and R. Kuntzman. 1965. *J. Pharmacol. Exptl. Therap.* 150: 84.
Schonheyder, F. and K. Volqvartz. 1945. *Act. Physiol. Scand.* 10: 62.
Schur, H. and A. Low. 1928. *Wien. Klin. Wochsch.* 41: 225.
Scow, R.O., F.A. Stricker, T.Y. Pick and T.R. Clary. 1965. *Ann. N.Y. Acad. Sci.* 131: 288.
Shapiro, B. 1957. *Progr. Chem. Fats Lipids* 4: 178.
Shapiro, B., I. Chowers and G. Rose. 1957. *Biochim. Biophys. Acta* 23: 115.
Steinberg, D., M. Vaughan and S. Margolis. 1960. *J. Biol. Chem.* 235: PC38.
Stock, K. and E. Westermann. 1966. *Arch. Exptl. Pathol. Pharmakol.* 253: 86.
Strand, O., M. Vaughan and D. Steinberg. 1964. *J. Lipid Res.* 5: 554.
Touabi, M. and B. Jeanrenaud. 1969. *Biochim. Biophys. Acta* 173: 128.
Vaughan, M. 1960. *J. Biol. Chem.* 235: 3049.
Vaughan, M. 1961. *J. Biol. Chem.* 236: 2196.

Vaughan, M. 1961. *J. Lipid Res.* 2: 293.
Vaughan, M. and J. Barchas. 1966. *J. Pharmacol. Exptl. Therap.* 152: 298.
Vaughan, M., J.E. Berger and D. Steinberg. 1964. *J. Biol. Chem.* 239: 401.
Vaughan, M. and D. Steinberg. 1963. *J. Lipid Res.* 4: 193.
Vaughan, M. and D. Steinberg. 1965. In *Handbook of Physiol.*, *Sect. 5: Adipose Tissue*, A.E. Renold and G.F. Cahill, Jr., eds. (Washington: Am. Physiol. Soc.), p. 239.
Walaas, O. and E. Walaas. 1950. *J. Biol. Chem.* 187: 769.
Wenke, M. 1966. *Advan. Lipid Res.* 4: 69.
Weiss, B. and R.P. Maickel. 1968. *Intern. J. Neuropharmacol.* 7: 393.
Wertheimer, E. and B. Shapiro. 1948. *Physiol. Rev.* 28: 451.
White, J.E. and F.L. Engel. 1958. *Proc. Soc. Exptl. Biol. Med.* 99: 375.
Wieland, O. and M. Suyter, 1957. *Biochem. Z.* 329: 320.
Winegrad, A.I. and A.E. Renold. 1958. *J. Biol. Chem.* 233: 267.
Wirsen, C. 1965. In *Handbook of Physiol.*, *Sect. 5: Adipose Tissue*, A.E. Renold and G.F. Cahill, Jr., eds. (Washington: Am. Physiol. Soc.), p. 197.
Yanagi, I., H. Okuda and S. Fujii. 1968. *J. Biochem.* 63: 249.
Yanagi, I., H. Okuda, H. Nakano, Y. Yamanouchi and S. Fujii. 1967. *J. Biochem.* 62: 599.

Selected Topics on the Function of Biogenic Amines in Exocrine Organs with Special Reference to Histamine in Stomach

Michael A. Beaven

National Heart Institute
Bethesda, Maryland

This paper is intended to be a brief survey of the function of biogenic amines in exocrine organs. In addition to this general survey, our own studies on the role of histamine in exocrine organs will be discussed. The biogenic amines that affect exocrine secretion include the catecholamines, epinephrine and norepinephrine; indole alkylamines; and histamine. These amines may act at several different sites: the secretory cells, the vascular system, or the contractile or structural elements.

Secretory cells are of various types, for example, the mucous and serous cells in the salivary glands. The balance of the secretions from these different cells determines the composition of the glandular secretion. Thus, a biogenic amine may affect the composition of the exocrine secretion by acting on one or more of these secretory cells. A biogenic amine may also affect the rate of secretion by altering the blood supply. A catecholamine for example, might inhibit glandular secretion through vasoconstriction. Little is known about the action of biogenic amines on the contractile elements such as the myoepithelial cells (Zimmerman, 1927). Such structural elements must exist within an exocrine gland so that secretory cells can function and yet, at the same time, contain the enormous pressures which must prevail in cells and ducts during periods of rapid secretion. If the contractile elements are capable of adjusting their tone and so adapting to changes within the gland, they may be a further site of action for biogenic amines. The expression of a few drops of saliva commonly observed after the injection of catecholamines may be due to stimulation of such contractile elements.

A biogenic amine may act directly on the exocrine gland, for example, histamine acts on the glandular cells of stomach to produce gastric secretion (Popielski, 1920). Or, a biogenic amine may act in concert with another system, for example, catecholamines promote prostatic secretion only in conjunction with cholinergic stimulation (Smith, *et al.*, 1966).

The actions of the three types of biogenic amines are quite diverse. Catecholamines, epinephrine and norepinephrine, are located in the sympathetic nervous system and have an important role in the control of exocrine secretion, influencing both the rate and composition of the secretions. Serotonin is found mainly in the gastrointestinal (GI) tract where it is presumed to have a role in gastric motility. Histamine, the first known of the biogenic amines, has proven the most difficult to study. The ubiquitous distribution of this amine does not help to localize its function, although a major emphasis in research on histamine has been its possible role as a mediator of gastric secretion.

Catecholamines

In the GI tract the catecholamines appear to be located exclusively in neuronal stores in sympathetic nerve terminals and, to a smaller extent, in the neurons and nerve cells. The evidence for this is well established and has been approached in several ways:

1. The histochemical fluorescence technique, developed by Hillarp, Falck, and their coworkers (Falck, 1962; Carlsson, et al., 1962; Falck and Owman, 1966), has shown that the sympathetic nervous system in various tissues consists of a dense mesh or plexus of sympathetic nerves containing many nerve varicosities or nerve terminals with large amounts of catecholamines. This work has told us much about the structure of the sympathetic nervous system.
2. The technique of Axelrod and coworkers (Axelrod, et al., 1959; Whitby, et al., 1961) has demonstrated selective labeling of catecholamine stores at sympathetic nerve endings by intravenous injection of tracer amounts of ^{3}H-epinephrine or ^{3}H-norepinephrine. With this technique it has been possible to study the release and turnover of these stores under various conditions.
3. Studies of the various drugs that affect the catecholamine stores have shown that intact catecholamine stores are required for proper function of sympathetic nerves (for example, see the review by Brodie and Beaven, 1963). An excellent description of the nature of the catecholamine stores in the sympathetic nervous system is given by Iverson (1967).

The role of the catecholamines in exocrine glands is to mediate sympathetic nervous activity which maintains a vasoconstrictor tone in the glands. A classical experiment of Claude Bernard (1858) showed that in dog salivary gland a permanent vasoconstrictor tone existed and that this tone could be removed by section of the sympathetic trunk and then restored by stimulation of the peripheral stump of the nerve. Excessive stimulation decreased or even arrested blood flow through the gland. The histochemical work of Falck (1962) has demonstrated that blood vessels, especially arteries and arterioles, are heavily supplied with sympathetic nerves. The action of the sympathetic nervous system on secretory cells is less clear. Glands that are well supplied with sympathetic nerves, for example, the cat submaxillary glands, produce profuse secretion in response to sympathetic stimulation or systemic administration of epinephrine (Burgen and Emmelin, 1961). The effect is particularly pronounced when epinephrine is given in small enough doses so that excessive vasoconstriction is avoided. In glands in which sympathetic innervation is not so complete, for example, cat parotid gland, epinephrine produces a small and transient flow of fluid (Stromblad, 1955). In rabbit, the parotid, but not the submaxillary gland, responds to epinephrine or sympathetic nerve stimulation (Nordenfelt and Ohlin, 1957). Therefore, the picture varies from species to species. In addition, the catecholamines act indirectly on the secretory cells by restricting the blood supply. The dryness of mouth usually associated with administration of sympathomimetic agents is due in large part to the general vasoconstriction produced by these drugs. It is known that a reduction of blood flow leads to a comparable reduction in the rate of secretion in a number of exocrine glands, for example, salivary gland (Burgen and Emmelin, 1961) and stomach (Martinson, 1965). It would seem that at high rates of secretion the supply of water from blood is a limiting factor on secretory rates. In our own studies saliva accounted for 25% of the plasma delivered to the salivary gland in cat, and the maximum rates of salivation during stimulation of chorda tympani or infusion of carbachol were reduced by restricting the blood supply (Jacobson, Severs and Beaven, unpublished observations).

Serotonin

It is generally assumed that this amine is located in the argentaffin cells in the intestine and in the pyloric region of the stomach (Barter and Everson-Pearse, 1953; Erspamer, 1954; and see discussion by Garattini and Valzelli, 1965); the best evidence for this is the high production of serotonin in carcinoid syndrome in which there is a proliferation of the argentaffin cells (Lembeck, 1953). Another possible source of serotonin is the "argentaffin-like" cells in stomach as described by Håkanson, *et al.* (1967). These cells form serotonin from 5-hydroxytryptophan but do not store this amine. A point to note is that serotonin stores in stomach and intestine are relatively resistant to the depleting action of reserpine (Carlsson, 1966). Apparently stores of serotonin in the GI tract have different properties from the neuronal stores of serotonin in the central nervous system. Little else is known about the site or mechanism of storage of serotonin in the GI tract.

The general outline of the synthetic pathway for serotonin is now established, and recent interest in the synthesis of serotonin has been concerned with the hydroxylation of tryptophan by tryptophan hydroxylase to 5-hydroxytryptophan, the immediate precursor of serotonin. Evidence for tryptophan hydroxylating activity in the mucosa of the GI tract was obtained by Håkanson and Hoffman (1967). They examined a number of tissues and found activity in the stomach and, to a lesser extent, in small intestine. Lovenberg and Besselaar (personal communication) have found tryptophan hydroxylase activity in isolated intact intestinal tissue. The hydroxylation step is of interest because it is the rate limiting enzymatic step in serotonin synthesis (Lovenberg, *et al.*, 1968), and may be the most vulnerable point for inhibition of serotonin synthesis by a specific inhibitor. Such an inhibitor, para-chlorophenylalanine, is available, and has been used to study the function of serotonin in the GI tract. Present evidence points to serotonin having a role in gastric and intestinal motility (see review by Bülbring, 1961). This evidence, apart from the direct action that serotonin has on the intestine, is the diarrhea observed in patients with the carcinoid syndrome. The diarrhea can be prevented by the administration of paraochlorophenylalanine in doses that inhibit the elevated serotonin production by 80 percent (Engelman, *et al.*, 1967).

Serotonin may also influence exocrine secretion by its action on the vascular system (Page, 1968). This action varies according to the level of systemic blood pressure. At low blood pressure the action of serotonin is stimulatory; at high blood pressure the action is inhibitory. These effects are reviewed by Haddy, *et al.* (1959) who suggest that serotonin has a function in modulating vasomotor tone. If serotonin has such a function, it must have appeared in evolution after the development of the nervous control of the vasomotor system. Recently Reite (1969) has attempted to trace the phylogenetic origin of the stimulatory and inhibitory actions of serotonin and histamine on smooth muscle in the vascular system. The specific actions of serotonin and histamine could not be detected along the phylogenetic line until the appearance of higher bony fishes. At this point of evolution, where the sympathetic nervous control of vascular smooth muscle appeared fully developed, an increase in blood pressure from 20 to 40-50 mm Hg (systolic) was also observed.

There is no indication that serotonin plays a direct role in control of exocrine secretion although Kim and Shore (1963) have suggested the possibility that serotonin may have an inhibitory role at least in gastric secretion. They base their conclusions on the finding that monoamine oxidase inhibitors, which block the metabolism of serotonin, partially inhibit the gastric secretion induced by reserpine or insulin.

Histamine

Role of histamine in gastric secretion. Histamine is located largely in the mast cells. The histamine stores in mast cells are relatively dormant with a turnover time of several weeks (Schayer, 1959). These stores do not appear to be essential for exocrine secretion, and exocrine function is unimpaired after depletion of the histamine stores of mast cells by compound 48/80. However, there exists a type of histamine store which is resistant to the action of compound 48/80 (Riley, 1959; Mongar and Schild, 1952; Paton, 1958). This hisatmine is thought to be of non-mast cell origin (Riley, 1959; Johnson, *et al.*, 1966). A classical example of non-mast cell histamine is that in stomach (Riley, 1959), although non-mast cell histamine is found to some extent in all exocrine organs (Brodie, *et al.*, 1966). Stores of non-mast cell histamine have a rapid turnover with a turnover time of one to two hours (Johnson, *et al.*, 1966; Beaven, *et al.*, 1967; Erjavec, *et al.*, 1967), and they may have a role in exocrine function. Non-mast cell histamine in stomach has long been considered a possible mediator of gastric secretion (see discussions by Code, 1965; Ivy and Bachrach, 1966). We have explored the possibility that non-mast cell histamine is a mediator of exocrine secretion in general. Our studies showed that non-mast cell histamine was released from salivary glands during salivation and that prolonged stimulation, for example, after pilocarpine, led to partial depletion of the glandular stores of histamine (Erjavec, *et al.*, 1967). Experiments with labeled histamine indicated that an acceleration in histamine turnover accompanied salivation. In addition, studies with radioactive L-histidine showed an apparent increase in histamine synthesis in salivary gland during periods of salivation (Erjavec, *et al.*, 1967). In these studies radioactive L-histidine was infused into the whole animal, and therefore the possibility existed that labeled histamine was synthesized elsewhere in the body and then transported to the salivary gland. The problem was approached in another way by perfusing radioative L-histidine directly into isolated salivary glands. In these later studies we found that little of the L-histidine was decarboxylated to form histamine even though stimulation of the isolated gland produced a copious flow of saliva (Jacobsen, *et al.*, 1970). Hence, the role of histamine in salivation still remains to be proven.

The evidence for histamine as a mediator in gastric secretion is more substantial. For example, an increase in gastric histidine decarboxylase levels has been observed by Kahlson and his collaborators (Kahlson, *et al.*, 1964) after administration of gastrin, the physiological regulator of gastric secretion. These observations gave rise to the idea that the histidine decarboxylase levels adapted according to physiological needs. The findings that gastrin (Haverback, *et al.*, 1965) or various drugs (Shore, 1965), such as reserpine and insulin, reduced histamine levels in gastric mucosa by 20 to 40%, further suggested that gastric secretion is mediated through release of histamine. However, the evidence to date is not completely satisfactory. The changes in histidine decarboxylase levels are significant but the values obtained show considerable spread. In addition, much of the information concerning the role of histamine in gastric secretion is obtained with rat, a species having considerable histidine decarboxylase activity in stomach. In species other than rodents histidine decarboxylase activity in stomach is very low (Waton, 1956). Also, we were troubled by the lack of consistency in the histidine decarboxylase activity and histamine levels in stomachs of rats. Over a period of six years, the histamine levels in Sprague-Dawley rats bred at the National Institutes of Health have declined from an average 26 μg histamine/g stomach (Johnson, *et al.*, 1966) to 20 to 24 μg/g (Beaven, *et al.*, 1968), to 19 μg/g (Beaven, *et al.*, 1969), and lately to 16 to 17 μg/g (Beaven, Severs and Horakova, unpublished observations). Recently we observed great differences in stomach histidine decarboxylase levels of rats housed in different locations at the NIH. In one case gastric histidine decarboxylase was almost non-existent but appeared on rehousing the rats in a new location. There appeared to be some unknown factors which were contributing to these inconsistencies; therefore we decided to study the nature of the differences in the histidine decarboxylase in rat stomach.

Bacterial origin of histamine in stomach. It was apparent that there were two histidine decarboxylase activities in rat stomach, one associated with the insoluble tissue debris obtained after a low speed centrifugation of stomach homogenate, and a soluble enzyme which remains in the supernatant fraction (Beaven, *et al.*, 1969). The soluble enzyme had a pH optimum of 7 and appeared similar to the mammalian histidine decarboxylase activity described in earlier studies (Weissbach, *et al.*, 1961). A major part of the histidine decarboxylase activity in rat stomach was associated with the insoluble material and was of bacterial origin. The present paper reviews our studies with this insoluble enzyme. These studies are described in detail elsewhere (Beaven, *et al.*, 1969).

The insoluble enzyme was quite different from the soluble mammalian histidine decarboxylase. It could be resuspended in buffer and reisolated by a low speed centrifugation with no loss of activity. The ease with which this insoluble enzyme was sedimented explained our failure to detect this enzyme in earlier studies. The enzyme had a pH optimum of 5.0 to 5.5 and did not require pyridoxal phosphate; it was heat stable and was not completely destroyed unless boiled for 20 minutes. These properties appeared similar to those of histidine decarboxylase of various bacteria: *Lactobacillus 30a* (Rosenthaler, *et al.*, 1965), *Clostridium welchii* (Epps, 1945), and *Escherichia coli* (Gale, 1946). The enzyme was specific for L-histidine; it decarboxylated L- but not D-histidine, it did not decarboxylate L-tryptophan or L-dihydroxyphenylalanine (DOPA) nor was it inhibited by α-methylDOPA.

The insoluble histidine decarboyxlase activity was located largely in the upper membranous part of the rat stomach. This part of the stomach is devoid of cells producing digestive enzymes and is distinct from the lower glandular part of the stomach. (For review of the anatomy of the rat stomach see Lambert, 1965). The amount of insoluble histidine decarboxylase in the stomachs was appreciable in all rats. However, there were extreme differences in the levels of enzyme ranging from activities of 40 to over 5000 mμmoles histidine decarboxylated/g tissue/hr (Table 1).

Care had to be taken in the measurement of histidine decarboxylase levels since the amount of enzyme itself increased with time. For example, when the intact tissue or the insoluble enzyme material was incubated at 37° and samples removed from the incubations periodically to be assayed for histidine decarboxylase activity, an increase in enzyme activity was observed. Table 2 shows this increase in one typical experiment. This increase in enzyme activity resembled a bacterial growth curve.

Investigation of the bacterial flora of normal rat stomachs showed a heavy growth of *Lactobacillus* in the upper membranous part of the stomach. As mentioned above, *Lactobacillus* contains a specific histidine decarboxylase with properties similar to the insoluble

TABLE 1

Histamine level and histidine decarboxylase activity in normal and germ free rat stomach*

	Stomach part					
	Glandular			Membranous		
Rats	Histamine[†]	Histidine decarboxylase[¶] pH 5.0	Histidine decarboxylase[¶] pH 6.8	Histamine[†]	Histidine decarboxylase[¶] pH 5.0	Histidine decarboxylase[¶] pH 6.8
Normal (n = 12)	19 ± 4	78 ± 61	72 ± 21	27 ± 18	2125[§] (40-5201)	642[§] (10-2372)
Germ free (n = 16)	13 ± 5	0	26 ± 20	2.5 ± 1.5	0	0

*Normal and germ free male rats (8 wk, Sprague-Dawley) were obtained from the Division of Research Services, National Institutes of Health. Stomachs were separated into two parts, the upper membranous and lower glandular parts (see text), and both parts were assayed for histamine levels by an enzymatic procedure and for histidine decarboxylase activity at pH 5.0 or pH 6.8 by a radioactive assay using L-histidine-^{14}C (carboxyl label) as described by Beaven, et al., 1969. This data has been abstracted from data published by Beaven, et al., 1969.

[†] μg Histamine/g stomach ± S.D.

[¶] mμmoles $^{14}CO_2$ formed/g stomach/hr ± S.D.

[§] (range)

TABLE 2

Rise of the insoluble histidine decarboxylase activity during incubations of homogenate of the upper membranous part of rat stomach at pH 5.0*

Incubation time before assay	Insoluble histidine decarboxylase activity
min	mμmoles $^{14}CO_2$ formed/ ml homogenate/hr
0	3.4
30	5.7
45	7.0
60	8.6

*The upper membranous (see text) parts of three rat stomachs were homogenized in 9 volumes 0.1 M citrate-sodium phosphate buffer, pH 5.0, containing 2.5 X 10^{-4} M L-histidine and pyridoxal phosphate, 10^{-5} M. One ml samples of the homogenate were incubated at 37° for various time periods and then assayed for histidine decarboxylase activity after the addition of 100 mμC L-histidine-^{14}C (carboxyl label). Histidine decarboxylase activity was measured at pH 5.0. The assay procedure is described by Beaven, et al., 1969.

rat stomach enzyme (Rosenthaler, et al., 1965). *Lactobacillus* and the enzyme were also found in the watery feces passed by the rats. In view of the coprophagic activity of this species, it was not surprising to find that *Lactobacillus* resided in the stomach as well as in the feces. In germ-free rats the insoluble histidine decarboxylase was absent (Table 1), thus confirming our suspicion that the enzyme was a bacterial one.

On re-examining the data from individual rats, it was apparent that the histamine level in the upper membranous part of the stomach, and to some extent in the lower glandular part, reflected the levels of the bacterial enzyme (Beaven, et al., 1969). It was also apparent from the low histidine decarboxylase activity observed at pH 6.8 in germ-free rats (Table 1) that the bacterial enzyme contributes to the activity measured at pH 6.8 in normal rats. Therefore, in normal rats a portion of the histidine decarboxylase and of histamine in stomach is of bacterial origin. Little or none of the histamine of bacterial origin is likely to reach other tissues. Our studies showed that 99.9% of the histamine infused into the portal vein is destroyed and does not reach the systemic circulation (Erjavec, Beaven and Brodie, unpublished observations). This would account for the finding that the urinary excretion of histamine and metabolites was the same in normal and germ-free rats (Gustaffson, *et al.*, 1957).

Effect of drugs and of starvation on bacterial histamine. The above results now make it possible to explain a number of earlier observations. First, the histamine that is found in appreciable quantities in gastric juice (Beaven, *et al.*, 1968) may be of bacterial origin. An interesting question is whether this histamine in gastric juice has a role in stimulating gastric secretion. An available source of substrate for the bacterial histidine decarboxylase is the histidine in saliva or gastric juice. Interestingly, saliva promotes gastric secretion in rats (Levine, 1965). The presence of bacterial histidine decarboxylase in stomach might account also for the changes in histamine levels observed in rats after reserpine treatment (Shore, 1965; Isaacs, *et al.*, 1970) or administration of cholinergic agents (Kahlson, *et al.*, 1964). For example, in our studies reserpine depleted the rat stomach of the insoluble histidine decarboxylase within 30 minutes of administration of the drug (Isaacs, *et al.*, 1970). Histamine levels then declined to levels normally found in germ-free rats, around 13 µg/g stomach. A similar decline in the insoluble histidine decarboxylase occurred after large doses of metacholine. These changes were observed in fed rats only, where, for example, histidine decarboxylase activity declined from a value of 38 to 19 m moles of histidine decarboxylated/g stomach/hr within ten min of administration of metacholine (Beaven, Severs and Horakova, unpublished observations). However, in starved rats such decreases were not observed; the levels of histidine decarboxylase remained minimal, between 3 to 6 mµmoles of histidine decarboxylated/g stomach/hr. Reserpine or cholinergic agents will induce rapid evacuation of the gastric contents and expose the upper part of the stomach, where the bacteria reside, to gastric acid as well as proteolytic enzymes. Starvation of rats will lead to a more gradual reduction of gastric contents and a gradual exposure of the upper part of the stomach to the gastric juice. Table 3

TABLE 3

Disappearance of the insoluble histidine decarboxylase activity in the upper membranous part of rat stomach during starvation*

Hours of starvation	Insoluble histidine decarboxylase activity
	$m\mu moles\ ^{14}CO_2\ formed/g$ $stomach/hr \pm S.D.\ (n)$
0	140 ± 55 (6)
2	90 ± 60 (4)
6	65 ± 25 (5)
8	37 ± 15 (4)
18	33 ± 8 (6)
24	22 ± 9 (6)

*Rats were kept in metabolic cages for the various times indicated. The animals had access to water but not food. Histidine decarboxylase activity was measured at pH 5.0 using L-histidine-^{14}C (carboxyl label) as described in the legend to Table 2.

shows that there is a slow decline in bacterial histidine decarboxylase levels after withholding food from rats. In other words, bacterial histidine decarboxylase activity is normally high in fed rats; this activity disappears slowly during starvation and more rapidly after administration of drugs that promote copious gastric secretion. In starved rats such drugs have little apparent effect on histidine decarboxylase activity since levels of the bacterial enzyme

are already low.
From our results it would appear that rat and possibly other rodents are not ideal animals in which to study the role of mammalian histidine decarboxylase of stomach. Germ-free rats might be suitable providing that care is taken to prevent inoculation of these rats with *Lactobacillus* from other rats. The advantage of studying gastric histamine in dogs, cats, or other carnivors is that the high content of gastric acid throughout the stomach maintains a sterile environment.

REFERENCES

Axelrod, J., H. Weil-Malherbe and R. Tomchick. 1959. *J. Pharmacol. Exp. Ther.* 127: 251.
Barter, R. and A.G. Everson-Pearse. 1953. *Nature (London)* 172: 810.
Beaven, M.A., Z. Horakova, H.L. Johnson, F. Erjavec and B.B. Brodie. 1967. *Fed. Proc.* 26: 233.
Beaven, M.A., Z. Horakova and W.B. Severs. 1969. *Europ. J. Pharmacol.*, in press.
Beaven, M.A., Z. Horakova, W.B. Severs and B.B. Brodie. 1968. *J. Pharmacol. Exp. Ther.* 161: 320.
Bernard, C. 1958. *C.R. Acad. Sci.* 47: 245.
Brodie, B.B. and M.A. Beaven. 1963. *Med. Exp.* 8: 320.
Brodie, B.B., M.A. Beaven, F. Erjavec and H.L. Johnson. 1966. In *Mechanism of Release of Biogenic Amines*. U.S. von Euler, S. Rosell and B. Uvnas, eds. Wenner-Gren Center International Symposium Series, Vol. 5. (Oxford: Symposium Publications Division, Pergamon Press), pp. 401-416.
Bülbring, E. 1961. In *Regional Neurochemistry: The Regional Chemistry, Physiology, and Pharmacology of the Nervous System.* S. Kety and J. Elkes, eds. (New York: Symposium Publications Division, Pergamon Press), pp. 437-441.
Burgen, A.S.V. and N.G. Emmelin. 1961. *Physiology of the Salivary Glands*. Monographs of the Physiological Society (London), No. 8. (London: Edward Arnold Publishers Ltd.), p. 121.
Carlsson, A. 1966. In *Handbook of Experimental Pharma-*

cology, XIX. 5-Hydroxytryptamine and Related Indolealkylamines. V. Erspamer, ed. (New York: Springer-Verlag), p. 532.
Carlsson, A., B. Falck and N.-Å. Hillarp. 1962. *Acta Physiol. Scand.* 56 (Suppl. 196): 1.
Code, C.F. 1965. *Fed. Proc.* 24: 1311.
Engelman, K., W. Lovenberg and A. Sjoerdsma. 1967. *New England J. Med.* 277: 1103.
Epps, H.M.R. 1945. *Biochem. J.* 39: 42.
Erjavec, F., M.A. Beaven and B.B. Brodie. 1967. *Fed. Proc.* 26: 237.
Erspamer, V. 1954. *Pharmacol. Rev.* 6: 425.
Falck, B. 1962. *Acta Physiol. Scand.* 56 (Suppl. 197): 1.
Falck, B. and C. Owman. 1966. In *Mechanism of Release of Biogenic Amines*. U.S. von Euler, S. Rosell and B. Uvnas, eds. Wenner-Gren Center International Symposium Series, Vol. 5. (Oxford: Symposium Publications Division, Pergamon Press), pp. 59-70.
Gale, E.F. 1946. *Adv. Enzymol.* 61: 1.
Garattini, S. and L. Valzelli. 1965. *Serotonin*. (New York: Elsevier Publishing Co.), pp. 54-55.
Gustafsson, B., G. Kahlson and E. Rosengren. 1957. *Acta Physiol. Scand.* 41: 217.
Haddy, F.J., P. Gordon and D.A. Emanuel. 1959. *Circ. Res.* 7: 123.
Håkanson, R. and G.J. Hoffman, 1967. *Biochem. Pharmacol.* 16: 1677.
Håkanson, R., B. Lilja and C. Owman. 1967. *Europ. J. Pharmacol.* 1: 188.
Haverback, B.J., M.I. Stubrin and B.J. Dyce. 1965. *Fed. Proc.* 24: 1326.
Isaacs, L., A.K. Cho and M.A. Beaven. 1970. Manuscript in preparation.
Iverson, L.L. 1967. *The Uptake and Storage of Noradrenaline in Sympathetic Nerves*. (Cambridge, Eng.: Cambridge,University Press).
Ivy, A.C. and W.H. Bachrach. 1966. In *Handbook of Experimental Pharmacology. XVIII/1. Histamine: Its Chemistry, Metabolism and Physiological and Pharmacological Actions*. O. Eichler and A. Farah, eds. (New York: Springer-Verlag), pp. 810-891.
Jacobsen, S., W.B. Severs and M.A. Beaven. 1970. Manu-

script in preparation.
Johnson, H.L., M.A. Beaven, F. Erjavec and B.B. Brodie. 1966. *Life Sci.* 5: 115.
Kahlson, G., E. Rosengren, D. Svahn and R. Thunberg. 1964. *J. Physiol. (London)* 174: 400.
Kim, K.S. and P.A. Shore. 1963. *J. Pharmacol. Exp. Ther.* 141: 321.
Lambert, R. 1965. *Surgery of the Digestive System in the Rat.* (Springfield, Ill.: Charles C. Thomas).
Lembrack, F. 1953. *Nature (London)* 172: 910.
Levine, R.J. 1965. *Life Sci.* 4: 959.
Lovenberg, W., E. Jequier and A. Sjoerdsma. 1968. *Adv. Pharmacol.* 6A: 21.
Martinson, J. 1965. *Acta Physiol. Scand.* 65: 300.
Mongar, J.L. and H.O. Schild. 1952. *J. Physiol. (London)* 118: 461.
Nordenfelt, I. and P. Ohlin. 1957. *Acta Physiol. Sc Scand.* 41: 12.
Page, I.H. 1968. *Serotonin.* (Chicago: Year Book Medical Publishers, Inc.), pp. 44-50.
Paton, W.D.M. 1958. *Progr. in Allergy* 5: 79.
Popielski, L. 1920. *Pflügers Arch. gesamte Phsyiol.* 178: 214.
Reite, O.B. 1969. *Acta Physiol. Scand.* 75: 221.
Riley, J.F. 1959. *The Mast Cells.* (Edinburgh: E. & S. Livingstone, Ltd.), pp. 132-136.
Rosenthaler, J., B.M. Guirard, G.W. Chang and E.E. Snell. 1965. *Proc. Natl. Acad. Sci.* 54: 152.
Schayer, R.W. 1959. In *Mechanism of Hyper-sensitivity. Henry Ford Hospital International Symposium.* (Boston: Little, Brown & Co.), pp. 227-233.
Shore, P.A. 1965. *Fed. Proc.* 24: 1322.
Smith, E.R., C. Ilievski and Z. Hadidian. 1966. *J. Pharmacol. Exp. Ther.* 151: 59.
Stromblad, R. 1955. *Acta Physiol. Scand.* 33: 83.
Waton, N.G. 1956. *Brit. J. Pharmacol.* 11: 119.
Weissbach, H., W. Lovenberg and S. Udenfriend. 1961. *Biochim. Biophys. Acta* 50: 177.
Whitby, L.G., J. Axelrod and H. Weil-Malherbe. 1961. *J. Pharmacol. Exp. Ther.* 132: 193.
Zimmerman, K.W. 1927. In *Handbuch der Mikrosckopischen Anatomie des Menschen.* Mollendorff, ed. (Berlin: Springer), Vol. 1: 161.

Histamine Formation as Related to Growth and Protein

Georg Kahlson and Elsa Rosengren

University of Lund
Lund, Sweden

Information on histamine is so scanty, with rare exceptions, that one may wonder whether the subject deserves the space accorded it at this symposium. Histamine research had a flying start under the aegis of the pioneers Henry Dale, Thomas Lewis, Wilhelm Feldberg, Carl Dragstedt and Jack Gaddum. Here was a body constituent seemingly designed to play an essential part in the machinery of the body. Yet, in spite of persistent search no place could be found for histamine in normal physiology. As late as 1950, at a symposium at the Royal Society of London, histamine was relegated to the sphere of pathology, designated as "not strictly physiological, and liberated when cells are injured" (Burn, 1950). Recent evidence, however, suggests that histamine does have a place in normal physiology, as a regulator of the microcirculation (Schayer, this volume), as a regulator of certain secretory processes (Beaven, this volume), and perhaps, as a regulator of growth and repair processes, as discussed in the present chapter.

Origin of Tissue Histamine

Until recently, the prevailing view was that histamine was only partly formed in the tissues, the rest being absorbed from the gut where histamine is produced by bacterial decarboxylation of histidine. With the advent of germ-free rats it became possible to test this concept. We investigated the urinary excretion of histamine and the distribution of this amine in various tissues of germ-free and ordinary rats, both groups being fed a histamine-free diet (Gustafsson, *et al.*, 1957). The two groups were similar in the histamine content of various tissues and in urinary histamie excretion. These findings indicated an endogenous origin for both tissue and urinary histamine. It was also found that the female rat excreted free histamine in amounts which could be easily assayed on isolated ileum preparations of the guinea pig. Thus the rate of formation and mobilization of endogenous histamine can be followed continuously day after day in the female rat by determining the urinary excretion of histamine. This method, still used to reveal overall changes in histamine formation, led to the discovery that histamine is formed at high rates by the pregnant rat.

Histamine formation in foetal tissues. In the rat abundant production of histamine starts on the 15th day of pregnanacy, as indicated by the rise in urinary excretion (Fig. 1). A peak is reached about a day before term and then, within about 24 hr, histamine output subsides to the pre-pregnancy level (Kahlson, *et al.*, 1958a). If the foetuses are removed surgically, without interfering with the other products of pregnancy, the rapid histamine formation promptly ceases (Kahlson, *et al.*, 1958b).

To establish which foetal tissue was responsible for histamine formation, we assayed the histidine decarboxylase activity of various foetal tissues *in vitro*, using the radioactive assay procedure developed

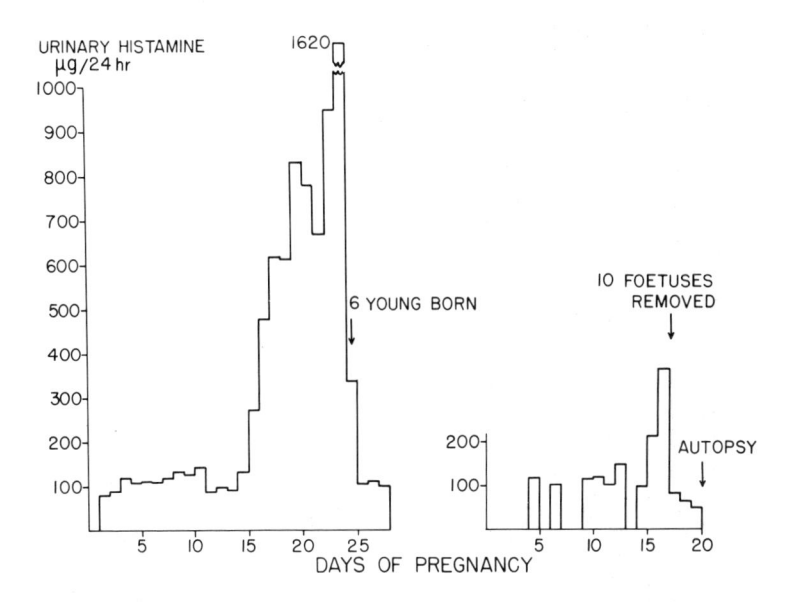

Figure 1. Urinary excretion of histamine in undisturbed pregnancy (left side) and in a rat where the foetuses were removed at the 17th day of pregnancy (right side). (From Kahlson, et al., 1958; by permission of J. Physiol.).

by Schayer and described in detail by Kahlson, et al. (1963b). The foetal liver produced histamine at an enormously high rate: Its histamine forming capacity (HFC) exceeded that of the new born or adult rat by 1000 times or more (Table 1). Other foetal tissues investigated also had much higher histidine decarboxylase activities than the same tissues from young or adult animals (Kahlson, et al., 1960). In foetal tissues the rates of histamine formation and turnover are high but the content of histamine is low. The histamine of foetal tissues is not formed or contained in mast cells. Non-mast cell histamine associated with rapid tissue growth we refer to as "nascent histamine," to emphasize the distinction from released histamine.

High rates of foetal histamine formation have been found in the mouse (Rosengren, 1963) and hamster (Rosengren, 1965). The human foetus also forms histamine

TABLE 1

Rate of histamine formation in 3 hr in rat foetuses, new-born, and adults, expressed as counts/min/g tissue. 1 µg ^{14}C-histamine gives 600 counts/min.

	Foetus 19 days	Foetus 20 days	Foetus 21 days	New-born < 3 hr	Young 2-8 days	Adult pregnant
Whole body	–	3,440	340	860	15	–
Liver	14,000 (23,000)*	8,730	2,280	2,240	5-40	5
Stomach	2,800	–	–	–	–	370-980[†]
Lungs	1,900	930	150	60	5	90-150[†]
Heart	–	–	140	30	0	–[†]
Kidneys	–	–	190	–	0	< 5[†]
Stomach + intestines	–	1,250	150	–	–	–
Brain	10	5	20	5	0	–
Rest of body (muscle, skin, bone)	130 (220)*	360	70	60	30	–

*Figures in brackets are the means of determinations in three litters; the other figures in this column are from a single litter. All the results in the second column are from one litter, and so also those in the third and fourth.

[†]Figures quoted from Kahlson, et al., 1958.

(Kahlson, *et al.*, 1959), although here histidine decarboxylase activity is at a rather low level in all tissues so far investigated (Lindberg, *et al.*, 1963).

Reparative growth and collagen formation. The presence of high histidine decarboxylase activity in rapidly growing foetal tissues suggested that histamine formation might be associated with rapid tissue growth. A study of rapid reparative growth in healing skin wounds appeared particularly attractive because in the skin histamine formation can be lowered or raised artificially. In rats we made incisions in the skin of the back and followed histamine excretion during the course of formation of granulation and wound tissue. Urinary histamine increased on the day of wounding, reached a peak on the fifth day, and returned to nor-

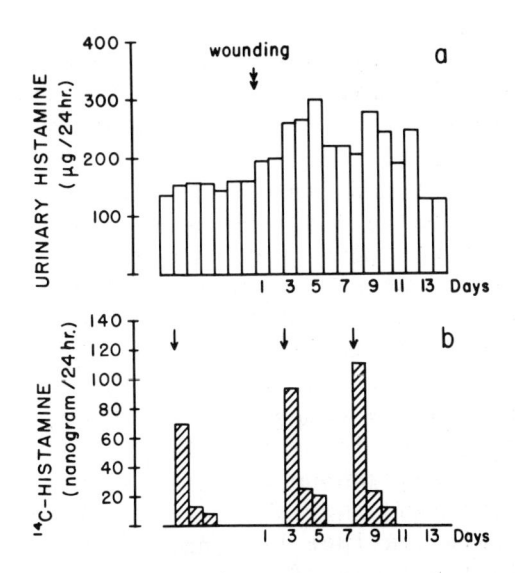

Figure 2. *a*. Daily urinary excretion of histamine in a female rat. *b*. Arrows indicate subcutaneous injection in the same rat of radioactive histidine (110 µg per injection); crosshatched columns represent urinary excretion of radioactive histamine resulting from these injections during the three subsequent days. (From Kahlson, *et al.*, 1960; by permission of *The Lancet*.)

mal on the 13th day, when healing was more or less complete (Fig. 2). Figure 2 also shows that the radioactive method gave the same results; more ^{14}C-histamine was formed from injected ^{14}C-histidine and excreted during healing than previously.

If histamine formation was causally related to the rate of repair of healing skin wounds, then inhibition of histamine formation should reduce the rate of healing. By use of semicarbazide in conjunction with a pyridoxine-deficient diet (*vide infra*), it was possible to inhibit histamine formation in wounded rats. The rate of healing was assessed by determining the tensile strength of the wound, i.e., the force just sufficient to disrupt the wound. When histamine formation was inhibited by 80%, healing was greatly retarded (Fig. 3). On discontinuing the inhibition treatment a period of rebound or "overshoot" ensued during which histamine formation in the skin increased 2-4-fold. During this phase of elevated histidine decarboxylase activity, which may last for more than a week, the rate of healing was considerably enhanced—by about 50% in the eight rats in Fig. 4 (Kahlson, *et al.*, 1960). During the phase of increased histamine formation the rate of collagen formation is also enhanced by about 50% (Sandberg, 1964). Experiments were also performed in which the healing wound tissues were irrigated with histamine diffusing from "long acting histamine" stores deposited under the skin. Extracellular histamine from this source did not promote healing or collagen formation; nor did antihistamine compounds retard healing or collagen formation. We interpret this to mean that the very process of endogenous histamine formation is related to growth, and that this action of nascent histamine is strictly distinct from any of the conventional pharmacological actions of histamine.

Inhibition of histamine formation. Studies on inhibition *in vivo* became feasible after the discovery that histamine excretion in the female rats reflects the rate of endogenous histamine formation and the introduction of isotopic methods of assay of histidine decarboxylase activity. Very strong inhibition, 80-90%, has been achieved by use of semicarbazide in conjunc-

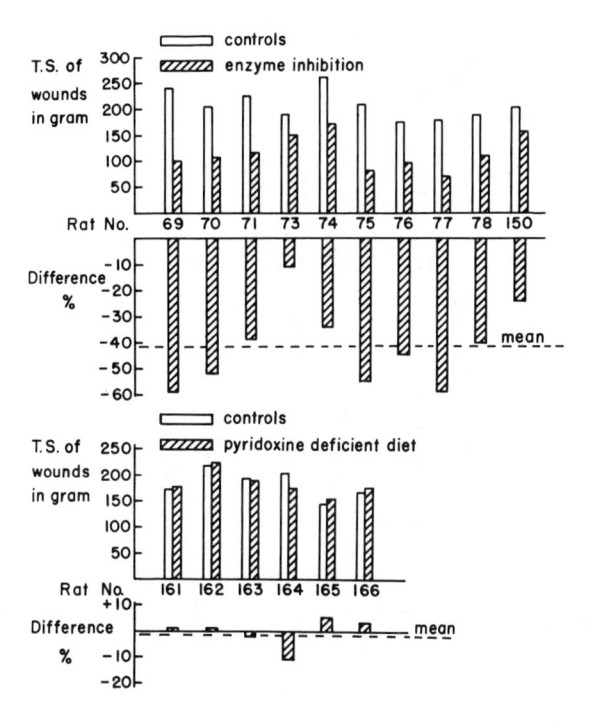

Figure 3. Each pair of columns represents the tensile strength (TS) of the control wound and the test wound in a group of ten rats. *a*. Healing under the influence of enzyme inhibition with semicarbazide superimposed on a pyridoxine-deficient diet. *b*. Effect of the deficient diet alone. The effect of this treatment on the TS is expressed as difference (per cent) against the control. (By permission of *The Lancet.*)

tion with a pyridoxine-deficient diet, as mentioned above. No other compound so far studied equals this procedure in its power to inhibit histamine formation *in vivo*. In our extensive inhibition studies (reviewed by Kahlson and Rosengren, 1968) we demonstrated that α-methylhistidine is a powerful inhibitor of histidine decarboxylase *in vitro* and fairly strong *in vivo*. Because α-methylhistidine is presumably a specific inhibitor of histidine decarboxylase, we have employed

230   Histamine Formation

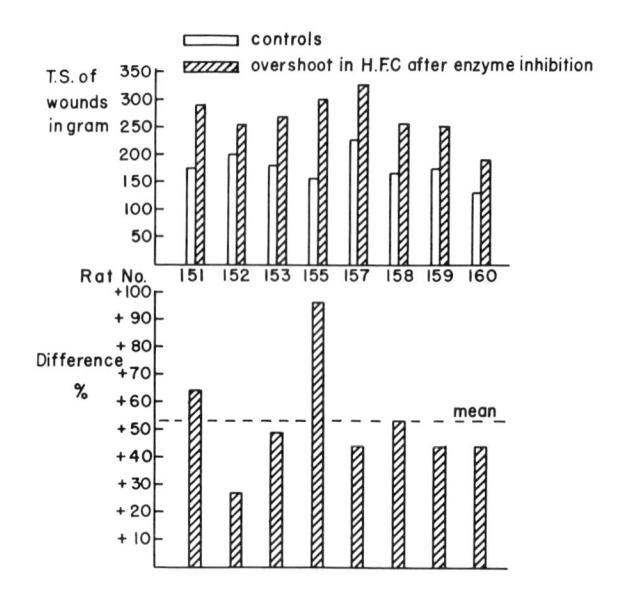

[bar chart with legend: controls; overshoot in H.F.C after enzyme inhibition. T.S. of wounds in gram, values 50-350, Rat No. 151 152 153 155 157 158 159 160. Lower chart: Difference %, +10 to +100, mean line around +50]

Figure 4. Healing during the phase of rebound, "overshoot." (By permission of *The Lancet*.)

it in our more recent experiments on protein synthesis, as described below.

Malignant growth. After a high HFC had been recognized as a concomitant of certain kinds of rapid tissue growth, it appeared likely that a similar phenomenon might also be found in malignant tissues, and this was found to be the case. Riley and his colleagues found that female rats bearing a subcutaneously implanted hepatoma had markedly elevated urinary histamine excretion levels which returned to normal immediately after removal of the tumor (Mackay, *et al.*, 1960). In our laboratory we have investigated various tumors. Results for only a few can be presented here. In the Ehrlich ascites tumor in mice we found high histidine decarboxylase activity in the early phase of growth, when the mitotic index was high, and low activity at later stages with low rates of cell division (Fig. 5). The histidine decarboxylase of the tumor was identical with that of rat foetal tissue on all biochemical cri-

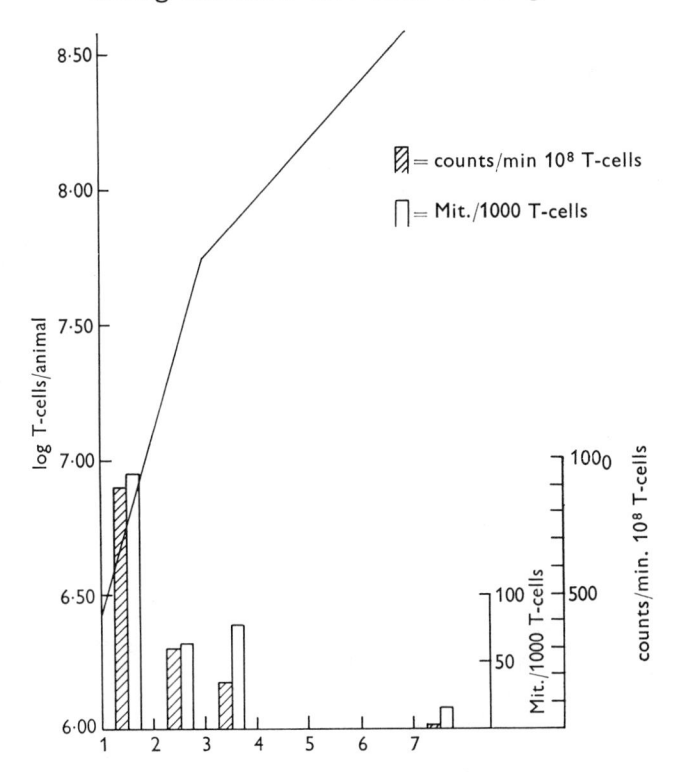

Figure 5. Measurements in 16 mice simultaneously inoculated with Lanschütz I ascites tumor cells. The continuous line represents growth in terms of total number of tumor cells per mouse. The hatched columns stand for amount of ^{14}C-histamine formed and the unhatched for frequency of tumor cell mitosis at 1, 2, 3, and 7 days after inoculation. (By permission of *J. Physiol.*)

teria investigated (Kahlson, *et al.*, 1963a).

A transplantable virus-induced rat sarcoma also had a high rate of histamine formation, and there was a tendency for the highest rate of histamine formation to be associated with a high mitotic index at early growth stages (Fig. 6). In the tumor-bearing host the liver became enlarged and had a HFC several times that of corresponding control livers (Ahlström, *et al.*, 1966).

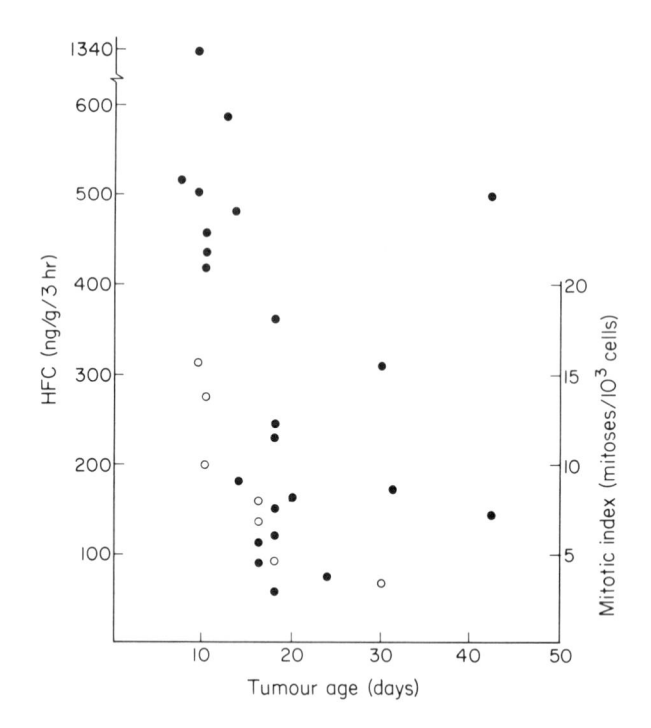

Figure 6. The variation of HFC (●) and mitotic index (○) with tumor age.

Our colleague Marian Johnston studied histamine formation in rats bearing a Walker mammary carcinosarcoma. Histamine excretion increased as the tumor grew (Fig. 7), and isotopic determinations *in vitro* showed that this tumor tissue formed histamine at a high rate, even higher than wound or granulation tissue.

In the tumor-bearing host the liver was enlarged and, as in rats bearing virus-induced sarcoma, had an elevated HFC (Johnston, 1967).

Histamine Formation and Protein Synthesis

Determinations of histidine decarboxylase activity in excised wound tissue 24 hr after wounding showed that histamine formation was 50-60 times the level in control skin, and then progressively fell to normal (Fig. 8; cf. Fig. 2). If plastic sponges are implanted

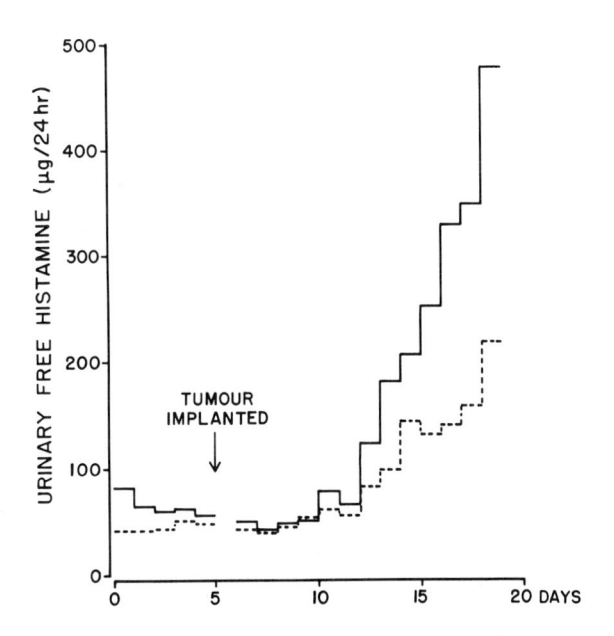

Figure 7. Urinary histamine excretion patterns in two female rats implanted at the arrow with the Walker carcinosarcoma.

in a wound, granulation tissue, consisting mainly of collagen, is formed. We observed that the formation of collagen was enhanced by artificially increasing histamine formation, and we also found that in granulation tissue histidine decarboxylase activity was high, with a maximum around the fifth day of healing, the day the gap of the wound closes (Fig. 8). These observations suggested a relation between HFC and protein synthesis.

The mammary gland is specifically designed to manufacture protein. Our colleagues Lilja and Svensson (1967) found that the lactating gland has a substantial elevation of HFC and that the increased histamine formation persists throughout lactation.

Foetal rat liver also synthesizes protein at a high rate and, as mentioned above, has a HFC about 1000-fold higher than that of liver from young rats. Protein synthesis was determined by measuring the rate of incorporation of ^{14}C-leucine into liver slices and then

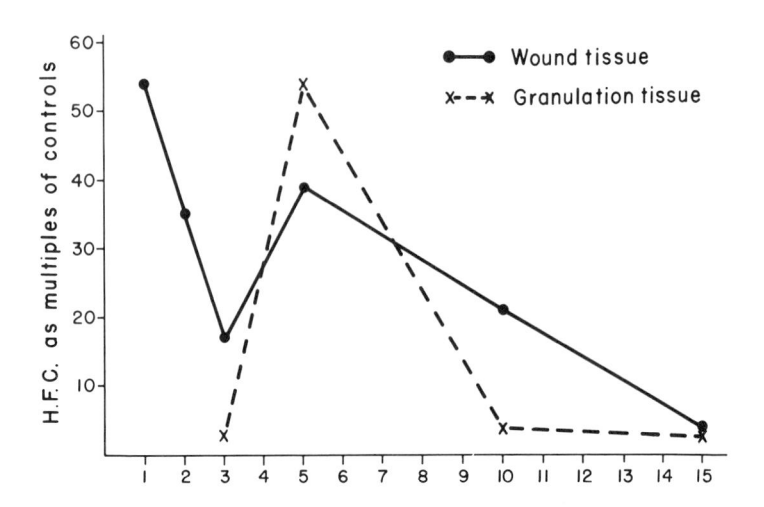

Figure 8. Rate of histamine formation (HFC) of excised wound tissue and granulation tissue expressed as multiples of the activity in whole skin from the same rat. Each figure is the mean of determinations in two rats. (From Kahlson, et al., 1960; by permission of *The Lancet*.)

fractionating the protein into a soluble fraction and a structural fraction (Table 2). In 19-day foetal liver, where HFC is at peak level, the rate of ^{14}C-leucine incorporation into each fraction was about three times that of five-day post-natal liver, where HFC was low. Addition of α-methylhistidine inhibited histamine formation about 80% and substantially reduced ^{14}C-leucine incorporation into foetal liver (53% of control in the soluble fraction, 71% of control in the structural protein fraction), but had no significant effect on five-day post-natal liver. We did not investigate whether higher concentrations of α-methylhistidine would further inhibit protein synthesis.

The observation (Table 2) that α-methylhistidine had no effect on ^{14}C-leucine incorporation into liver slices from five-day-old rats indicates that the histidine analogue *per se* did not interfere with incorporation of the leucine, suggesting that the α-methylhistidine effect in the foetal livers was directly related

TABLE 2

Incorporation of ^{14}C-L-leucine into rat liver slices at two age stages during 90-min incubation period

Tested compounds in final concentrations	19-Day rat foetal liver				5-Day rat young liver	
	Soluble protein fraction	%	Structural protein fraction	%	Soluble protein fraction	Structural protein fraction
Control	30.04 ± 1.46 (10)	100	19.96 ± 1.15 (10)	100	10.21 ± 1.34 (8)	6.76 ± 1.02 (8)
DL-α-methylhistidine, 25 mM	15.91 ± 0.47 (9) P < 0.001	53	14.16 ± 0.55 (9) P < 0.001	71	8.76 ± 1.09 (8) N.S.	6.18 ± 0.95 (8) N.S.
Histamine, 7.5 µg base/ml AMG, 10^{-4} M DL-α-methyl histidine, 2.5 mM	18.83 ± 1.04 (7) P < 0.001	63	16.81 ± 1.03 (7) 0.05 < P < 0.1	84		
L-histidine, 3.0 mM	29.68 ± 1.43 (5) N.S.	99	20.32 ± 1.51 (5) N.S.	102	7.65 ± 1.55 (8) N.S.	5.15 ± 1.02 (8) N.S.
Puromycin diHCl, 200 µg/ml	0.41 ± 0.03 (4)	1.4	0.45 ± 0.02 (4)	2.3	0.07 ± 0.01 (4)	0.11 ± 0.02 (4)

Proteins were separated into soluble and insoluble fractions as described by Grahn, et al. (1969). Figures are mean values (cpm · 1000 per mg protein) ± S.E. of mean. Figures in brackets are the number of observations. The levels of significance are referred to control values. P-values greater than 0.1 are considered not significant.

to the inhibition of histidine decarboxylase. Addition of exogenous histamine to the foetal tissue did not however, remove the inhibition of ^{14}C-leucine incorporation (Grahn, et al., 1969). Here, as in wound healing, we encounter the inability of extracellular histamine to fulfill the role of nascent histamine.

Protein synthesis and its inhibition were also studied in several malignant tissues. For brevity, only results with Rous virus sarcoma in rats will be presented here. The HFC of the tumor tissue was about the same as that of granulation tissue of healing skin wounds, i.e., about 50-fold higher than that of intact rat skin. The rate of incorporation of ^{14}C-leucine into minced tumor tissue was determined without and with inhibitors of histamine formation. Either of the two inhibitors used, α-methylhistidine and NSD-1055 (4-bromo-3-hydroxybenzyloxyamine; trade name, Brocresine) strongly inhibited histamine formation in minced tumor tissue; and inhibited ^{14}C-leucine incorporation by about 50% into the two protein fractions investigated (Table 3), (Grahn and Rosengren, 1969).

We have not yet investigated whether α-methylhistidine or NSD-1055 would retard tumor growth *in vivo*. Judging from our *in vitro* experiments, rather high concentrations of inhibitor would be required. This seems difficult to attain, as α-methylhistidine is destroyed *in vivo* by decarboxylation (Kahlson, et al., 1963b) and NSD-1055 is toxic (Levine, et al., 1965; Johnston and Kahlson, 1967).

The process of rapid tissue growth is not always associated with high histidine decarboxylase activity. We have recorded several exceptions. Raina and Janne found increased ornithine decarboxylase activity in regenerating rat liver following partial hepatectomy, and increased formation of spermidine in Ehrlich ascites tumor cells in mice and in the developing chick embryo (see Article in this volume). Russel and Snyder (1968), confirmed, as have others, our findings of high HFC in foetal rat liver and in some rat tumors. They also reported elevated ornithine decarboxylase activity in some rapidly growing tissues such as the regenerating rat liver, chick embryo and in the STAT-1 rat sarcoma. The product of ornithine decarboxylase activity, putrescine, is a diamine, as is histamine. On the assumption that in certain rapidly growing tissues putres-

TABLE 3

Incorporation of 1-^{14}C-L-leucine into rat Rous virus sarcoma tissue in 90 min

Expt. no.	Soluble fraction inhibitor			Insoluble fraction inhibitor		
	No inhibitor	DL-α-methyl-histidine, 2.5 mM	NSD-1055, 0.5 mM	No inhibitor	DL-α-methyl-histidine, 2.5 mM	NSD-1055, 0.5 mM
1	10.7 (3)	4.7 (3)		7.0 (3)	3.3 (3)	
4	13.8 (4)	8.1 (4)		11.9 (4)	7.9 (4)	
5	10.1 (5)	4.4 (5)		14.0 (5)	6.3 (5)	
6	8.7 (2)	4.4 (2)	4.8 (2)	10.0 (2)	5.7 (2) 6	6.3 (2)
7	9.6 (2)		5.6 (4)	11.4 (2)		6.7 (4)
8	7.5 (2)	4.1 (2)	4.3 (4)	9.7 (2)	5.7 (2)	5.1 (4)

The figures are mean cpm X 10^3 per mg protein in samples of minced tissue. The number of determinations in each experiment are given in parentheses. Molarities are final concentrations.

cine plays a role similar to that of nascent histamine, we suggest that our original hypothesis be broadened to state that high rates of intracellular diamine formation are associated with and presumably essential to certain types of rapid growth.

REFERENCES

Ahlström, C.G., M. Johnston and G. Kahlson. 1966. *Life Sci.* 5: 1633.
Burn, J.H. 1950. *Proc. Roy. Soc. (London)* 137: 281.
Grahn, B., R. Hughes, G. Kahlson and E. Rosengren. 1969. *J. Physiol. (London)* 200: 677.
Grahn, B. and E. Rosengren. 1969. To be published.
Gustafsson, B., G. Kahlson and E. Rosengren. 1957. *Acta Physiol. Scand.* 41: 217.
Johnston, M. 1967. *Experientia* 23: 152.
Johnston, M. and G. Kahlson. 1967. *Brit. J. Pharmacol. Chemother.* 30: 274.
Kahlson, G., K. Nilsson, E. Rosengren and B. Zederfeldt. 1960. *Lancet* 279: 230.
Kahlson, G. and E. Rosengren. 1968. *Physiol. Rev.* 48: 1.
Kahlson, G., E. Rosengren and C. Steinhardt. 1963a. *J. Physiol. (London)* 169: 487.
Kahlson, G., E. Rosengren and R. Thunberg. 1963b. *J. Physiol. (London)* 169: 467.
Kahlson, G., E. Rosengren and H. Westling. 1958a. *J. Physiol. (London)* 143: 91.
Kahlson, G., E. Rosengren, H. Westling and T. White. 1958b. *J. Physiol. (London)* 144: 337.
Kahlson, G., E. Rosengren and T. White. 1959. *J. Physiol. (London)* 145: 30 P.
Kahlson, G., E. Rosengren and T. White. 1960. *J. Physiol. (London)* 151: 131.
Levine, R.J., T.L. Sato and A. Sjoerdsma. 1965. *Biochem. Pharmacol.* 14: 139.
Lilja, B. and S.E. Svensson. 1967. *J. Physiol. (London)* 190: 261.
Lindberg, S., S.E. Lindell and H. Wesling. 1963. *Acta Obstet. Gynecol. Scand.* 42, Suppl. 1: 49.
Mackay, D., P.B. Marshall and J.F. Riley. 1960. *J. Physiol. (London)* 153: 31 P.
Rosengren, E. 1963. *J. Physiol. (London)* 169: 499.
Rosengren, E. 1965. *Proc. Soc. Exptl. Biol. Med.* 118: 8
Russel, D. and S.H. Snyder. 1968. *Proc. Natl. Acad. Sci.* 60: 1420.
Sandberg, N. 1964. *Acta Chir. Scand.* 127: 9.

Biogenic Amines and Microcirculatory Homeostasis

Richard W. Schayer*

Research Center
Rockland State Hospital
Orangeburg, New York

The fundamental function of the circulatory system is to provide an adequate blood flow through the capillaries thus permitting exchange of gases, nutrients and waste products between blood and tissue cells. The heart and larger blood vessels contribute to microcirculatory blood flow by establishing pressure gradients between end points of different circuits (Folkow, *et al.*, 1965).

Certain biogenic amines are of major importance in regulating the larger blood vessels. Since for maintenance of homeostasis, the microcirculation must adapt to changes in the "macrocirculation," some biogenic amines will affect the small vessels indirectly (Folkow, *et al.*, 1965; Zweifach, 1961). However, the microcirculation possesses its own regulatory mechanisms,

---

*Supported by United States Public Health Service Grant AM-10155.

independent of the large vessels and of nervous control, and this paper will consider only the direct effects of biogenic amines on the small vessels.

In evaluating the possible physiological role of certain biogenic amines in microcirculatory homeostasis, their pharmacological effects will first be considered. Bradykinin, although usually classified as a peptide, is also a biogenic amine and is included.

Pharmacological Effects of Biogenic
Amines on the Microcirculation

A. Catecholamines: at the level of the smallest vessels they are, insofar as tests have been made, invariably constrictors (Zweifach, 1961, and personal communication).
B. Serotonin: effects are highly variable; it may be a constrictor or a dilator depending on species and vascular bed. In the rat and mouse it increases microvascular permeability (Zweifach, 1961).
C. Acetylcholine: dilator.
D. Bradykinin: dilator; increases permeability (Lewis, 1960); may, at least in some cases, act through release of histamine.
E. Histamine: invariably a dilator; increases permeability.

Physiological Effects of Biogenic
Amines on the Microcirculation

A. Catecholamines: during physiological release in "stress" states, they are constrictors; under normal conditions they appear to be of relatively little importance (Altura and Zweifach, 1965a).
B. Serotonin: in rat and mouse, but not in other species, released serotonin causes hyperemia and increased permeability at the site of local injury. It is an unattractive condidate as a microcirculatory regulator because of its variable effects on vascular beds, the failure of anitiserotonin compounds to produce changes in blood flow, and the fact that its precursor, 5-hydroxy-

tryptophan, cannot normally be detected in blood.
C. Acetylcholine: there is no substantial evidence of general participation in microcirculatory control; parasympathetic stimulation produces hyperemia in certain glands.
D. Bradykinin: since this is the only substance, excepting histamine, which has been suggested as a general microcirculatory regulator (Rocha e Silva, 1963) it will be considered more fully after the observed activities of the microcirculation are described.
E. Histamine: this amine has been the primary subject of the author's research, its involvement in microcirculatory homeostasis will be the main thesis of this paper.

The Microcirculation and its Control

Included in the microcirculation are the minute precapillary arterioles, the capillaries, and the minute post capillary venules (Wiedeman, 1963). Metarterioles, preferential arteriovenous channels from which capillaries are distributed, are present in some tissues. A precapillary sphincter is the last smooth muscle cell along the branch of a terminal arteriole; its contraction or relaxation determines whether the capillary is closed or open to blood flow. Normally only a relatively small number of capillaries (20-30% in some tissues) are open.

To a considerable extent, the responses of the microcirculation to environmental changes can be explained in terms of intrinsic constrictor-dilator mechanisms built into the microvascular smooth muscle cell. The constrictor mechanism, which imparts "tone" to these cells, is unidentified; since the best available evidence suggests that it does not involve catecholamines or any other biogenic amine (Barcroft, 1963; Altura and Zweifach, 1965a) it will not be considered further. The nature of the intrinsic dilator mechanism of the microcirculation will be the major topic of this paper.

Autonomous activities of the microcirculation which participate in homeostasis include the following:

242   Microcirculatory Homeostasis

a) Vasomotion: the normal, periodic, unsynchronized opening and closing of precapillary sphincters. This phenomenon provides the most precise adjustment of capillary blood flow to meet nutritive requirements of tissues.
b) Reactive hyperemia: the immediate hyperemia observed when circulation is restored to a region to which it had been previously blocked by mechanical means. Presumably this response is of value in overperfusing a tissue which has been, to some degree, deprived of blood, as in tissues compressed during sitting or lying on a rigid surface. The process of *autoregulation* appears to be a form of reactive hyperemia.
c) Post-exercise hyperemia: the immediate hyperemia observed when exercise is initiated in skeletal muscle. This response is essential for providing oxygen and nutrients required for muscular work.
d) Slowly-developing vasodilatation in systemic "stress": the gradual, not immediate, opening of the microvasculature following application of severe stressors of almost any kind. This phenomenon is of key importance to the gradual pooling of blood observed in shock.
e) Slowly-developing vasodilatation at the site of localized injury: the gradual opening of the microvasculature in the region affected by a localized irritant stimulus of almost any kind. This phenomenon is of key importance to the slowly-developing phase of inflammation.

Dilator Metabolites

In current textbooks, the dilator responses listed above are attributed to a variety of "dilator metabolites" formed in other tissues. While it is undoubtedly true that such metabolites exist, and play some role in control of blood flow, there are major reasons why their importance should not be overestimated.

1. It seems inconceivable that "dilator metabolites" formed by other tissues in a variety of distinct circumstances would provide microcirculatory re-

sponses adequate for homeostasis.
2. The "dilator metabolite" concept presupposes that nature did not provide the microcirculation with its own indwelling regulatory mechanisms. In view of the many remarkable homeostatic controls known to exist, the postulate of such a crude control for the microcirculation is not acceptable. Preoccupation with the "dilator metabolite" concept impedes development of a unified picture of microcirculatory control by intrinsic processes.
3. The microcirculation must be capable of a continuous adaptation, for example, to the normal daily variations in temperature. Such adaptation could be simply achieved by variable rates of formation of an intrinsic dilator, catalyzed by action of an inducible enzyme system.
4. The "dilator metabolite" concept requires that the dilator be carried in the blood; yet a mass of evidence indicates that no blood-borne substance can even approximately mimic the natural activities of the microcirculation; further, experiments have recorded failure to detect dilator activity in blood or lymph draining a dilated tissue (Folkow, 1949; Alexander, 1963).
5. If post exercise hyperemia were primarily due to products of muscle metabolism, how can one explain why warming a tissue, e.g., by placing an arm in warm water, produces a similar hyperemia; presumably here the output of metabolites would be much different than in vigorous exercise.
6. Vasomotion is unsynchronized; how could two closely spaced precapillary sphincters open independently if the dilator were blood-borne?

The Histamine Theory of
Microcirculatory Regulation

It is proposed that a simple mechanism, for which experimental support will be given, can to a considerable degree unify the autonomous dilator activities of microcirculation. This view holds first, that microvascular smooth muscle cells contain an inducible enzyme system which continuously produces a dilator, and that the rate of dilator synthesis can be increased or decreased to meet homeostatic requirements through alterations in enzyme activity; second, that the dilator acts primarily on intracellular "intrinsic" receptors.

The inducible enzyme is believed to be histidine decarboxylase; the product, histamine, the dilator.

Histamine formed continuously within microvascular smooth muscle cells would diffuse through the walls giving the following distribution: a) *intracellular histamine*: some molecules remain in the cytoplasm, being free to stimulate intrinsic receptors favoring relaxation of the smooth muscle cell; b) *loosely bound histamine*: as histamine molecules pass through the cell wall, some become loosely bound; c) *extracellular histamine*: histamine molecules reaching the lumen of the vessel are washed away when blood is flowing, but accumulate when blood flow is blocked. Histamine molecules entering the blood stream are of little further significance for dilution, inactivation and loose binding occur; also, there is evidence that higher concentrations of histamine are required to affect extrinsic dilator receptors.

From this logical distribution of intrinsically-formed histamine into three categories, the autonomous dilator responses may be interpreted as follows:

a) Vasomotion: in the precapillary sphincter of a closed capillary, newly-formed histamine diffuses out, and accumulates extracellulary since blood is not flowing. Then histamine "backs up" filling binding sites in the cell wall and then accumulating intracellularly and occupying intrinsic receptor sites. When sufficient dilator receptors are stimulated so that the constrictor force, mainly "tone," is overcome, the sphincter relaxes and blood enters the capillary. Then extracellular histamine washes away and the intracellular concentration gradually decreases. When constrictor forces again predominate the sphincter closes and the cycle is complete. Owing to the inducible characteristics of histidine decarboxylase, whose activity may either be increased or decreased relative to "normal," the essential process of vasomotion can adapt to environmental needs by readjustment of the rate of histamine production.

b) Reactive hyperemia: when blood flow is obscured, histamine molecules accumulate extracellulary, then in the cell wall, and finally at intrinsic receptor sites. When the obstruction is removed the muscle immediately relaxes in rough proportion to the time of occlusion (and hence to the extent of intracellular histamine accumulation). As the flowing blood washes away extracellular histamine, the intracellular histamine level, and hence the rate of blood flow, gradually return to normal.

Autoregulation may involve a similar process. In an isolated perfused tissue a moderate increase in perfusion pressure, by increasing flow through precapillary sphincters, may wash out histamine, cause some sphincters to close, thus producing the observed increase in resistance to flow. Conversely, a moderate drop in perfusion pressure, by decreasing the rate of histamine washout, may permit some sphincters to stay open for a longer than normal interval and thus reduce resistance to flow.

c) Post-exercise hyperemia: the onset of muscular exercise is accompanied by a rapid production of heat. There is an almost immediate increase in local temperature causing an increase in intrinsic histamine synthesis. Precapillary sphincters open, and through the process of "conducted vasodilatation," the arterial side opens (Hilton, 1962). The vasodilatation produced by moderate warming of a limb could be due to the same mechanism. Moderate cooling, presumably by reducing the rate of histamine formation, causes vasoconstriction. These rapid responses are due to temperature effects on enzymes, a slight increase causing improved catalytic action; this is not attributable to inductive changes in histidine decarboxylase which is a slow process.

d) Slowly-developing dilator responses: for maintenance of homeostasis the production of histamine must have adaptive features. If small changes in histidine decarboxylase activity are normally required for adjustment to minor environmental changes, then marked increase in histamine production, in response to drastic local, or systemic stimuli, could lead to the microvascular dilatation observed in inflammation and shock, respectively. Thus unity in microvascular behavior was recognized years ago by Lewis (1927) who emphasized that responses of small vessels to the mildest stimuli passed gradually to responses sufficiently intense to threaten life, by a simple transition, the differences being only in quantity, not quality.

Following a localized irritant stimulus, histidine decarboxylase activation first becomes detectable in about one hour, and near maximal in about four hours. This time sequences corresponds closely to vasodilatation in the slow phase of inflammation in which a latent period of roughly one hour is also observed. In systemic stress, histidine decarboxylase activation also corresponds closely with microvascular behavior; in this case however, the constrictor effects of released catecholamines are superimposed.

Evidence Favoring a Microvascular Regulatory Function for Histamine

1. Exogenous histamine is invariably a dilator of the smaller blood vessels and can initiate reflex dilatation of larger vessels; it has these actions in every mammalian species tested. There is no apparent reason why histamine formed within sphincter cells should not have similar effects.
2. The mediator of microvascular dilatation acts continuously and could not be a tachyphylaxis-producing substance. Histamine shows little tendency to tachyphylaxis in any of its pharmacological activities.
3. Histidine decarboxylase is inducible to activi-

ties ranging from small changes up to roughly 20 times normal. Changes in activity are initiated by a wide variety of stimuli, some stressful, or injurious, but others non-injurious and physiological (Schayer, 1963; Kahlson and Rosengren, 1968). Under certain conditions, e.g., placing an animal in a warm environment, histidine decarboxylase activity of some tissues may drop to below normal levels (Schayer, 1963a). Insofar as can be measured, the rate of change of histidine decarboxylase activity is roughly parallel to the rate of slowly-developing microcirculatory adaptation.

4. Histidine decarboxylase is unique among known enzymes in that its activity can be increased in every species, and in every tested tissue of the commonly studied species, mice and rats (Schayer, 1962). This finding clearly indicates a widespread requirement and is compatible with a microcirculatory role. It has been shown that the author's histidine decarboxylase assay gives results which correspond closely to those obtained from studies on histamine formation *in vivo* (Reilly and Schayer, 1968; 1969). The inducible form of histidine decarboxylase is not associated with mast cells (Schayer, 1963a).

5. In contrast to catecholamines, serotonin and a number of other vasoactive substances, histamine is formed by a single-step decarboxylation of a dietary amino acid, L-histidine, which is available in free form, in adequate concentrations, in all cells and body fluids.

6. Although the smallest vessels cannot be isolated and tested directly for histidine decarboxylase activity, this enzyme does occur in larger vessels and can be activated in them by various stimuli (Schayer, 1962; Kahlson, *et al.*, 1966).

7. Histidine decarboxylase activation and microcirculatory activity are both autonomous. Neither can be blocked by removal of, or interference with, any nervous or endocrine regulatory mechanism (Schayer, 1962; 1963).

8. The fact that histidine decarboxylase induction

occurs to a greater degree in mice and rats than in other tested species, e.g., man, cat and guinea pig, is consistent with a microcirculatory function; the small vessels of the latter group are highly sensitive to histamine, while in mice and rats they are less sensitive.
9. Inhibitors of protein synthesis, e.g., cycloheximide, puromycin and tenuazonic acid, are the only known substances capable of blocking activation of histidine decarboxylase. Although means are not available for measuring the effect of these drugs on capillary blood flow, they have been found to be extraordinarily powerful anti-inflammatory agents when used in systemic doses capable of blocking histidine decarboxylase activation. In contrast, actinomycin D, which inhibits synthesis of RNA and indirectly blocks activation of most known inducible enzymes, fails to block histidine decarboxylase activation or inflammation (Schayer and Reilly, 1968). In recent studies on inflammation of mouse lung caused by intranasal instillation of endotoxin, we have found that addition of actinomycin D to the endotoxin greatly enhances both inflammation and histidine decarboxylase activation in lung tissue (Reilly and Schayer, 1969).
10. Antihistamines, irrespective of structure, when applied topically to microcirculatory preparations, cause constriction of smooth muscle cells (Haley and Andem, 1950; Conrad, 1951; Altura and Zweifach, 1965). This catecholamine-like action is not shown by antiserotonin or anti-acetylcholine drugs unless they are also antihistamines. Although antihistamines are known to cause histamine-like effects, presumably by histamine release, microcicrulatory constriction appears to be the only known pharmacodynamic action of these drugs which is opposite that of histamine. This constrictor effect of antihistamines is consistent with the existence of intrinsic histamine; when the dilator action is partially blocked by drug molecules reaching receptors, the tone-force causes the cell to contract. The alternate ex-

planation, a direct vasoconstrictor effect of antihistamines, seems highly implausible; it implies that a large group of drugs of diverse structures, having no group identity other than ability to block histamine actions, also possess a histamine-unrelated ability to constrict capillaries. Relatively high concentrations of antihistamine are required to produce constriction but this might be expected for antagonism of intrinsic histamine.

11. A point of indirect evidence, but one which I consider to have considerable force, is the ability of the histamine-microcirculation concept to permit a simple, unified interpretation of the major physiological and pharmacological effects of the glucocorticoid hormones. Since this thesis has been published in detail (Schayer, 1964; 1967) no further comments will be made.

12. There is no other known substance suited to be a generalized microcirculatory dilator. Although bradykinin has been proposed there are serious deficiencies in its qualifications: a) it evidently cannot be formed intrinsically since its protein precursor is found in plasma, not intracellularly; b) methods for its detection are said to be fraught with hazards; good correspondence between *in vitro* and *in vivo* findings has not been obtained; c) there is no evidence that it is formed by an inducible enzyme system; d) experiments with drugs which inhibit bradykinin formation, or which destroy it and thus block its activities, reveal no evidence of physiological involvement in microcirculatory control (Webster, et al., 1967); e) bradykinin does have a tendency to induce a gradually diminishing response of the circulatory system (tachyphylaxis), particularly when large doses are given (Miles, 1961). Tachyphylaxis is often the result of gradual depletion of the true mediator, and there is evidence that to some extent, bradykinin activity may be indirect, and due to histamine release (Edery and Lewis, 1964); f) evidence that bradykining is the mediator of functional vasodilata-

tion of certain glands, findings considered so convincing that this mechanism has entered some textbooks as factual, is now considered of doubtful validity mainly because of experiments by Schachter (1969). For a more sympathetic treatment of bradykinin and its possible circulatory role, a recent review by Lewis (1968) should be consulted.

REFERENCES

Alexander, R.S. 1963. *Ann. Rev. Physiol.* 25: 213.
Altura, B.M. and B.W. Zweifach. 1965. *Am. J. Physiol.* 209: 545.
Altura, B.M. and B.W. Zweifach. 1965a. *Am. J. Physiol.* 209: 550.
Barcroft, H. 1963. In *Handbook of Physiology*. Section 2. Circulation. Vol. 2. (Washington, D.C.: American Physiological Society), pp. 1353-1385.
Conrad, V. 1951. *Compt. Rend. Soc. Biol.* 145: 1875.
Edery, H. and G.P. Lewis. 1964. *J. Physiol.* 169: 568.
Folkow, B. 1949. *Acta Physiol. Scand.* 17: 289.
Folkow, B., C. Heymans and E. Neil. 1965. In *Handbook of Physiology*, Section 2. Circulation. Vol. 3. (Washington, D.C.: American Physiological Society), pp. 1787-1824.
Haley, T.J. and M.R. Andem. 1950. *J. Pharmacol. Exp. Therap.* 100: 393.
Hilton, S.M. 1962. *Physiol. Rev.* 42 (Suppl. 5, pt. 2): 265.
Kahlson, G. and E. Rosengren. 1968. *Physiol. Rev.* 48: 155.
Kahlson, G., E. Rosengren and R. Thunberg. 1966. *Lancet* I: 782.
Lewis, T. 1927. *The Blood Vessels of the Human Skin and Their Responses*. (London: Shaw).
Lewis, G.P. 1960. *Physiol. Rev.* 40: 647.
Lewis, G.P. 1968. In *Recent Advances in Pharmacology*, 4th Ed., J.M. Robson and R.S. Stacy, eds. (London: Churchill), pp. 213-246.
Miles, A.A. 1961. *Fed. Proc.* 20 (No. 2, part III): 141.
Reilly, M.A. and R.W. Schayer. 1968. *Brit. J. Pharma-*

col. 34: 551.
Reilly, M.A. and R.W. Schayer. 1969. *Brit. J. Pharmacol.* In press.
Rocha e Silva, M. 1963. *Ann. N.Y. Acad. Sci.* 104: 190.
Schachter, M. 1969. *Physiol. Rev.* 49: 509.
Schayer, R.W. 1962. *Am. J. Physiol.* 202: 66.
Schayer, R.W. 1963. *Progr. Allergy* 7: 187.
Schayer, R.W. 1963a. *Ann. N.Y. Acad. Sci.* 103: 164.
Schayer, R.W. 1964. *Persp. Biol. Med.* 8: 71.
Schayer, R.W. 1967. *Persp. Biol. Med.* 10: 409.
Schayer, R.W. and M.A. Reilly. 1968. *Am. J. Physiol.* 215: 472.
Webster, M.E., N.S. Skinner, Jr. and W.J. Powell, Jr. 1967. *Am. J. Physiol.* 212: 553.
Wiedeman, M.P. 1963. In *Handbook of Physiology*, Section 2, Vol. II, (Washington, D.C.: American Physiological Society), pp. 891-934.
Zweifach, B.W. 1961. *Functional Behavior of the Microcirculation.* (Springfield, Ill.: Thomas).

Binding and Metabolism of Gamma-Aminobutyric Acid and Other Physiologically Active Amino Acids in the Brain

K. A. C. Elliott
McGill University
Montreal

Several classes of compounds are known to regulate function and metabolism in various organisms and organs; this symposium is limited to the class of amines —which alone is an ample field. My own main interest is in mammalian neurobiology and I would note that, as far as I can remember, *all* the natural organic substances that directly affect neuronal activity and all neurotropic drugs, are nitrogenous compounds, mostly amines. The best known natural synaptic transmitter substance, acetylcholine, is a quaternary amine; discussion of this partciular amine could, of course, overwhelm this symposium. The group of substances that have most recently come to the fore as neuro-regulators are free amino acids, that is, amino acids not peptide-bound in proteins or smaller peptides.

That one of these amino acids, L-glutamic acid, is of special interest in the metabolism of the brain became apparent long ago when it was found that it, alone among the amino acids, could serve as substrate for the oxygen uptake by brain slices (Quastel and Wheatley, 1932; Weil-Malherbe, 1936). The development of paper and resin chromatographic techniques disclosed the selection of amino acids that are free in the brain (e.g., Tallan, 1962) and led to the simultaneous discovery in 1950 by Awapara and Roberts, with their co-workers, of the presence of γ-aminobutyric acid (GABA) in brain and virtually only in brain in mammals. These investigators also worked out the enzyme systems involved in the production and removal of GABA. Five years later we, in Montreal, discovered the inhibitory effect of GABA on the stretch receptor neuron of the crayfish. This observation started energetic work on effects of GABA on the mammalian brain in many laboratories. Since then an immense number of biochemical, neurophysiological and neuropharmacological studies have been made on the effects of these amino acids on mammalian brain, on invertebrate nervous organs and, by micro electrode studies, on single neurons in brain and spinal cord. References can be found in reviews (e.g., Elliott and Jasper, 1959; Elliott, 1965; Roberts, 1962; Curtis and Watkins, 1965; Krnjevic, 1965; Lovell, 1969). Since I have been concerned mainly with biochemical aspects, I shall only cursorily summarize the pertinent neurophysiological aspects of the subject.

Neurophysiological aspects. Free glutamic acid occurs in relatively large amount in brain, around 10 μmole/g in mammals. It can support brain tissue respiration but glucose, if present, is consumed preferentially. When iontophoretically injected extracellularly in the neighborhood of cortical or spinal cord neurons, it excites these neurons, apparently by increasing permeability to sodium ions with consequent depolarization. Aspartic acid, which occurs in lesser amount, around 2.5 μmole/g, is also excitatory. Brain from various species contains about 2 μmole of GABA/g and this substance inhibits the excitation, by glutamate or electrical stimulation, of all cortical neurons and of a number of invertebrate neural systems. GABA also inhibits the activity of Renshaw cells in the cat spinal cord. In the crayfish the inhibitory effect of GABA is prevented by the convulsant picrotoxin, but not by strychnine; the crayfish is susceptible to picrotoxin but not to strychnine. Strychnine, the convulsant action of which is mainly due to action on the spinal cord in mammals, prevents the effect of inhibitory neurons in the spinal cord. These neurons seem to act through a transmitter that is not GABA and may be glycine (Werman, *et al.*, 1968; Curtis, *et al.*, 1968). Cat spinal cord contains a relatively large amount of glycine, up to 7 µmole/g in grey matter (Graham, *et al.*, 1966). There may be another, picrotoxin sensitive, inhibitor in the cord.

The inhibitory effects of GABA have been explained as due to a re-polarization or stabilization of the potential of neurons, caused by increased permeability of relevant membranes to chloride and perhaps to potassium ions. GABA affects a number of peripheral nervous mechanisms in the mammal; these effects are difficult to study because of wide species variations and very marked tachyphylaxis which almost prevents repeating an observation on any one object (Hobbiger, 1958; Elliott and Hobbiger, 1959). Intravenous injection into ourselves produced very strange, short-lived sensations. These were probably due to peripheral action because GABA passes the blood-brain barrier rather sparsely (van Gelder and Elliott, 1958) unless the barrier system is damaged (Strasberg, *et al.*, 1967). What does pass seems to be absorbed specifically by neurons (Ford, personal communication).

Other omega-amino acids. Other omega-amino acids besides GABA, and certain other amino acids, have been shown to be able to exert neuroinhibitory effects though they do not all occur in brain. As already mentioned, glycine is a likely inhibitory transmitter substance particularly in the spinal cord. We find only about 0.7 µmole of glycine/g in rat brain. Hayashi (1958, 1959, 1966) has reported remarkable anticonvulsant effects of γ-amino-β-hydroxybutyric acid (GABOB) in dogs and in epilectic human subjects. This substance does have neuroinhibitory effects but, in a rather thorough study, my co-workers, Y. Yoshino and

F.V. DeFeudia, have not been able to identify any in brain. β-Alanine is present in small amount, about 0.05 μmole/g in the rat, and this substance has neuroinhibitory properties. Micro-injection studies show that taurine, also an ω-amino acid, exerts a weak neuroinhibitory action. It occurs in brain usually in greater amount than GABA but it is present also in other tissues. Like GABA it does not appear in protein. Its structure, α-aminoethanesulfonic acid, is unusual and I believe somebody will find a very interesting role for it but it has not been seriously considered as a transmitter substance. It might be noted that serotonin and the catecholamines behave rather like GABA in micro-injection studies (Krnjevic, 1965).

At least two compounds of GABA exist in brain and spinal cord in very small amount, namely homocarnosine (γ-aminobutyryl histidine) and γ-aminobutyryl choline. That GABA itself is an inhibitory synaptic or neuromuscular transmitter has been more or less proven in the lobster (Otsuka, *et al.*, 1967) and the case for it in the mammalian brain is now very strong (Mitchell and Srinivasan, 1969).

Relation of glucose metabolism to active free amino acids. The excitatory amino acids, glutamic and aspartic acid, and GABA the inhibitor, differ fundamentally in metabolic pattern from the catecholamines and serotonin. The latter are derived from the aromatic amino acids and are finally disposed of largely by enzymes that produce specific end products that are excreted. The carbon atoms of the active free amino acids, on the other hand, are derived much more immediately from glucose, the main substrate of energy metabolism in the brain. They are produced in branch reactions from the keto-acids produced in the normal pathway of glucose catabolism; the carbon atoms are ultimately returned to this pathway. Glutamic acid and GABA, with succinic semialdehyde, provide an alternative pathway to part of the Krebs cycle and this is referred to as "the GABA shunt." Figure 1 illustrates the inter-relations between the main free amino acids in brain and glucose metabolism. α-Oxoglutarate is reduced while accepting ammonia to give glutamate.

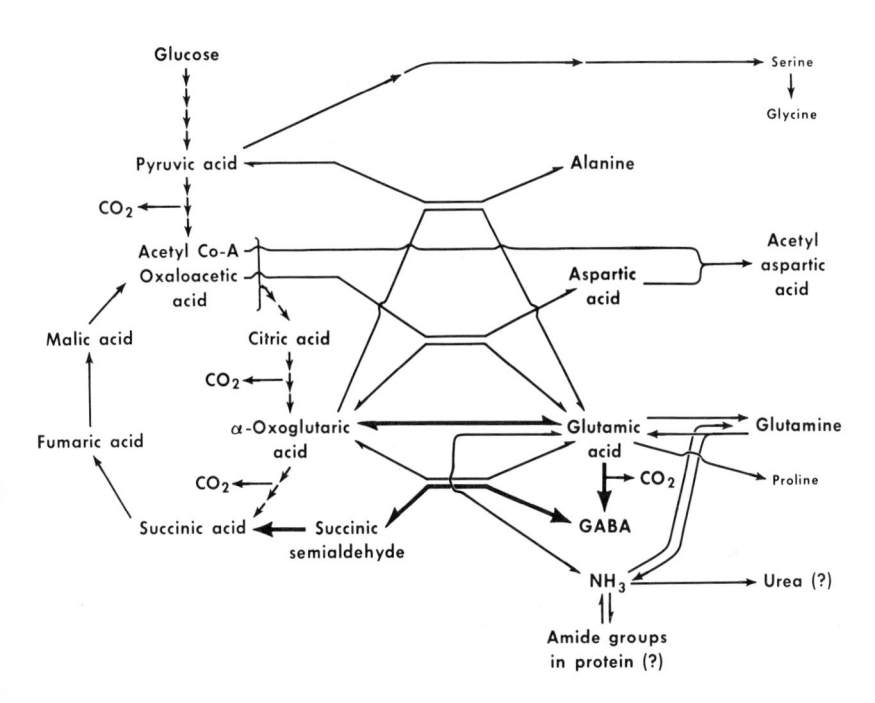

Figure 1. Interrelations of glucose metabolism and free amino acids in brain.

Glutamate is decarboxylated to yield GABA. GABA transaminates with α-oxoglutarate to give succinic semialdehyde and this is dehydrogenated to give succinic acid and we are back to the Krebs cycle. This "GABA shunt" has been estimated (Elliott, et al., 1965; Machinyama, et al., 1965; Myles and Wood, 1969) to be the pathway followed by 10 to 17% of the total glucose carbon consumed in the brain. Glutamate also transaminates with pyruvate to give alanine and with oxaloacetic acid to give aspartic acid. Aspartic acid, with acetyl CoA, yields N-acetyl aspartic acid. At the expense of metabolic energy in the form of ATP, glutamate can take up more ammonia to give glutamine; the latter can be broken down again by a different enzyme.

Most, perhaps all, of the amino substances involved in this system are, or may be, concerned in the regulation of neuronal activity. Glutamate and aspartate are, or can be, excitatory and GABA inhibitory. So far a neurological role for alanine has not been stressed. Glycine, which is derived indirectly from pyruvate, is a likely inhibitor, at least in the spinal cord. No role for acetyl aspartic acid has been shown; it may constitute a reservoir of active acetyl groups, perhaps for the production of acetylcholine. Ammonia has long been known to be produced in nervous activity and is considered to be toxic; the α-oxoglutarate-glutamic acid-glutamine system constitutes a kind of ammonia buffer system which could control the concentration of ammonia.

The source of the ammonia is uncertain (see Weil-Malherbe, 1962). It seems likely that it arises from hydrolysis of amide groups in proteins (Vrba, 1955). In an attempt to prove this my colleague, Y. Yoshino, estimated the amide groups in normal rat brain protein and during convulsions but found no significant difference. The problem is that careful work by Richter and Dawson (1948) had shown extra production of free ammonia in brain during convulsions but the amount produced was very small, about 0.12 μmole/g, and this is less than the experimental error involved in the determination, by hydrolysis, of the very large amount of amide groups, about 50 μmole/g of brain in mixed brain proteins. Any ammonia finally liberated from the free amino acid system presumably enters the urea-producing process. Apparently all the enzymes for this are present in brain except those that produce citrulline, which may have to be provided from outside the brain (Buniatian and Davtian, 1966).

The decarboxylase that produces GABA and all the transaminases, including the one which disposes of GABA, are dependent on pyridoxal phosphate as also are a number of other reactions. There are numerous indications of a strong relation between reduced levels of pyridoxal phosphate and convulsions (e.g., Bilodeau, 1965). But understanding of the effects of drugs that react with the aldehyde group of pyridoxal or affect its phosphorylation, require consideration of the firmness of attachment of the coenzyme to the relevant apoenzymes and the subcellular locations of these enzymes.

K.A.C. Elliott   259

Occlusion, binding and absorption. In our earliest
studies on GABA, when we still knew this active substance as Florey's Factor I, we observed that it could
not all be extracted from brain tissue with saline
solution. We now know that all the neuro-active substances—acetylcholine, the catecholamines, serotonin,
and GABA—exist in tissue mainly in bound or occluded
situations, presumably storage conditions in which they
are neither active themselves nor subject to the action
of the mechanisms that destroy them. In our laboratory
(for references see Strasberg and Elliott, 1967) we find
the GABA in the brain in three states as illustrated in
Fig. 2. When brain is homogenized in 0.32 M sucrose

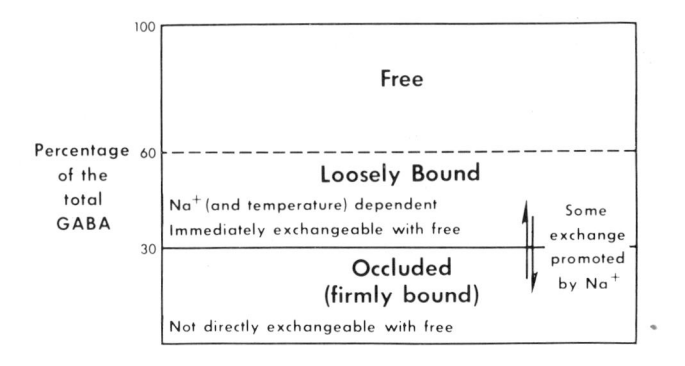

```
                    100 ┌──────────────────────────────┐
                        │                              │
                        │            Free              │
                        │                              │
      Percentage     60 ├──────────────────────────────┤
         of the        │        Loosely Bound          │
         total         │ Na⁺ (and temperature) dependent │  Some
         GABA          │ Immediately exchangeable with free │ exchange
                    30 ├──────────────────────────────┤ promoted
                        │          Occluded            │ by Na⁺
                        │       (firmly bound)         │
                        │ Not directly exchangeable with free │
                        └──────────────────────────────┘
```

Figure 2. Bound forms of GABA and the effect of sodium ions in brain suspensions.

and centrifuged, about 40% of the GABA is found occluded in the particulate sediment. This GABA does
not readily exchange with radioactive free GABA in the
medium. It is evidently contained in sealed-off fractions of cells or subcellular particles, the bounding
membranes of which are rather impermeable to it. These
containers are fragile since the amount found in this
state decreases with any extra manipulation of the suspension. If the homogenization is done in a Ringer-
type solution, or if sodium chloride is added to the
sucrose solution, more of the total GABA is found in
the sediment. The extra GABA thus bound is freely
exchangeable with radioactive GABA in the suspending

medium. We believe it is adsorbed onto surfaces. The presence of sodium also promotes some exchange of the occluded GABA with radioactive GABA in the medium.

My co-worker, Dr. Zehava Gottesfeld, has shown that this sodium-dependent binding is blocked by very low concentrations of ouabain, the sodium pump-ATPase inhibitor, or of protoveratrine. Tetrodotoxin, which blocks the sodium influx involved in action potentials, does not affect the binding. However, it reverses the block caused by protoveratrine but not that caused by ouabain (Fig. 3). We do not understand the meaning of these observations. The sodium-potassium-magnesium-activated ATPase is undoubtedly concerned. Calcium ions are probably involved but we have not been able to show clear effects of calcium except for some decrease in binding in its absence.

We believe that the free GABA that we find is largely artefact. Some free GABA is probably naturally present and varies with the physiological activity of the tissue. It may represent GABA in transit from the occluded or storage form to the sodium-dependent binding site where, perhaps, it exerts its inhibitory action. The other free amino acids of brain can also be bound (Gaitonde, *et al.*, 1965; Elliott, *et al.*, 1965) but whether their binding is similar to that of GABA has not yet been established. I feel that it is likely that the transport into cells of the whole variety of substances which is dependent on sodium and is inhibited by ouabain, may be divisible into a sodium-dependent adsorption followed by entry into the cell.

Near sites where acetylcholine is liberated, cholinesterase is present and can dispose of the acetylcholine almost instantly, though there is evidence (Elliott and Henderson, 1951) that some can be reoccluded. The catecholamines and serotonin can be inactivated by two enzymes, monoamine oxidase and O-methyltransferase, and can also be, and very probably are, re-absorbed into a storage state. GABA can be disposed of by transamination with α-oxoglutarate but there is no indication that this is a very rapid process. If brain slices are incubated in the presence of oxygen and glucose, with or without exogenous GABA, there is virtually no net change in the total amount

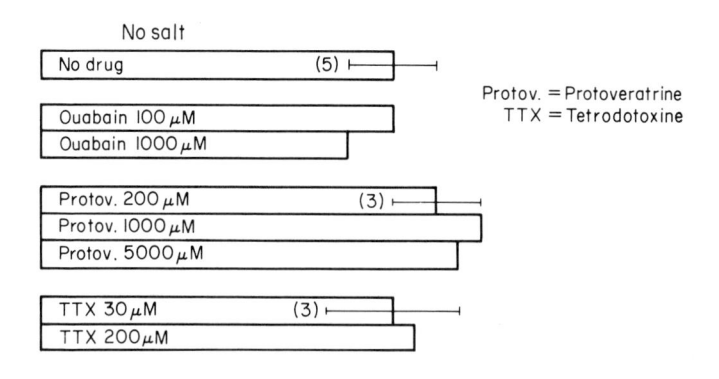

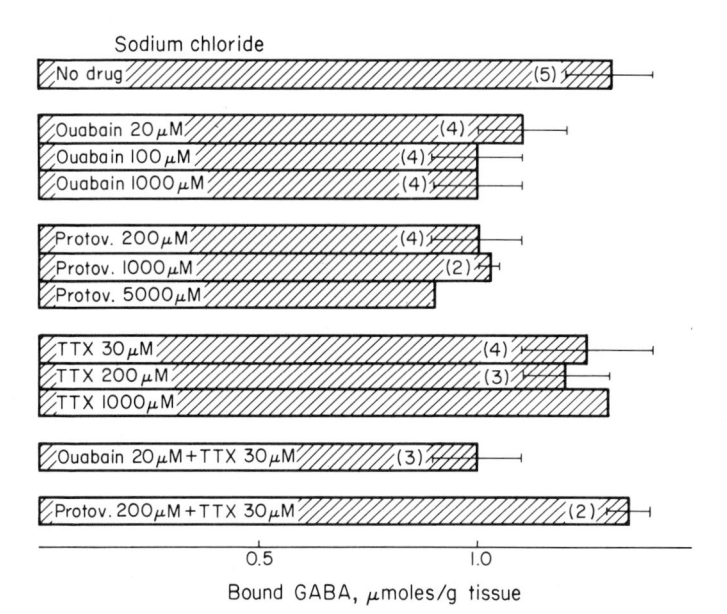

Figure 3. The effects of sodium chloride ouabain, protoveratrine and tetrodotoxin on the binding of GABA in brain suspensions. (Z. Gottesfeld and K.A.C. Elliott, to be published.)

of GABA. It is produced and removed at equal rates. Anaerobically, or in the decapitated head, the total amount of GABA increases. But brain slices absorb GABA from the medium very strongly. (Slices also absorb other amino acids but not nearly to such a high apparent concentration gradient.) Pre-absorption of GABA is probably the main physiological method of disposal of free active GABA. This is a relatively slow process and I believe that the physiological actions of GABA are of a relatively long lasting, regulatory, kind rather than almost instantaneous like those of acetylcholine. I guess the actions of the other amines are intermediate.

As might be expected, Dr. Gottesfeld finds that the uptake of GABA by slices is inhibited by ouabain and protoveratrine but not by tetrodotoxin. Tetrodotoxin reverses the action of protoveratrine but not of ouabain (Fig. 4).

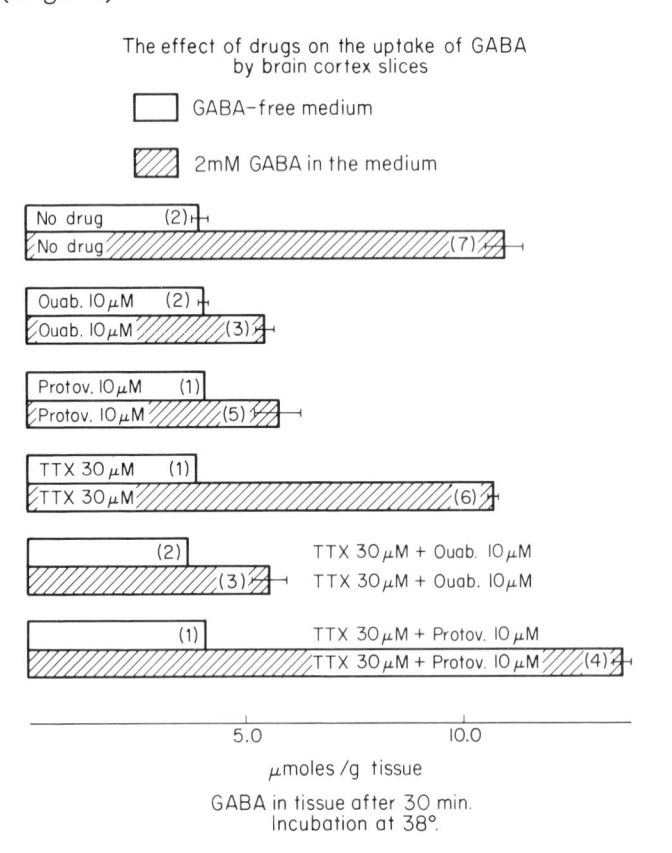

Figure 4. The effects of ouabain, protoveratrine and tetrodotoxin on the uptake of GABA by slices of cerebral cortex. (Z. Gottesfeld and K.A.C. Elliot, to be published.)

Entry of glucose carbon into free and protein-bound amino acids. Labeled carbon atoms from parenterally administered radioactive glucose appear very rapidly and in large amount in free amino acids in brain (Vrba, et al., 1962; Gaitonde, et al., 1965). Lately, Dr. Yoshino and I have confirmed and extended this work. Uniformly labeled glucose was given intravenously and the amounts and radioactivity in all the free amino acids were determined by means of an automatic amino acid analyser coupled, by means of a flow cell, to a scintillation counter. Figure 5 shows the distribution

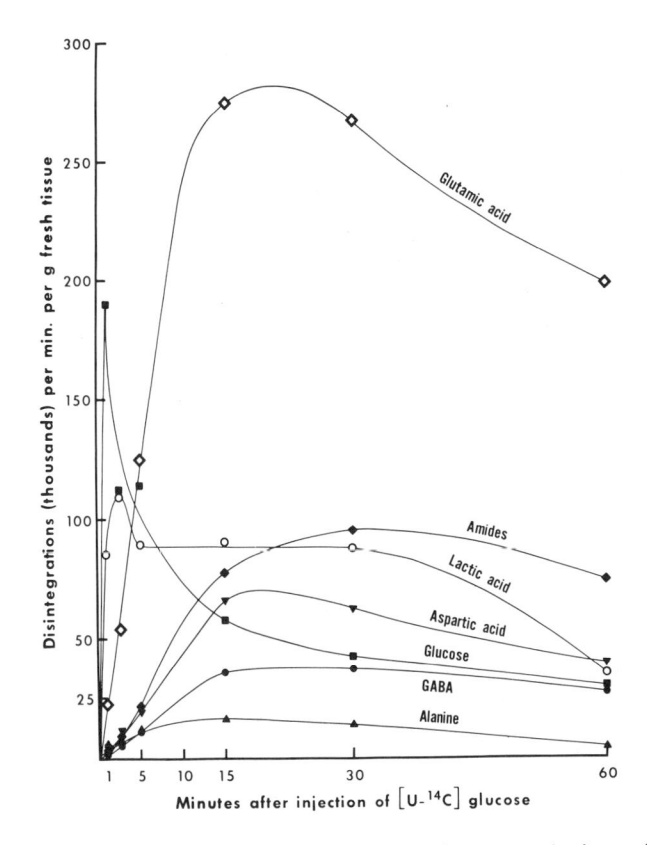

Figure 5. The amounts of radioactivity in glucose, lactic acid and free amino acids in rat brain after intravenous injection of U-^{14}C glucose. (Y. Yoshino and K.A.C. Elliott, to be published.)

264   GABA and Active Amino Acids in Brain

of radioactivity in the brain. Within a minute of the
injection radioactive carbon appears in by far the
greatest amount in glutamic acid, as must be expected,
but considerable amounts appear in the amides, mostly
glutamine, and in aspartic acid, GABA and alanine.
Figure 6 shows the time course of changes in the spe-

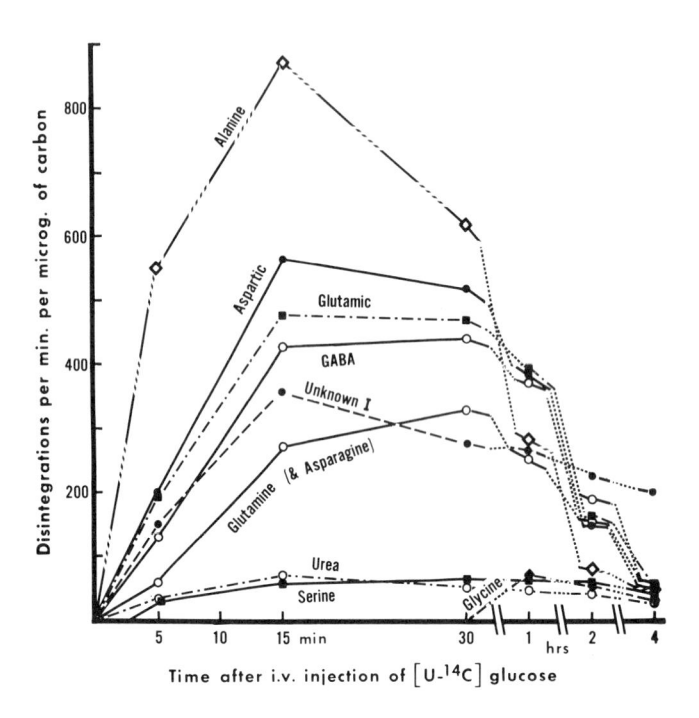

Figure 6. The specific radioactivity of
free ninhydrin-positive substances in rat
brain after intravenous injection of U-^{14}C
glucose. (Y. Yoshino and K.A.C. Elliot,
to be published.)

cific activities. Alanine becomes labeled fastest, as
might be expected, since its parent substance pyruvate,
is the first keto acid produced in glucose metabolism.
Aspartic and glutamic acids, GABA, and glutamine become labeled a little more slowly as also does an unknown compound which is present in very small amount.
Low but definite radioactivity appears in serine and
urea and, later, in glycine. By four hours after the

injection, radioactivity has almost completely disappeared from all the amino acids. Pentobarbital anaesthesia caused changes which were in accord with the well known decrease in the oxidative metabolism of glucose in brains of anaesthetized animals; radioactivity increased in alanine but decreased in the other amino acids. Hypoxia caused the changes which were to be expected from decreased oxidative metabolism and increased anaerobic glucolysis. There was a marked increase in the amount of alanine and a small increase in GABA and generally decreased specific activities in all the amino acids. Convulsions, induced by picrotoxin or pentylenetetrazol, were accompanied by changes very similar to those occurring in hypoxia.

As I mentioned earlier, the neuro-active amino acids differ from the other active amines in the way their production and removal are interlocked with the main energy metabolism. They differ, it seems to me, also from the ordinary run of amino acids in that they exert functions as free substances whereas we commonly think of the functions of amino acids as parts of peptide chains, determining the special conformations of these large molecules. However, with the exception of GABA, these amino acids do enter into protein molecules. Protein turnover in brain has been observed by many workers (e.g., Gaitonde and Richter, 1956; Lajtha, et al., 1957; Roberts, et al., 1959; Vrba, et al., 1962). Dr. Yoshino and I feel that a truly physiological picture can be obtained only with glucose as the source of carbon atoms, reaching the brain via the natural circulation from blood containing a natural concentration of glucose. Figure 7 shows the appearance of radioactivity in the amino acids hydrolyzed from mixed proteins prepared from rat brain after intravenous injection of radioactive glucose. Radioactivity was detectable after 5 minutes in protein-bound alanine and in glutamic and aspartic acids (which included their amides), and somewhat later and to considerably less extent, in serine, proline and glycine. No radioactivity was detected at any time in any other protein-bound amino acid. Very similar pictures were obtained with protein from various regions of the brain. Figure 8 shows the extended time course for

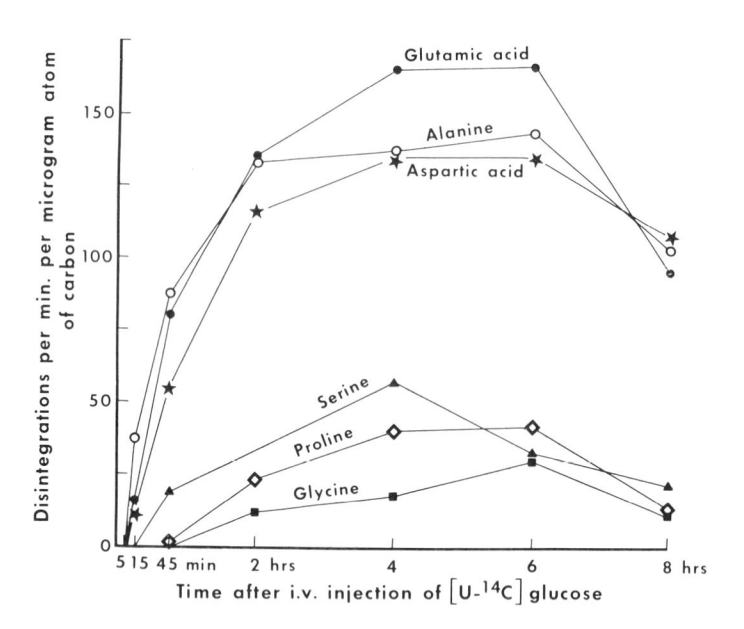

Figure 7. Appearance of radioactivity in protein-bound amino acids in rat brain after intravenous injection of U-^{14}C glucose. (Y. Yoshino and K.A.C. Elliott, to be published.)

protein-bound glutamic acid; the picture was similar for alanine and aspartic acid. There appear to be two main groups of proteins, one of which has a relatively rapid turnover and another, possibly larger, group with a slower turnover.

Release of amino acids from brain protein. In normal animals, at 24 hr after the injection of radioactive glucose, little or no radioactivity remains in free amino acids but the activity in protein is still considerable. Pentobarbital narcosis at this time has no effect but, as Table 1 shows, protein breakdown with the re-appearance of radioactive free amino acids, including GABA, occurs during drug-induced convulsions or under oxygen at high pressure or during insulin hypoglycemia or hypoxia.

Neither the studies on initial labeling of the free amino acids, nor on the appearance of radioactivity in

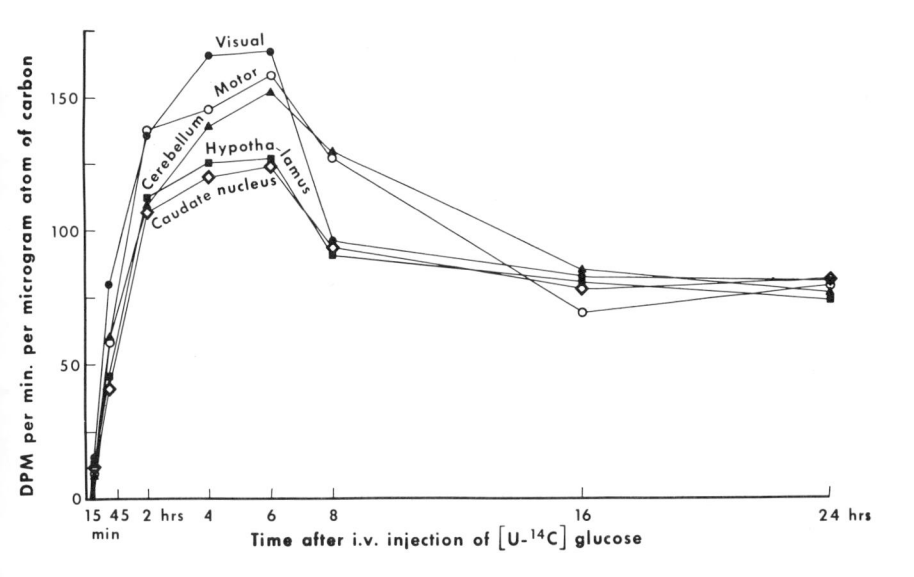

Figure 8. Appearance of radioactivity in protein-bound glutamic acid (including glutamine) in various regions of brains of rats after intravenous injection of U-^{14}C glucose. (Y. Yoshino and K.A.C. Elliott, to be published.)

protein and the later breakdown of labeled protein give us any indication of a primary role of the amino acids in narcosis or convulsions. They do, however, suggest that the effects of insulin shock therapy or elecroconvulsive therapy, may be connected with breakdown of proteins due to lack of substrate or to insufficient oxygen supply to satisfy the excessive energy demand in hyperactive brain. Controlled hypoxia might perhaps work as well as the more drastic therapeutic procedures.

GABA and convulsions. Though GABA does not pass the blood brain barrier very freely, perhaps because it is enzymatically destroyed during passage (van Gelder, 1965), administration of GABA has, under some circumstances, been found to inhibit convulsive activity in some animals. Many studies have been made on the relation of the amount of GABA in the brain to

TABLE 1

Reappearance of radioactivity in free amino acids

dpm/µg-atom of carbon

| | Normal | Narcosis | | Convulsions | | | High pressure oxygen 60 min | Hypoglycemia | Hypoxia |
		Pentobarbital	Pentylenetetrazol	Picrotoxin early	Picrotoxin later			Insulin 3-4 hr	5% O$_2$ in N$_2$ 30 min
Alanine	0	0	80	0	145		45	0	*390*
Glutamic acid	20	30	60	90	*180*		*125*	*140*	*260*
Aspartic acid	0	0	50	*130*	*180*		*110*	*140*	*290*
Glutamine and asparagine	15	35	25	45	90		85	90	*150*
Serine	0	0	65	0	*190*		50	85	70
GABA	0	0	20	0	*160*		25	60	*120*

Rats had received 50 µC of U-^{14}C glucose 24 hr previously.

Averages: all italicized figures are highly significantly different from normal.

K.A.C. Elliott   269

the onset of convulsions, variously induced (see Lovell, 1969). Wood, *et al.* (1969), for instance, find a decrease in the GABA content of the brains of animals subjected to oxygen at high pressure (OHP) and a correlation between the extent of this decrease and the susceptibility of the animals to OHP-induced seizures. In general, however, I believe that there is no such relation and that the total amount of GABA in the brain is usually irrelevant. The effective GABA is probably not the bulk of the GABA in storage but the amount of GABA in the free, active state. I do not know how this could be measured. An illustration of the situation arose early in studies with van Gelder (Elliott and van Gelder, 1960). We confirmed the earlier observation by Killam and Bain (1957) that treatment of rats with thiosemicarbazide caused convulsions and lowered the total GABA found postmortem in the brain. We found that several monoamine oxidase inhibitors, including Iproniazid, tended to raise GABA levels. Iproniazid would counteract the decrease in GABA, but *not the convulsions*, caused by thiosemicarbazide.

Other roles of GABA. There seem to be inter-relations between the systems concerned with GABA and those with catecholamines and serotonin. Recently Popov and Matthies (1969) have shown that several MAO inhibitors, but not all, can inhibit GABA transaminase, the enzyme responsible for removing GABA, thus causing an increase in GABA levels. Some authors (Fujita, 1959; Elliott and van Gelder, 1960; Bose, *et al.*, 1963) have found increases in GABA levels in brains of rats treated with reserpine; other workers found no effect in rats (Tallan, 1962; Mussini and Marcucci, 1962) and decreases have been reported in mice (Palm, *et al.*, 1962), in rats (Marshall and Yockey, 1968) and in monkeys (Singh and Malhortra, 1964). It seems that reserpine can exert varying effects on GABA levels depending on species or strain. It is likely that the effects are somehow connected with the known release of biogenic amines by reserpine. Administration of compounds which raise levels of serotonin, norepinephrine or GABA in the brains of mice has been found to protect against audiogenic seizures (Schlesinger, *et al.*, 1968). Various treatments which lower the concentrations of catecholamines and serotonin in the brains of rabbits have been found to be accompanied by increases in the activity of pyridoxal kinase and *vice versa* (Ebadi, *et al.*, 1968). Since pyridoxal phosphate is concerned in both the production and removal of GABA, it seems that the roles of the amino acids and the other amines must be inter-related.

Since the symposium a report by Yessaian, Armenian and Buniatian (1969) has appeared which shows that intraperitoneal injections of GABA to rats, or administration of amino-oxyacetic acid which increases brain GABA, lowered the level of norepinephrine in brain, heart and spleen, but not in adrenals, and raised the level of serotonin in brain. With brain slices, GABA in the medium was found to release norepinephrine and to inhibit the release of serotonin. No effects on dopamine were observed nor on the activities of 5-hydroxytryptophan or dihydroxyphenylalanine decarboxylases or monamine oxidase. These results certainly suggest interplay between the omega-amino acid and the monoamines.

Other observations indicate that there are aspects of the free amino acid story which have only begun to be explored and may be of high significance. Recently Tewari and Baxter (1969) have reported that GABA and glycine, also an "ω-amino acid," have a remarkable effect on a ribosomal system from immature rat brain. GABA, which does not enter proteins itself, and glycine, seemed to be stimulating the synthesis or turnover of specific proteins. We should remember that the biological production of GABA was first observed with bacteria. GABA has long been known to be present in plants. Warburg and co-workers (Warburg, *et al.*, 1957; Warburg, 1958) claimed that GABA and glutamic acid are involved in photosynthesis in *Chlorella*. I feel sure that free amino acids have some special functions of general biological significance apart from their roles in neurobiology or as substances to be synthesized into protein molecules.

The organizer of this symposium, Dr. Blum, in studies on *Tetrahymena*, has indicated that catecholamines may have a general function in the regulation of metabolism. Serotonin is present in venoms and in bananas and other fruit. Acetylcholine is known to be produced by some bacteria and to be present in large amount in such interesting, nerveless, locations as potatoes, stinging nettles, and royal jelly of bees. Evidently neurobiology cannot divorce itself from General Physiology.

REFERENCES

Bilodeau, F. 1965. *J. Neurochem*. 12: 671.
Bose, B.C., R. Vijayvargiya, A.Q. Safi and S.K. Sharma. 1963. *Arch. int. Pharmacodyn*. 146: 114.
Buniatian, H. Ch. and M.A. Davtian. 1966. *J. Neurochem*. 13: 743.
Curtis, D.R. and J.C. Watkins. 1965. *Pharmacol. Rev*. 17: 347.
Curtis, D.R., L. Hosli, G.A.R. Johnston and I.H. Johnston. 1968. *Exp. Brain Res*. 5: 235 and 6: 1.
Ebadi, M.S.R., R.L. Russel and E.E. McCoy. 1968. *J. Neurochem*. 15: 659.
Elliott, K.A.C. 1965. *Brit. Med. Bull*. 21: 70.
Elliott, K.A.C. and N.M. van Gelder. 1960. *J. Physiol*. 153: 432.
Elliott, K.A.C. and F. Hobbiger. 1959. *J. Physiol*. 146 146: 70.
Elliott, K.A.C. and H.H. Jasper. 1959. *Physiol. Rev*. 39: 383.
Elliott, K.A.C. and N. Henderson. 1951. *Am. J. Physiol*. 165: 365.
Elliott, K.A.C., R.T. Khan, F. Bilodeau and R.A. Lovell. 1965. *Can. J. Biochem*. 43: 407.
Ford, D. 1968. Personal communication.
Fujita, S. 1959. *Okayama Igakkai Zasshi*. 70: 1337.
Gaitonde, M.K. and D. Richter. 1956. *Proc. Roy. Soc. B*. 145: 83.
Gaitonde, M.K., D. Dahl and K.A.C. Elliott. 1965. *Biochem. J*. 94: 345.
Gelder, N.M. van. 1965. *J. Neurochem*. 12: 239.
Gelder, N.M. van and K.A.C. Elliott. 1958. *J. Neurochem*. 3: 139.
Graham, L.T., Jr., R.P. Shank, R. Werman and M.H.

Aprison. 1967. *J. Neurochem.* 14: 465.
Hayashi, T. 1958. *Nature (London)* 182: 1076.
Hayashi, T. 1959. *Neurophysiology and Neurochemistry of Convulsion.* (Tokyo: Dainihon-Tosho).
Hayashi, T. 1966. In *Enzymes and Mental Health.* G.J. Martin and B. Kirsch, eds. (Philadelphia: Lippincott), p. 160.
Hobbiger, F. 1958. *J. Physiol.* 142: 147 and 144: 349.
Killam, K.F. 1957. *J. Pharmacol.* 119: 263.
Killam, K.F. and J.A. Bain. 1957. *J. Pharmacol.* 119: 255.
Krnjevic, K. 1965. *Brit. Med. Bull.* 21: 5.
Lajtha, A., S. Furst, A. Gerstein and H. Waelsch. 1957. *J. Neurochem.* 1: 289.
Lovell, R.A. 1969. In *Handbook of Neurochemistry.* A. Lajtha, ed. (New York: Plenum), Vol. 6, in press.
Machiyama, Y., R. Balazs and T. Julian. 1965. *Biochem. J.* 96: 68P.
Marshall, F.D., Jr. and W.C. Yockey. 1968. *Biochem. Pharmac.* 17: 640.
Mitchell, J.F. and V. Srinivasan. 1969. *J. Physiol.* Proceedings, in press.
Mussini, E. and F. Marcucci. 1962. In *Amino Acid Pools.* J.T. Holden, ed. (Amsterdam: Elsevier), p. 486.
Myles, M.S. and J.D. Wood. 1969. *J. Neurochem.* 16: 685.
Otsuka, M., E.A. Kravitz and D.D. Potter. 1967. *J. Neurophysiol.* 30: 725.
Palm, D., H. Balzer and P. Holtz. 1962. *Int. J. Neuropharamcol.* 1: 173.
Popov, N. and H. Matthies. 1969. *J. Neurochem.* 16: 899.
Questel, J.H. and A.H.M. Wheatley. 1932. *Biochem. J.* 26: 725.
Richter, D. and R.M.C. Dawson. 1948. *J. Biol. Chem.* 176: 1199.
Roberts, E. 1962. In *Neurochemistry, 2nd edit.* K.A.C. Elliott, I.H. Page and J.H. Quastel, eds. (Springfield, Ill: Charles C. Thomas), p. 636.
Roberts, R.B., J.B. Flexner and L.B. Flexner. 1959. *J. Neurochem.* 4: 78.
Schlesinger, K., W. Boggan and D.X. Freedman. 1968. *Life Sciences* 7 (Part I): 437.
Singh, S.I. and C.L. Malhortra. 1964. *J. Neurochem.* 11: 865.

Strasberg, P. and K.A.C. Elliott. 1967. *Can. J. Biochem.* 45: 1795.
Strasberg, P., K. Krnjevic, S. Schwartz and K.A.C. Elliott. 1967. *J. Neurochem.* 14: 755.
Tallan, H.H. 1962. In *Amino Acid Pools*. J.T. Holden, ed. (New York: Elsevier), pp. 465-471.
Tewari, S. and C.F. Baxter. 1969. *J. Neurochem.* 16: 171.
Vrba, R. 1955. *Nature (London)* 176: 1258.
Vrba, R. 1957. *Review of Czechoslovak Medicine* III-2: 1.
Vrba, R., M.K. Gaitonde and D. Richter. 1962. *J. Neurochem.* 9: 465.
Warburg, O. 1958. *Science* 128: 68.
Warburg, O., H. Klotsch and G. Krippahl. 1957. *Z. Naturforschung* 12b: 622.
Weil-Malherbe, H. 1936. *Biochem. J.* 30: 665.
Weil-Malherbe, H. 1962. In *Neurochemistry, 2nd Edition*. K.A.C. Elliott, I.H. Page and J.H. Quastel, eds. (Springfield, Ill: Charles C. Thomas), p. 321.
Werman, R., R.A. Davidoff and M.H. Aprison. 1968. *J. Neurophysiol.* 31: 81.
Wood, J.D. 1969. *J. Neurochem.* 16: 281.
Wood, J.D., W.J. Watson and G.W. Murray. 1969. *J. Neurochem.* 16: 281.
Yessaian, N.H., A.R. Armenian and H. Ch. Buniatian. 1969. *J. Neurochem.* 16: 1425.

Do Polyamines Play a Role in the Regulation of Nucleic Acid Metabolism?*

A. Raina and J. Janne

Department of Medical Chemistry
University of Helsinki
Helsinki, Finland

Although aliphatic polyamines have long been known to be normal tissue constituents, it is only during the past fifteen years that new data have revealed the universal distribution of putrescine, spermidine and spermine in all living organisms. The compelling question is then, what is the physiological function of these compounds? Some of the effects they exert in biological systems are fairly well established. For example, they are essential growth factors for certain microorganisms, and they are able to stabilize nucleic acids, viruses, membrane structures, etc. Among some other possibilities studied is whether they play a role in the synthesis of macromolecules. It is reasonable to believe that *in vivo* they are important for the stabil-

---
*This work was supported by grants from the National Research Council for Medical Sciences, Finland, and from the Sigrid Jus'elius Foundation.

ity of ribosomes, thus sharing the function usually assigned to the Mg^{++} ion. There are indications that polyamines may have other sites of action in protein synthesis, e.g., in the binding of aminoacyl transfer RNA to ribosomes by lowering the requirement for Mg^{++} (Tanner, 1967; Takeda, 1969). Although it has been demonstrated with dialyzed preparations that polyamines may not be essential for *in vitro* protein synthesis, but can be replaced by Mg^{++}, this does not prove that polyamines are not important, for example, in modifying the rate of protein synthesis *in vivo*. For further discussion of various aspects of the physiology of polyamines not directly related to the present topic, the reader is referred to some recent reviews (Tabor and Tabor, 1964; Cohen and Raina, 1967; Jänne, 1967).

During our studies on polyamine synthesis in animal systems, we found that to a great extent changes in polyamine synthesis and accumulation paralleled those occurring in ribonucleic acid. We detected, for example, a correlation between total polyamine content and RNA content in rat tissues of different ages and established that polyamine synthesis increased sharply under conditions in which RNA synthesis was stimulated, e.g., in the regenerating rat liver and after treatment with various hormones and drugs. Furthermore, we were able to show that in bacterial systems spermidine, under certain conditions, increases the accumulation of newly synthesized RNA. These observations, which have been discussed in greater detail elsewhere (Cohen and Raina, 1967), led us to suggest that polyamines may play a part in the regulation of ribosomal RNA synthesis. In the following discussion the possibility that polyamines are concerned in RNA metabolism will be considered in the light of data relating to animal systems, although some comments on results obtained in work on bacteria will also be made. DNA metabolism is largely left out of this discussion.

The following main topics will be considered here: 1) correlation in accumulation and synthesis between polyamines and RNA; 2) effect of polyamines on the accumulation of newly synthesized RNA in whole cells; and 3) effect of polyamines on the accumulation of newly synthesized RNA in nuclear and nucleolar preparations.

Correlation in Accumulation and Synthesis Between Polyamines and RNA

When studying polyamine metabolism in the developing chick embryo (Raina, 1963), we noticed that during the development of the embryo the peak in the concentrations of polyamines coincided with that previously reported for nucleic acids. A closer analysis made later by Caldarera, et al. (1965) confirmed this observation. As we shall see, the changes taking place in the rate of accumulation and synthesis of polyamines very closely parallel those observed in RNA both during physiological growth, in the regenerating liver, and after treatment with various hormones and drugs.

We have shown earlier that marked changes occur in the concentration of spermidine in various rat tissues during the active growth period, whereas that of spermine remains fairly constant (Jänne, et al., 1964). In spite of the great changes recorded in spermidine content, the ratio of total polyamine-N to RNA-P remains fairly constant in a particular tissue, as shown in Fig. 1 (Ic'en and Raina, unpublished). No clear correlation was observed between the concentrations of polyamines and DNA. In this context I would like to recall that in bacterial systems a close correlation has been observed between the accumulation of cellular spermidine and RNA (Raina, et al., 1967).

Regenerating liver, as a rapidly growing tissue, seemed to be particularly suitable for studying the relationship of polyamines to rapid tissue growth and to RNA metabolism in particular. It is well known that in the regenerating liver RNA synthesis is activated almost immediately after operation (e.g., Fujioka, et al., 1963), there being a detectable net accumulation of RNA some 16 to 20 hr after operation. Figure 2 compares the accumulation of polyamines and RNA in regenerating rat liver. The net amount of spermidine increased as early as 16 hr postoperatively. A significant increase in RNA was seen somewhat later. The ratio of total polyamine-N to RNA-P remained constant during the regneration period (Raina, et al., 1965, 1966). Similar results were reported by Dykstra and Herbst (1965). Using ^{14}C-methionine as a precur-

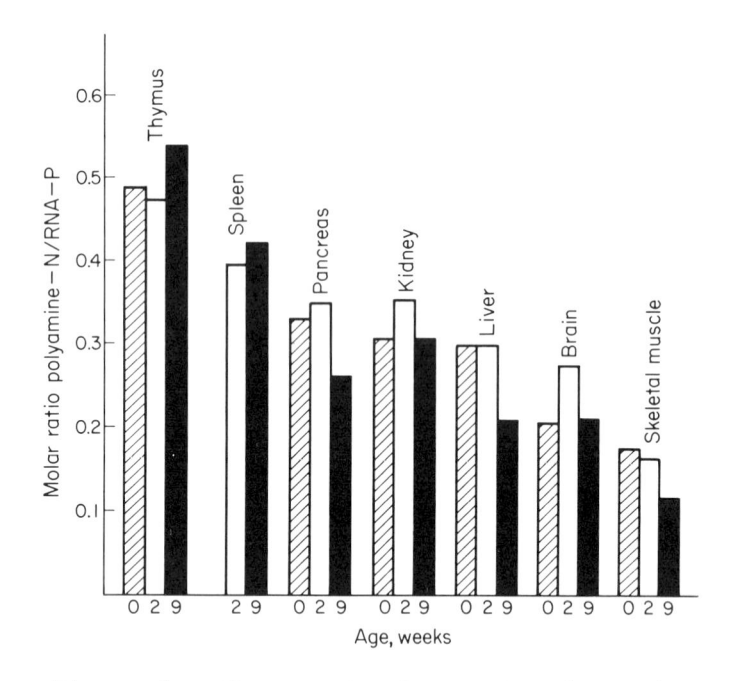

[Figure: bar chart]

Figure 1. Changes in the ratio of total polyamine nitrogen to RNA phosphate in various rat tissues in relation to age.

sor we were able to show that the synthesis of spermidine was markedly increased as early as 4 to 8 hr postoperatively. Another observation of significant importance was the discovery (Jänne, 1967) that putrescine synthesis *in vivo* was strongly stimulated immediately after partial hepatectomy with resultant marked accumulation of endogenous putrescine (Fig. 3 and Fig. 4). These observations were then confined by the demonstration, independently in our laboratory (Jänne and Raina, 1968) and by Russell and Snyder (1968a, b), that the activity of ornithine decarboxylase (ODC), an enzyme catalyzing the formation of putrescine from ornithine, was markedly stimulated in the regenerating liver.

The above data indicate that the increased synthesis and accumulation of putrescine resulting from increased ODC activity is an event taking place very early in the regenerating liver, and that this leads to increased synthesis of liver spermidine (Jänne and Raina, 1968).

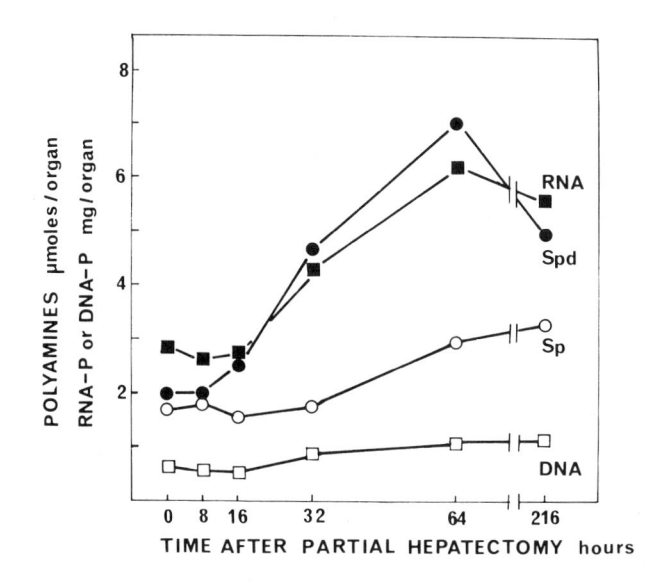

Figure 2. Total polyamine and nucleic acid contents in rat liver after partial hepatectomy. Spd, spermidine; Sp, spermine. (From Raina, et al., 1966).

The latter notion is supported by the observation that putrescine *in vitro* stimulates S-adenosylmethionine decarboxylase, spermidine being the ultimate reaction product (Pegg and Williams-Ashman, 1968a). Ornithine decarboxylase may play a crucial role in the regulation of the rate of polyamine synthesis. Therefore, later in this discussion we will compare the changes taking place in ODC activity with those found in RNA polymerase activity.

Several hormones are known to affect the rate of synthesis of RNA and protein and have also proved to be useful in studies on the relationship between polyamines and RNA metabolism. For example, one of the earliest effects of growth hormones on liver metabolism is the stimulation of RNA polymerase (Widnell and Tata, 1964). Growth hormone also causes accumulation of liver spermidine in hypophysectomized rats (Kostyo, 1966) and stimulates the synthesis of putrescine and spermidine in normal rat liver (Jänne, 1967). Figure 5 shows the marked increase in the incorporation of ^{14}C-methionine into liver spermidine after treatment

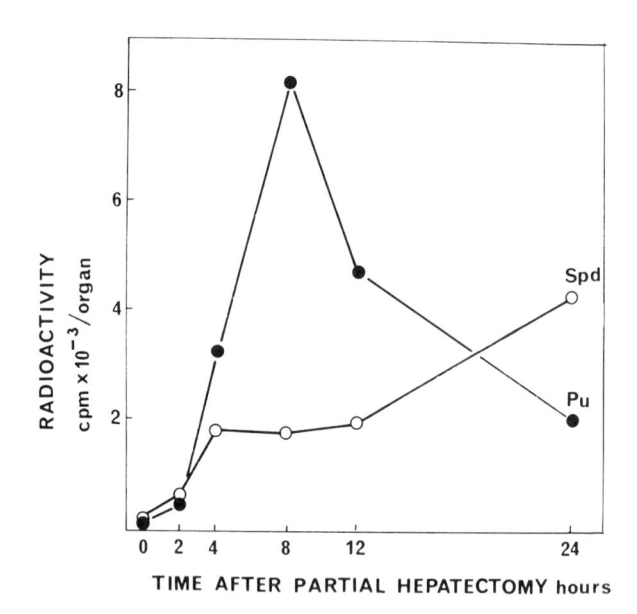

Figure 3. Incorporation of radioactivity from ^{14}C-ornithine into liver putrescine and spermidine after partial hepatectomy. DL-ornithine-5-^{14}C (4 μC/rat) was given intraperitoneally 2 hr before analysis. The activities are expressed as cpm per residual liver. (From Jänne, 1967).

with growth hormone and some stimulation in the synthesis of RNA and protein also. We have further observed that growth hormones promotes a sharp stimulation in liver ODC activity, with a simultaneous increase in RNA polymerase activity, as measured in intact nuclei (Fig. 6). Unlike some other enzymes that are "inducible" by hormones, such as tryptophan pyrrolase and tyrosine aminotransferase, the activities of liver ODC and RNA polymerase were increased by growth hormone even in adrenalectomized animals (Jänne and Raina, 1969). Growth hormone also enhanced ODC activity in the isolated perfused liver (Jänne, et al., 1969).

Some recent studies have further demonstrated a close correlation in the changes in polyamines and RNA in the prostate and seminal vesicle of the orchiectom-

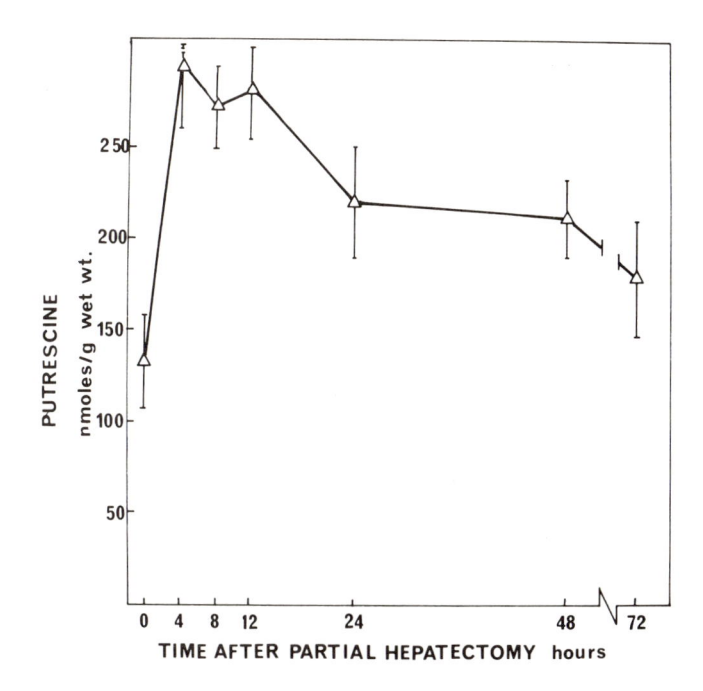

Figure 4. Accumulation of putrescine in regenerating rat liver. Five to six animals in each group. Vertical bars indicate standard deviation of the mean. (From Jänne, 1967).

ized rat. Castration produced a marked drop in the concentrations of polyamines and RNA in these organs. During subsequent testosterone treatment the two were accumulated in parallel fashion (Caldarera, *et al.*, 1968; Moulton and Leonard, 1969). A similar correlation was also seen in the corresponding enzyme activities, i.e., in RNA polymerase, ornithine decarboxylase and S-adenosylmethionine decarboxylase (Pegg and Williams-Ashman, 1968b).

Recently, we have followed the changes taking place in the synthesis of polyamines and RNA, as well as in their accumulation in rat tissues after treatment with various drugs. As seen in Fig. 7, after treatment of the rat with folic acid, a drug which is known to increase RNA synthesis in the kidney (Threlfall, *et al.*, 1967), the earliest change recorded was the stimulation

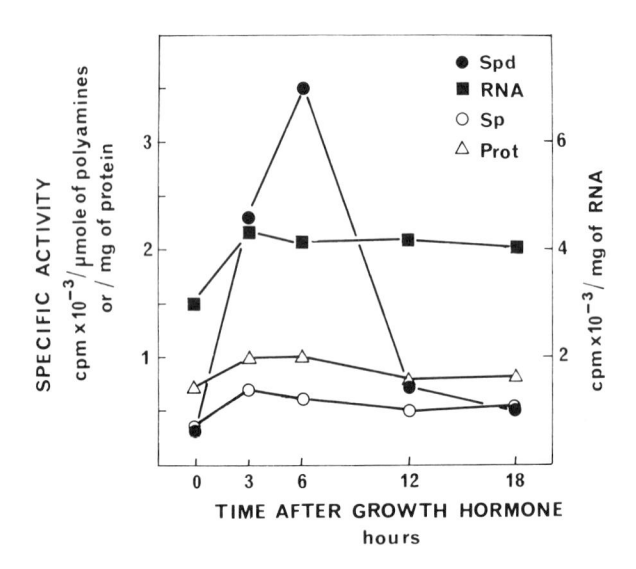

Figure 5. Effect of growth hormones on the synthesis of polyamines, RNA and protein in normal rat liver. Two-month-old rats, weighing about 120 g, received 7 I.U. of growth hormone at the times indicated and 6 µC of DL-methionine-2-^{14}C 60 min or 2 µC of orotic acid-6-^{14}C 30 min before sacrifice. (From Jänne, *et al.*, 1968).

of ornithine decarboxylase activity followed by accumulation of putrescine. An increase in RNA polymerase activity and accumulation of RNA and spermidine occurred between 6 and 12 hr after treatment. Another compound, thioacetamide is known to cause a characteristic accumulation and increased synthesis of RNA in liver nucleoli (e.g., Villalobos, *et al.*, 1964), and to stimulate liver ornithine decarboxylase activity (Fausto, 1969). As seen in Fig. 8, after thioacetamide treatment, putrescine synthesis was again the first process to be activated, and this was followed by an increase in RNA polymerase activity and accumulation of spermidine and RNA. Parallel changes in polyamines and nucleic acids have also been recorded in the chick embryo after iproniazid treatment (Caldarera, *et al.*, 1965). A similar correlation was observed between

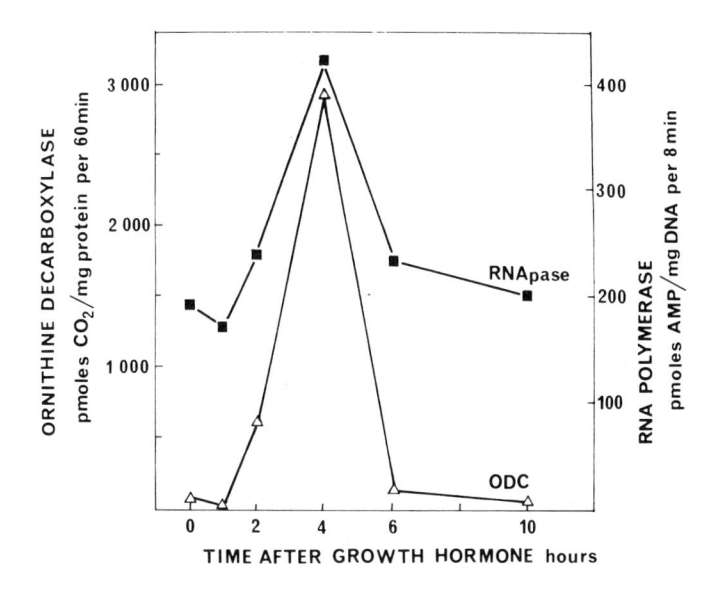

Figure 6. Time course of the effect of growth hormone treatment on RNA polymerase and ornithine decarboxylase levels in rat liver. 8 I.U. of growth hormone/100 g body weight was injected intraperitoneally at time zero and the animals were sacrificed at the times indicated. The controls (time 0) received solvent only 4 hr before sacrifice. (From Jänne and Raina, 1969).

spermidine and RNA in mouse liver during food deprivation and after various drug treatments (Seiler, et al., 1969).

One can now try to find an interpretation of the observations cited above. Does stimulation of RNA synthesis lead to increased synthesis and accumulation of polyamines or *vice versa*, or are these two processes perhaps stimulated independently? It seems probable that stimulation of polyamine synthesis is preceded by increased transcription of at least certain genes, because stimulation of putrescine synthesis after growth hormone treatment or partial hepatectomy can be abolished by actinomycin D (Jänne, et al., 1968; Russell and Snyder, 1969). The increased polyamine levels might then affect the rate of transcription of some other genes and/or affect the rate of degradation of the newly synthesized RNA. The results obtained from studying the effect of exogenous polyamines on RNA synthesis *in vivo* and *in vitro* and discussed in the following sections, might support this kind of hypothesis, although the evidence is by no means unequivocal.

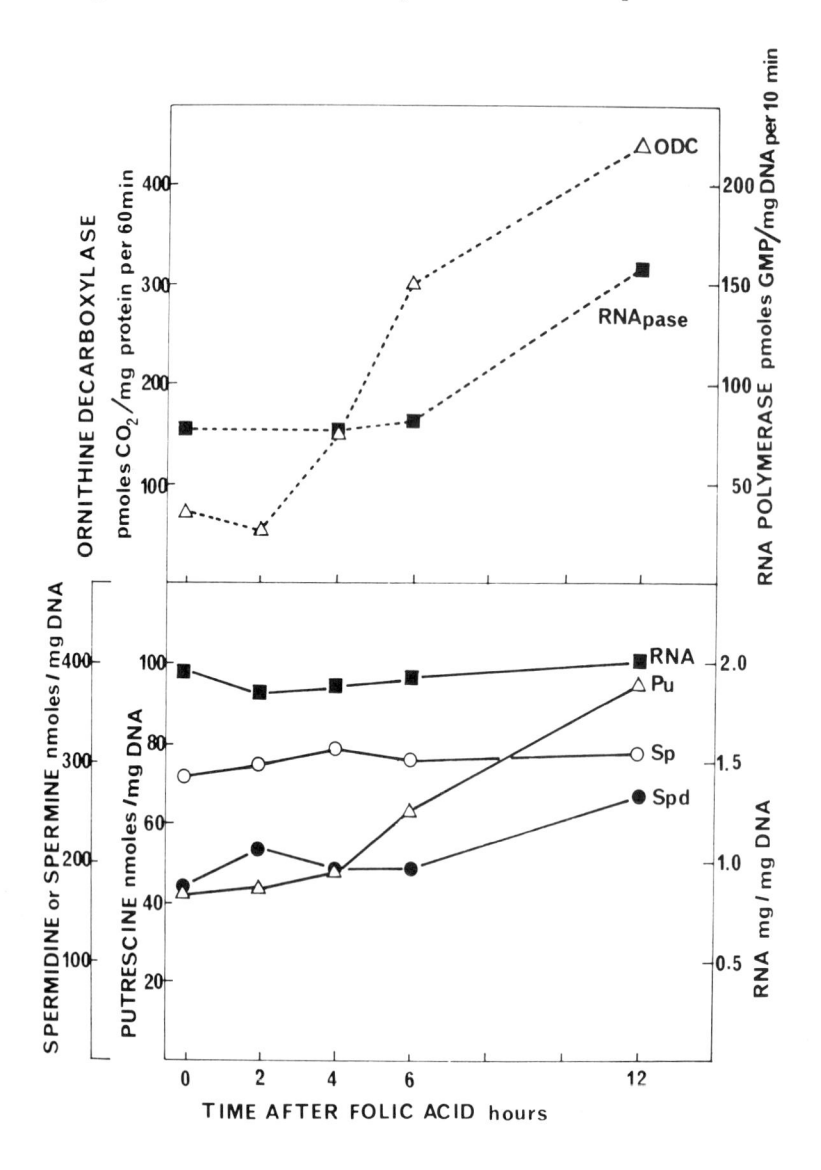

Effect of Polyamines on the Accumulation of Newly Synthesized RNA in Whole Cells

Several reports have indicated that exogenous polyamines can increase the accumulation of the label from radioactive precursors in whole cells and tissue cultures from different sources, although the mechanism(s) of their action is in most cases poorly characterized. For example, we have shown that exogenous spermidine can relax RNA synthesis in *Escherichia coli* 15 TAU during amino acid starvation. Possible mechanisms of the spermidine effect in this system were discussed in greater detail in an earlier paper (Raina, *et al.*, 1967). In this connection I would like to report just one experiment (Fig. 9) showing that the spermidine effect might partly be due to inhibition of the degradation of newly formed RNA (Ic'en and Raina, unpublished). The same observation has also been made by Boyle and Cohen (1968).

We have studied the effect of polyamines on the accumulation of newly synthesized RNA in Ehrlich ascites cells *in vitro*. As shown in Table 1, both spermidine and spermine are able to increase the incorporation of ^{14}C-orotic acid into RNA, whereas putrescine has little effect in this system. Our data further indicate that the stimulation produced by spermidine and spermine is

---

Figure 7. (Opposite page.) Changes in the activity of RNA polymerase and ornithine decarboxylase, as well as in the concentrations of RNA and polyamines in rat kidney after treatment with folic acid. The animals were injected intraperitoneally with 25 mg of folic acid/100 g body weight. The isolation of nuclei and other methods of analysis were as described earlier (Raina, *et al.*, 1967; Jänne and Raina, 1969), except that before the assay of RNA polymerase activity the nuclei were freed of most of the contaminating folic acid by centrifugation through concentrated sucrose (Threlfall, *et al.*, 1967). Extensive lysis occurred in the 2-hr sample of nuclei making the RNA polymerase assay impossible. At each time the kidneys of 5 animals were pooled for analysis.

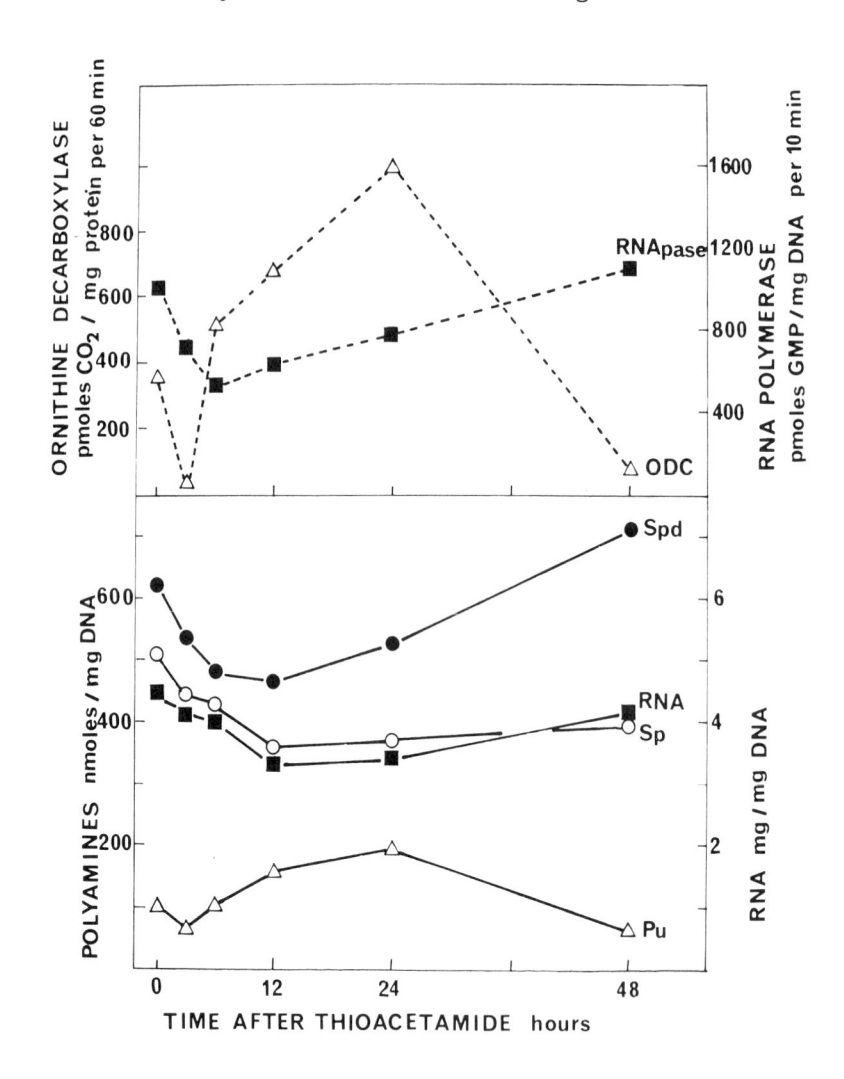

Figure 8. Changes in the activity of RNA polymerase and ornithine decarboxylase, as well as in the concentrations of RNA and polyamines in rat liver after treatment with thioacetamide. 5 mg of thioacetamide per 100 mg body weight was given intraperitoneally. For methods of analysis see the legend to Fig. 7.

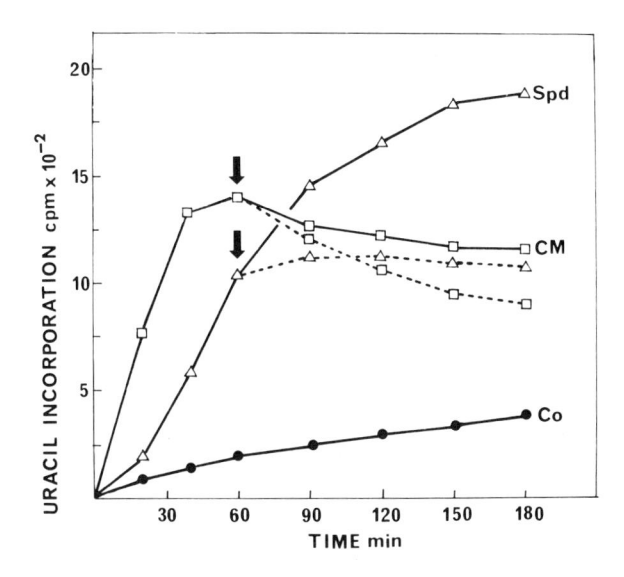

Figure 9. Effect of chloramphenicol and spermidine on the accumulation of ^{14}C-uracil in RNA in cultures of E. coli 15 Tau during arginine starvation. At 60 min two of the cultures (broken lines) were chased with 20-fold excess of cold uracil. For methods of analysis see Raina, et al. (1967). Co, control (thymine + ^{14}C-uracil); CM, chloramphenicol, 100 µg/ml; Spd, spermidine, 80 µmole/ml.

probably not due to increased permeability to the radioactive precursor used (Raina and Jänne, 1968). Larger variations in the radioactive precursor pool seem likely with radioactive uridine, because it is rapidly taken up by the cells.

Our next step was further exploration of the effect of polyamines on the accumulation of RNA in Ehrlich ascites cells. The cells were pulse-labeled with ^{14}C-uridine in the presence and absence of polyamines and chased with cold uridine and actinomycin D, after which the disappearance of the label from the acid-insoluble fraction was followed. As shown in Fig. 10, both spermidine and spermine considerably decreased the degradation of the labeled material. This result does

TABLE 1

Effect of polyamines on the accumulation of the label from orotic acid-^{14}C in RNA and in the acid-soluble fraction of Ehrlich ascites cells

Polyamine added	Conc. mM	RNA cpm	Change %	Acid-soluble fraction cpm	Change %
None	–	686	–	358	–
Putrescine	5	665	– 3.1	328	– 8.4
Spermidine	2	735	+ 7.1	390	+ 8.9
	5	805	+ 17.3	382	+ 6.7
Spermine	2	942	+ 37.3	370	+ 3.4
	5	1250	+ 82.2	474	+ 32.4

Cells were preincubated for 30 min in the presence of polyamines before addition of 0.1 µC/ml of orotic acid-^{14}C. Analysis 3 hr after addition of the label. (From Raina and Jänne, 1968).

not exclude the possibility that polyamines may also increase the rate of synthesis of labeled RNA in this system.

Other observations bearing on this topic will be discussed briefly. Results comparable with those cited above were reported by Goldstein (1965), who found that spermine increased the accumulation of ^{14}C-uridine in RNA in Walker carcinosarcoma cell cultures, but did not affect the incorporation of ^{14}C-thymidine into DNA, and inhibited protein synthesis. On incubation of salivary glands of *Drosophila melanogaster* in the presence of low concentrations of spermidine (0.5 to 6.4 X 10^{-4} M), Dion and Herbst (1967) obtained either inhibition or stimulation of ^{3}H-uridine incorporation, depending on the spermidine concentration used. In the chick embryo, spermidine and spermine injected onto the chorioallantoic membrane increased the incorporation of ^{14}C-formate into the ribosomes of

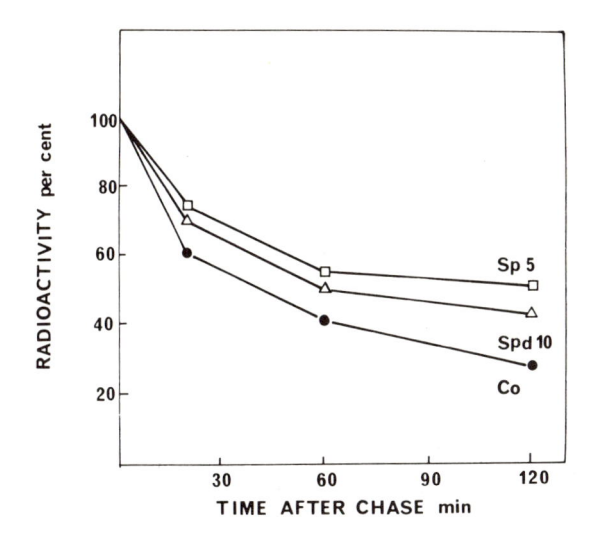

Figure 10. Effect of spermidine and spermine on the degradation of newly synthesized RNA in Ehrlich ascites cells *in vitro*. Cells were preincubated in the presence of polyamines for 60 min before the addition of 0.1 µC/ml of ^{14}C-uridine. After labeling for 15 min, 100-fold excess of unlabeled uridine and 5 µg/ml of actinomycin D were added to the cultures. The cultures were as follows: Co, incubated without polyamines, Spd and Sp, incubated in the presence of 10 mM spermidine or 5 mM spermine, respectively. (From Raina and Jänne, 1968).

chick embryo brain (Caldarera, *et al.*, 1969). Barros and Giudice (1968), working with sea urchin embryos, observed increased labeling of ribosomal RNA when embryos were developed in the presence of spermidine. These authors also state that the increased labeling is not attributable to an effect on permeability to the precursor.

It is clear that the above results should be interpreted with some caution until indirect effects, such as effects on precursor pools or effects at translational level, are excluded. In any case, the observations presented above should encourage a more detailed study of the effect of polyamines in cell and tissue cultures.

Effect of Polyamines on the Accumulation of Newly Synthesized RNA in Nuclear and Nucleolar Preparations *in vitro*

The stimulation of RNA polymerase by polyamines has been repeatedly demonstrated in a variety of systems utilizing RNA polymerase preparations from microbial, plant and animal sources (e.g., Fox, *et al.*, 1965; Krakow, 1966; Stout and Mans, 1967; Ballard and Williams-Ashman, 1966). In general, stimulation by polyamines is seen only when native double-stranded DNA is used as a primer. The exact mechanism by which the RNA polymerase reaction is stimulated by polyamines as well as by other cations is not fully understood at present. There is some evidence to suggest that polyamines may either affect the rate of release of the RNA product from the enzyme or prevent its binding to the template (Fox, *et al.*, 1965; Krakow, 1966). Apparently, the polyamine effect in this system is not due to inhibition of nucleases. Further insight into the mode of action of polyamines in the RNA polymerase reaction has been gained recently (Fuchs, *et al.*, 1967; Abraham, 1968; Petersen, *et al.*, 1968). Whatever the mechanism may be, the concentrations of polyamines able to stimulate RNA polymerase *in vitro* are within the physiologic range. It seems possible, therefore, that changes in the concentrations of polyamines may have an effect on the rate of transcription in physiologic conditions as well.

Fewer data have been reported concerning the effect of polyamines on RNA synthesis in nuclear and nucleolar preparations. In the former case the possibility that polyamines may change the permeability of the nuclear membrane to substrates makes interpretation of the results difficult. Furthermore, the possibility has to be borne in mind that polyamines may only penetrate the nuclear membrane slowly. Although exogenous polyamines help to preserve nuclear morphology during *in vitro* incubation, the stimulation of RNA synthesis obtained in some cases with polyamines in this system (MacGregor and Mahler, 1967) is probably not related to this phenomenon.

Interesting results have been reported by Caldarera, *et al.* (1968), who have shown that spermidine and spermine are able to stimulate RNA synthesis in nuclei isolated from prostates of castrated rats. As mentioned earlier, testosterone treatment stimulates both RNA polymerase activity and the synthesis and accumulation of polyamines. Therefore these authors suggested that the testosterone effect on RNA synthesis in the prostate is mediated by spermine or spermidine. Before the significance of these results can be fully evaluated, a more extensive study is needed and confirmation in other systems.

It has been demonstrated that in animal tissues a large fraction, perhaps 70 to 80%, of the newly synthesized RNA is rapidly degraded (Darnell, 1968). It seems very possible that, in addition to factors affecting the rate of synthesis, others affecting the rate of degradation of the newly synthesized RNA may be important in determining RNA accumulation in animal tissues. If such regulation exists, polyamines are one group of substances that might possibly be involved. Some evidence to support this concept has been obtained in studies on RNA synthesis in nuclear and nucleolar preparations and will be discussed below.

By pulse-labeling RNA *in vivo* and then incubating isolated nuclei in RNA polymerase assay conditions, it can be shown that the labeled RNA is rapidly degraded during incubation (e.g., MacGregor and Mahler, 1967). The same is true of RNA products synthesized in intact nuclei *in vitro*. As seen in Table 2, this degradation was largely prevented by addition of 5 mM spermidine or spermine at the time of the chase, and was completely prevented by addition of 0.4 M $(NH_4)_2SO_4$.

We have suggested earlier that polyamines may play a role in the regulation of the synthesis of ribosomal RNA (Cohen and Raina, 1967). That the nucleolus acts as a site for the synthesis of ribosomal RNA is well established. Therefore it seemed worth while to study the effect of polyamines on nucleolar RNA synthesis *in*

TABLE 2

Effect of polyamines and ammonium sulfate on the degradation of newly synthesized RNA in rat liver nuclei *in vitro*

Addition at 8 min	Radioactivity left acid-precipitable at 32 min	
	cpm	%
Chase (ATP + actinomycin D)	903	62.6
Chase + 5 mM spermidine	1082	74.9
Chase + 5 mM spermine	1225	84.8
Chase + Mn^{++} + (NH$_4$)$_2$SO$_4$	1441	99.8

Nuclei were isolated from rat liver and assayed for RNA polymerase activity as described earlier (Jänne and Raina, 1969), using ^{14}C-ATP as the labeled precursor. After incubation for 8 min, the samples were chased with 10-fold excess of cold ATP and 50 μg/ml of actinomycin D. Polyamines and MnCl$_2$ (0.6 mM) plus (NH$_4$)$_2$SO$_4$ (0.4 M, final concentration) were added at the time of chasing, and incubation was continued for an additional 24 min.

vitro. As seen in Table 3, both putrescine and spermidine stimulated nucleolar RNA polymerase activity. Spermine was found to be inhibitory in concentrations as low as 5 X 10^{-4} M. We soon discovered that the effect of spermidine in this system was dependent on Mg^{++} concentrations (Fig. 11), the highest stimulation being obtained at suboptimal Mg^{++} levels. However, spermidine could not totally replace the Mg^{++} ion. The inhibitory effect observed with spermine and, at higher levels, with spermidine and Mg^{++} is obviously due to microscopically visible aggregation of nucleoli under these conditions. In contrast to spermidine, putrescine enhanced the rate of RNA synthesis in a wide range of Mg^{++} concentrations (Fig. 12). We tried to gain further insight into the mode of action of polyamines in this system. As seen in Fig. 13, spermidine added late during incubation significantly increased the additional incorporation of the label and

TABLE 3

Effect of putrescine and spermidine on nucleolar RNA synthesis *in vitro*

Polyamine	mM	^{14}C-GMP incorporated nmoles/mg DNA/10 min	Stimulation
			%
None		10.86	-
Spermidine	0.1	12.25	12.7
	0.5	12.80	17.8
	3.0	12.89	18.6
	5.0	13.23	21.8
Putrescine	2.0	11.68	7.5
	5.0	14.82	36.4
	10.0	16.40	51.0
Putrescine + Spermidine	1.0 3.0	15.03	38.3

The nucleoli were isolated from rat liver by the sonication procedure described by Busch (1968) with slight modifications. RNA polymerase activity was assayed as described earlier (Jänne and Raina, 1969), using ^{14}C-GTP (0.1 mM) as the labeled precursor. The concentration of $MgCl_2$ in the incubation mixture was 2.6 mM.

also decreased the degradation of the RNA product after chasing. In a comparable experiment with putrescine (Fig. 14), this polyamine restarted the accumulation of the label in RNA, but did not significantly change the rate of degradation after chasing. It therefore seems probable that polyamines, spermidine in particular, may exert a dual effect on the accumulation of RNA in nucleolar preparations by increasing the RNA polymerase activity and inhibiting the degradation of the RNA product. These data, although suggestive, do not necessarily indicate that in physiological conditions polyamines have similar functions. For example, the nucleases responsible for the degradation of RNA

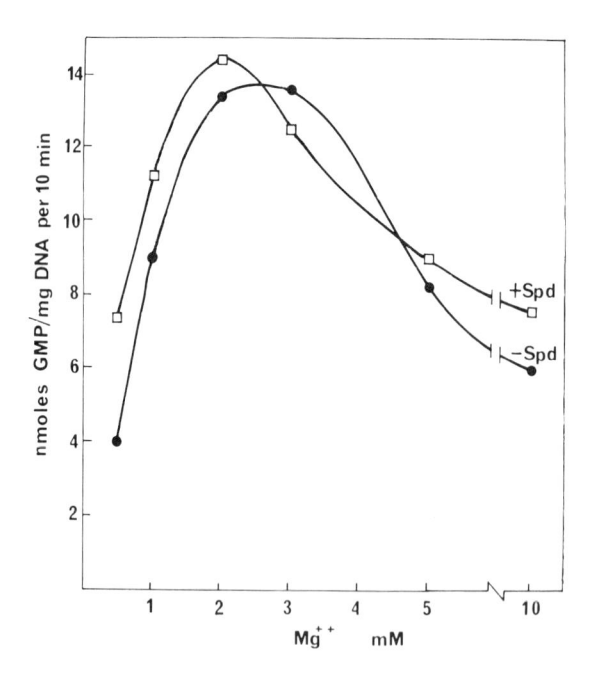

Figure 11. Effect of spermidine on nucleolar RNA synthesis at different Mg^{++} concentrations. For methods of analysis see the legend to Table 3. The concentration of spermidine was 2 mM.

during *in vitro* incubation may not function in physiological conditions. Further, the conditions under which RNA synthesis is stimulated may be far from physiological, e.g., in respect to Mg^{++} concentration. On the other hand, the aggregation of nucleolar particles at higher polyamine concentrations may mask the true effect of the latter. This should also be taken into account in studies on the effect of polyamines on other systems, e.g., on protein synthesis *in vitro*.

In this context it may be mentioned that polyamines may have other sites of action in RNA metabolism. As an example, one may refer to the preliminary report of Leboy (1968), indicating stimulation of RNA methylase by spermidine. It has also been reported that spermidine added to a system containing purified bacterial RNA polymerase is able to increase the asymmetry of

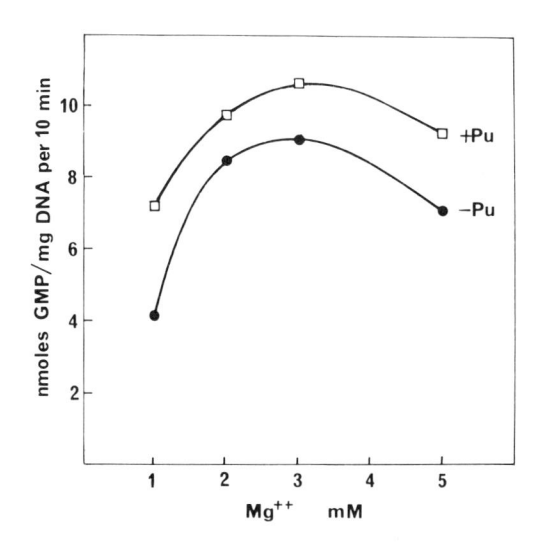

Figure 12. Effect of putrescine on nucleolar RNA synthesis at different Mg^{++} concentrations. The concentration of putrescine was 10 mM.

the RNA product, probably by eliminating some nonspecific starting points of the enzyme (Ascoli, 1968).

Conclusion and Comments

The results reveiwed here have shown that changes in the concentrations of polyamines and RNA, as well as the activation of the synthesis of these compounds, parallel each other in a variety of physiologic and experimental states such as those obtaining after partial hepatectomy or after treatment of the animals with various hormones and drugs. It has further been shown that polyamines can affect the accumulation of newly synthesized RNA in whole cell and tissue cultures, and *in vitro* systems whether nuclear or nucleolar preparations are used as the enzyme source or purified RNA polymerase is employed.

One may now ask how these data should be interpreted. Do polyamines play a role in the regulation of RNA metabolism? Regardless of the more or less provocative title of this paper, it is not possible to give a definitive answer to this question. We have

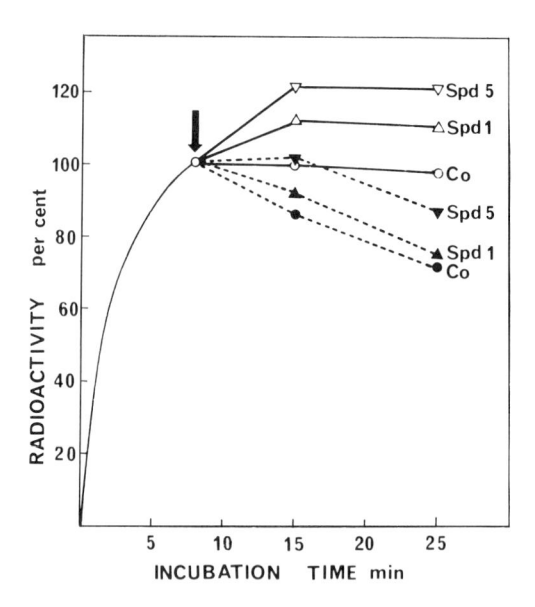

Figure 13. Effect of spermidine on the accumulation of newly synthesized RNA in nucleolar preparations *in vitro*. ^{14}C-UTP (0.1 mM) was used as the labeled precursor. At 8 min a part of the samples (broken lines) were chased with 10-fold excess of cold UTP and 50 µg/ml of actinomycin D ± spermidine. A part of the samples were treated with spermidine (1 and 5 mM) or with buffer only (Co). The Mg^{++} concentration in the incubation mixture was 3 mM.

fairly strong experimental evidence that a large fraction of cellular polyamines is physiologically associated with structures containing nucleic acids, such as ribosomes, and exerts a stabilizing effect on these structures. A further question is whether polyamines affect the rate of RNA synthesis. Although the activation of polyamine synthesis in sites such as the regenerating liver, is a very early event, there is little evidence to suggest that polyamines act as general gene activators. One may, however, speculate that changes in the concentrations of polyamines may affect the rate of transcription of certain genes, e.g., those serving as templates for ribosomal RNA.

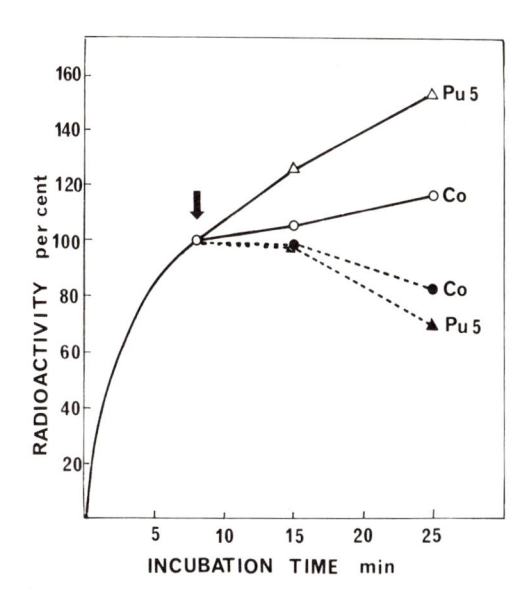

Figure 14. Effect of putrescine on the accumulation of newly synthesized RNA in nucleolar preparations *in vitro*. Incubation conditions and general procedure as indicated in the legend to Fig. 13. The putrescine concentration was 5 mM.

Furthermore, they may increase the accumulation of newly synthesized RNA by inhibiting its degradation, e.g., by changing the conformation of RNA into a form less accessible to the action of nucleases. A question of particular interest is whether polyamines can affect the rate of ribosome formation. At the present we know rather little about this process, i.e., how ribosomal proteins become associated with ribosomal RNA.

Although the question of the role of polyamines in RNA metabolism remains unanswered, the data that have accumulated during the past few years seem to warrant further discussion of this problem and will, I hope, by increasing interest in polyamine research, contribute towards a better understanding of the physiology of polyamines. Their possible regulatory role in cell metabolism should not be forgotten.

ACKNOWLEDGMENTS

We wish to thank Dr. Martti Siimes and Dr. Rainer Ic'en for collaboration in some parts of this investigation. The skillful technical assistance of Mrs. Sirkka Kanerva and Mrs. Riitta Leppänen is gratefully acknowledged.

REFERENCES

Abraham, K.A. 1968. *European J. Biochem.* 5: 143.
Ascoli, F. 1968. *Quad. "Ric. Sci."* 48: 126. Cited in *Chem. Abstr.* 1969. 70: 12.
Ballard, P.L. and H.G. Williams-Ashman. 1966. *J. Biol. Chem.* 241: 1602.
Barros, C. and G. Giudice. 1968. *Exptl. Cell Res.* 50: 671.
Boyle, S.M. and P.S. Cohen. 1968. *J. Bacteriol.* 96: 1266.
Busch, H. 1968. In *Comprehensive Biochemistry*, Vol. 23, M. Florkin and E.H. Stotz, eds. (Amsterdam: Elsevier Publishing Co.), p. 39.
Caldarera, C.M., B. Barbiroli and G. Moruzzi. 1965. *Biochem. J.* 97: 84.
Caldarera, C.M., M.S. Moruzzi, B. Barbiroli and G. Moruzzi. 1968. *Biochem. Biophys. Res. Commun.* 33: 266.
Caldarera, C.M., M.S. Moruzzi, C. Rossoni and B. Barbiroli. 1969. *J. Neurochem.* 16: 309.
Cohen, S.S. and A. Raina. 1967. In *Organizational Biosynthesis*, H.J. Vogel, J.O. Lampen and V. Bryson. eds. (New York: Academic Press, Inc.), p. 157.
Darnell, J.E., Jr. 1968. *Bacteriol. Rev.* 32: 262.
Dion, A.S. and E.J. Herbst. 1967. *Proc. Natl. Acad. Sci.* 58: 2367.
Dykstra, W.G., Jr. and E.J. Herbst. 1965. *Science* 149: 428.
Fausto, N. 1969. *Federation Proc.* 28: 366.
Fox, C.F., R.I. Gumbort and S.B. Weiss. 1965. *J. Biol. Chem.* 240: 2101.
Fuchs, E., R.L. Millette, W. Zillig and G. Walter.

1967. *European J. Biochem.* 3: 183.
Fujioka, M., M. Koga and I. Lieberman. 1963. *J. Biol. Chem.* 238: 3401.
Goldstein, J. 1965. *Exptl. Cell Res.* 37: 494.
Jänne, J. 1967. *Acta Physiol. Scand.* Suppl. 300: 1.
Jänne, J., M. Kekomäki, A. Raina and E. Hölttä. 1969. In *Abstracts, 6th Meeting of Federation of European Biochemical Societies*, Madrid, p. 195.
Jänne, J. and A. Raina. 1968. *Acta Chem. Scand.* 22: 1349.
Jänne, J. and A. Raina. 1969. *Biochim. Biophys. Acta* 174: 769.
Jänne, J., A. Raina and M. Siimes. 1964. *Acta Physiol. Scand.* 62: 352.
Jänne, J., A. Raina and M. Siimes. 1968. *Biochim. Biophys. Acta* 166: 419.
Kostyo, J.L. 1966. *Biochem. Biophys. Res. Commun.* 23: 150.
Krakow, J.S. 1966. *J. Biol. Chem.* 241: 1830.
Leboy, P.S. 1968. *J. Cell Biol.* 39: 78a.
MacGregor, R.R. and H.R. Mahler. 1967. *Arch. Biochem. Biophys.* 120: 136.
Moulton, B.C. and S.L. Leonard. 1969. *Endocrinology* 84: 1461.
Pegg, A.E. and H.G. Williams-Ashman. 1968a. *Biochem. Biophys. Res. Commun.* 30: 76.
Pegg, A.E. and H.G. Williams-Ashman. 1968b. *Biochem. J.* 109: 32P.
Petersen, E.E., H. Kroger and U. Hagen. 1968. *Biochim. Biophys. Acta* 161: 325.
Raina, A. 1963. *Acta Physiol. Scand.* 60:(Suppl. 218): 1.
Raina, A., M. Jansen and S.S. Cohen. 1967. *J. Bacteriol.* 94: 1967.
Raina, A. and J. Jänne. 1968. *Ann. Med. Exp. Biol. Fenniae* 46: 536.
Raina, A., J. Jänne and M. Siimes. 1965. In *Abstracts, 2nd Meeting of Federation of European Biochemical Societies*, Vienna. A 114.
Raina, A., J. Jänne and M. Siimes. 1966. *Biochim. Biophys. Acta* 123: 197.
Russell,D.H. and S.H. Synder. 1968a. *Federation Proc.* 27: 642.

Russell, D.H. and S.H. Snyder. 1968b. *Proc. Natl. Acad. Sci.* 60: 1420.
Russell, D.H. and S.H. Snyder. 1969. *Mol. Pharmacol.* 5: 253.
Seiler, N., G. Werner, H.A. Fischer, B. Knötgen and H. Hinz. 1969. *Z. Physiol. Chem.* 350: 676.
Stout, E.R. and R.J. Mans. 1967. *Biochim. Biophys. Acta* 134: 327.
Tabor, H. and C.W. Tabor. 1964. *Pharmacol. Rev.* 16: 245.
Takeda, Y. 1969. *Biochim. Biophys. Acta* 182: 258.
Tanner, M.J.A. 1967. *Biochemistry* 6: 2686.
Threlfall, G., D.M. Taylor, P. Mandel and M. Ramuz. 1967. *Nature* 215: 755.
Villalobos, J.G., Jr., W.J. Steele and H. Busch. 1964. *Biochim. Biophys. Acta* 91: 233.
Widnell, C.C. and J.R. Tata. 1964. *Biochem. J.* 93: 2P.

Norepinephrine and the Circadian Rhythm of Rat Hepatic Tyrosine Transaminase Activity

Ira B. Black

*Laboratory of Clinical Science
National Institute of Mental Health
Bethesda, Maryland*

Rhythmicity of biologic function has been identified in organisms throughout the plant and animal kingdoms. In addition to synchronizing the physiology and behavior of a species with a periodically varying environment, these biorhythms synchronize the individuals of a species with one another, and synchronize multiple phenomena within a single individual. Viewed in such a context, biological cycles are of obvious survival value and contribute to the evolutionary progression of a species.

Biological rhythms may be endogenous to the organism or may be driven by environmental stimuli. Although the driving oscillation of a cycle may reside within the organism, synchronization in time may depend on external entraining agents. Environmental cues which synchronize biological rhythms in time have been termed Zeitgeber (Aschoff, 1951). Persistence of a biological cycle under constant environmental conditions constitutes evidence for its endogenous character. Classically, light and temperature have been considered the critical geophysical variables. Both exogenous and endogenous biorhythms vary in period from minutes, as in the case of cyclic ribosomal aggregation (Kaempfer, *et al.*, 1968), to decades, as in the case of human sexual maturation and climacteric. Circadian rhythms, those with a period approximately 24 hours, have been studied most extensively and constitute the prime models of biological periodicity.

The nervous system appears to serve a pivotal function in the regulation of biorhythms. In the case of exogenous cycles, the nervous system, through the retina and optic pathways, for example, transduces environmental information into biochemical rhythmicity. In the case of endogenous rhythms, the role of the nervous system appears to have changed from that of transducer to that of generator, with driving oscillations incorporated into the neural character.

This communication reports investigations of the daily rhythm in rat hepatic tyrosine transaminase activity and its regulation by catecholamines. Tyrosine transaminase is the first of a sequence of enzymes which metabolize the amino acid tyrosine (Schepartz, 1951). Tyrosine transaminase turns over extremely rapidly, with a half-life of approximately 1.5 hr (Kenney, 1967), and thus constitutes a sensitive model for the study of regulatory biochemistry. The enzyme is subject to regulation by a number of hormones, and by substrate. Glucocorticoids (Lin and Knox, 1957) increase tyrosine transaminase activity through an actinomycin D (Greengard, *et al.*, 1963) sensitive process, due to increased synthesis of enzyme protein (Granner, *et al.*, 1968). Glucagon (Greengard and Baker, 1966), insulin (Holten and Kenney, 1967) and L-tryptophan (Rosen and Milholland, 1963) also induce tyrosine transaminase in the adrenalectomized rat, and growth hormone has been reported to increase (Ottolenghi and Cavagna, 1968) and decrease (Kenney, 1967) enzyme activity under different conditions.

Initial experiments in this laboratory defined the characteristics of the enzyme rhythm (Black and Axelrod, 1968a) and subsequent studies suggested regulation of the cycle by the nervous system through the mediation of norepinephrine (Black and Axelrod, 1968b, c; 1969). *In vitro* experimentation (Black and Axelrod, in press) provided information regarding the molecular locus of action of the catecholamine and indicated the mechanism by which norepinephrine may contribute to the generation of the tyrosine transaminase rhythm.

Characteristics of the Tyrosine Transaminase Rhythm

Rat hepatic tyrosine transaminase activity varies over a two- to four-fold range daily (Potter, *et al.*, 1966) and this rhythm persists in the absence of the pituitary or adrenal glands (Wurtman and Axelrod, 1967; Civen, *et al.*, 1967; Shambaugh, *et al.*, 1967). To define the relationship of this enzyme cycle to the environment, rats were exposed to normal diurnal lighting, constant light, or constant darkness for one week. In constant darkness the rhythm persisted essentially unchanged, while seven days of constant light resulted in unaltered period or phase, but decreased amplitude (Fig. 1) (Black and Axelrod, 1968a). Thus, although the rhythm is not generated by external light cues, constant light decreases either the amplitude of the driving oscillation and/or the responsiveness of the measured parameter (enzyme activity) to the oscillation.

To determine whether light and dark stimuli during the 24-hr day synchronize the cycle in time, groups of rats were exposed to the normal or a reversed lighting schedule. In those groups exposed to schedule-reversal, the rhythm was phase-shifted nearly 180° (Fig. 2), indicating that lighting entrains the enzyme cycle, and that both a light signal and a dark signal at critical times are necessary for synchronization (Black and Axelrod, 1968a). Thus, although the tyrosine transaminase rhythm is endogenous to the rat, lighting conditions serve to synchronize the enzyme rhythm in time. Additional studies have indicated that the rhythm persists in the fasted animal (Black and Axelrod, 1968a; Potter, *et al.*, 1968) and in the rat deprived of adrenal or pituitary glands.

304   Norepinephrine and Circadian Rhythm

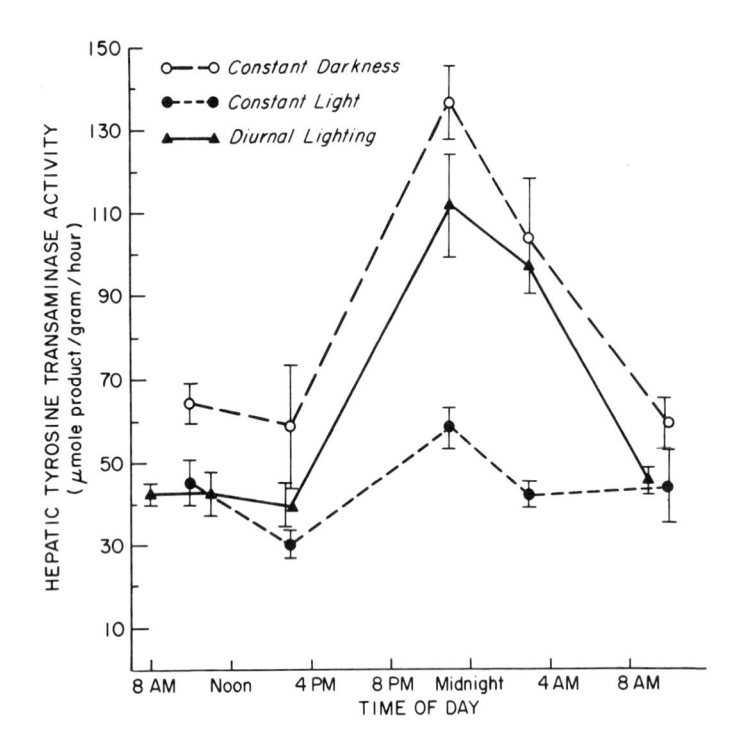

Figure 1. Effect of constant light or constant darkness on tyrosine transaminase activity. Controls were subjected to the normal diurnal light cycle, with lights on from 5 A.M. to 7 P.M. Experimental groups were exposed either to constant light or to constant darkness for 1 week. Each group at each time consists of six animals. Vertical bars indicate standard errors of the mean. In all groups the 11 P.M. value differs from the 3 P.M. value at $p < 0.001$. (Black and Axelrod, 1968a).

Role of Norepinephrine

The cycle may be intrinsic to the liver cell or may be driven by humoral or neural mechanisms. The last alternative was examined by the manipulation of specific neurotransmitters. Since glucocorticoids and amino acids increase tyrosine transaminase activity, only

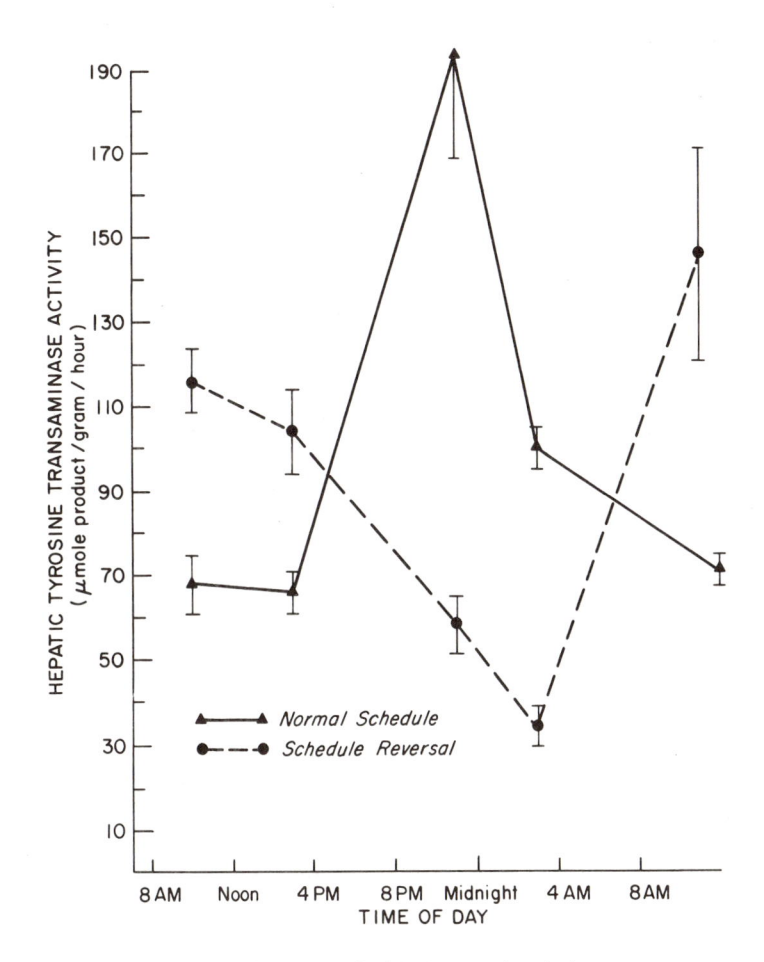

Figure 2. Effect of light schedule reversal on tyrosine transaminase rhythm. Control rats were exposed to the normal diurnal light-dark cycle (see Fig. 1). The experimental group was subjected to a reversed lighting cycle with lights on from 7 P.M. to 5 A.M. Vertical bars indicate standard errors of the mean. Both rhythms are statistically significant at $p < 0.001$. (Black and Axelrod, 1968a).

fasted, adrenalectomized rats were employed.

Reserpine, a depletor of biogenic amines, was administered to rats in a dose of 2.0 mg/kg. Within

8 hr brain norepinephrine exhibited greater than a 90% depletion and hepatic tyrosine transaminase activity increased three-fold (Black and Axelrod, 1968b). Doses of less than 2.0 mg/kg were associated with variable depletion of norepinephrine and inconstant increases in tyrosine transaminase activity. To determine whether increased enzyme activity was dependent upon the amine-depleting action of reserpine, the drug was administered after pretreatment with the monoamine oxidase inhibitor, Catron. This regimen prevents norepinephrine depletion. Reserpine administered alone was associated with a three-fold elevation in enzyme activity, whereas the Catron pretreated rats showed no depletion of brain norepinephrine and no significant rise in tyrosine transaminase activity (Black and Axelrod, unpublished observations).

To determine whether the elevated enzyme activity was dependent on the synthesis of new enzyme protein, puromycin, an inhibitor of protein synthesis, was administered to rats previously treated with reserpine. Reserpine caused the anticipated rise in tyrosine transaminase activity, whereas puromycin reduced activity to control levels in rats pretreated with reserpine (Fig. 3) (Black and Axelrod, 1968b).

Since reserpine depletes serotonin as well as norepinephrine, the effect of depletion of norepinephrine alone was studied. α-Methyl-para-tyrosine (α-m-t) inhibits tyrosine hydroxylase, the rate-limiting enzyme in norepinephrine biosynthesis, which catalyzes the conversion of $\underline{L}$-tyrosine to $\underline{L}$-DOPA. Norepinephrine is selectively depleted while the levels of serotonin are not primarily affected. Norepinephrine may be repleted by administration of precursors distal to the block in norepinephrine biosynthesis. $\underline{L}$-DOPA was employed for this purpose. α-Methyl-para-tyrosine increased tyrosine transaminase activity four-fold and reduced brain norepinephrine by approximately 90%. $\underline{L}$-DOPA increased brain norepinephrine in rats pretreated with α-m-t and lowered tyrosine transaminase activity to basal values (Table 1) (Black and Axelrod, 1968b).

Thus the depletion of norepinephrine is associated with a three- to four-fold, puromycin-sensitive rise

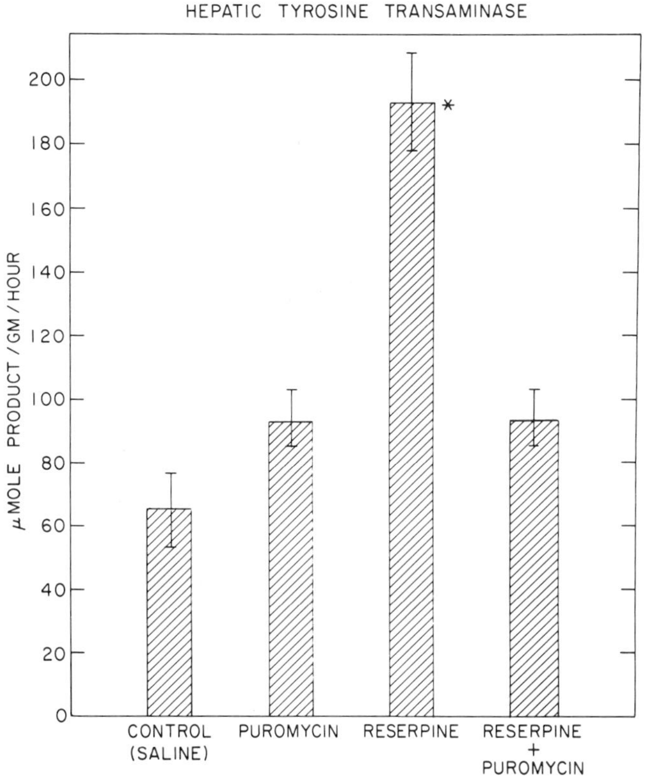
HEPATIC TYROSINE TRANSAMINASE
μMOLE PRODUCT / GM / HOUR
0
20
40
60
80
100
120
140
160
180
200
*
CONTROL (SALINE)
PUROMYCIN
RESERPINE
RESERPINE + PUROMYCIN

Figure 3. Effects of reserpine and puromycin on tyrosine transaminase activity. One group received reserpine 2 mg/kg at 12 midnight. Those treated with puromycin alone received 50 mg/kg at 8 A.M. and 10 A.M. Rats given the combination were treated with reserpine 2 mg/kg at 12 midnight and puromycin 50 mg/kg at 8 A.M. and 10 A.M. Controls were treated with saline at appropriate times. All injections were i.p. All rats were killed at 12 noon. Vertical bars indicate standard errors of the mean.

*Differs from control at $P < 0.001$, differs from reserpine + puromycin at $P < 0.001$. (Black and Axelrod, 1968b).

Norepinephrine and Circadian Rhythm

TABLE 1

The effect of depletion and repletion of norepinephrine on hepatic tyrosine transaminase activity*

Group	Hepatic tyrosine transaminase μmole product/ g/hour	Brain norepinephrine (NE) μg NE/g brain
Control	43.4 ± 12.3	0.458 ± 0.023
α-Methyl-para-tyrosine	190.1 ± 18.4†	0.056 ± 0.016¶
L-DOPA	34.6 ± 4.1	0.431 ± 0.037
α-Methyl-para-tyrosine + L-DOPA	51.1 ± 5.9	0.206 ± 0.034

Rats receiving α-methyl-para-tyrosine were injected with 200 mg/kg at 8 A.M. Those treated with L-DOPA alone received 100 mg/kg at 8 A.M. The group treated with the combination was injected with α-methyl-para-tyrosine 200 mg/kg at 12 midnight and 50 mg/kg at 8 A.M. and L-DOPA 100 mg/kg at 8 A.M. Controls were treated with 1 ml of saline at appropriate times. All rats were killed at 12 noon. All injections were intraperitoneal.

*Results expressed as mean ± standard error.

†Differs from control at $P < 0.001$, differs from α-methyl-para-tyrosine + L-DOPA at $P < 0.001$.

¶Differs from control at $P < 0.001$, differs from α-methyl-para-tyrosine + L-DOPA at $P < 0.05$.

in tyrosine transaminase activity. These observations are consistent with a mechanism in which norepinephrine, either directly or indirectly, decreases the number of tyrosine transaminase molecules in the liver cell. In the absence of the adrenal glands, norepinephrine must be acting via the central and/or peripheral nervous system(s) and may exert its effect through neural or blood-borne mediation. To examine the former possibility, peripheral neural impulses were interrupted by pharmacologic blockade of sympathetic ganglia. Administration of the ganglion blocking agent chlorisondamine was associated with a two-fold rise in tyrosine transaminase activity within six hours (Black and Axelrod, 1969). This increase in enzyme activity was reversed by the administration of norepinephrine, indicating a peripheral neural adrenergic mediation in the regulation of tyrosine transaminase (Black and Axelrod, 1969). Adrenalectomized rats, however, were found to be extremely sensitive to the toxic effects of chlorisondamine, and a dose of 15 mg/kg, necessary for ganglionic blockade, was associated with a finite mortality rate. Nerve impulses were, therefore, also interrupted by spinal cord transection immediately cephalad to C-7. This procedure resulted in a two-fold rise in tyrosine transaminase activity in six hours and this elevation was also reversed by the administration of norepinephrine (Black and Axelrod, 1969).

To determine the role of peripheral noradrenergic neural suppression of tyrosine transaminase activity in the regulation of the enzyme rhythm, chlorisondamine was administered to rats at times corresponding to basal and peak enzyme activities. Interruption of neural activity abolished the tyrosine transaminase rhythm at peak levels (Black and Axelrod, 1969).

Conversely, elevation of tissue norepinephrine should abolish the daily enzyme rhythm at basal levels. Norepinephrine was elevated by blocking intraneural degradation by monoamine oxidase. In groups killed 24 hr after the administration of β-phenylisopropylhydrazine (Catron) the enzyme rhythm was abolished, while the concentration of brain and heart norepinephrine increased significantly (Fig. 4) (Axelrod and Black, 1968c).

Since monoamine oxidase inhibitors elevate both endogenous serotonin and norepinephrine, the tissue catecholamine was selectively increased by the administration of large doses of L-DOPA. This precursor of norepinephrine caused a significant 15% increase in brain norepinephrine and suppressed the tyrosine transaminase activity rhythm by abolition of the normal evening peak (Fig. 5) (Axelrod and Black, 1968c).

DIURNAL TYROSINE TRANSAMINASE ACTIVITY RHYTHM
IN SALINE TREATED AND L-DOPA TREATED RATS

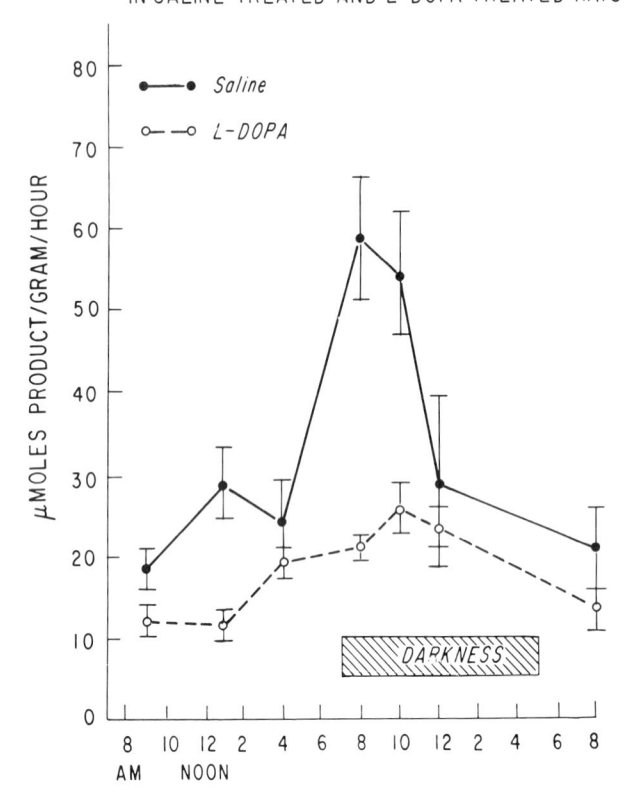

Figure 4. Effect of Catron on the diurnal rhythm of tyrosine transaminase activity. Each group of four to seven rats was given "Catron," 5 mg/kg intraperitoneally in a volume of 1 ml 24 hr before death. Controls were injected with saline at appropriate times. Vertical bars indicate standard errors of the mean. ●——●, saline; O---O, monoamine oxidase inhibitor. (Axelrod and Black, 1968c.)

Thus, two structurally dissimilar compounds, Catron and L-DOPA, which raise tissue norepinephrine by different mechanisms, suppress the characteristic 10 P.M. peak in hepatic tyrosine transaminase activity.

DIURNAL RHYTHM OF TYROSINE TRANSAMINASE
ACTIVITY IN SALINE TREATED AND MONOAMINE
OXIDASE INHIBITOR TREATED RATS

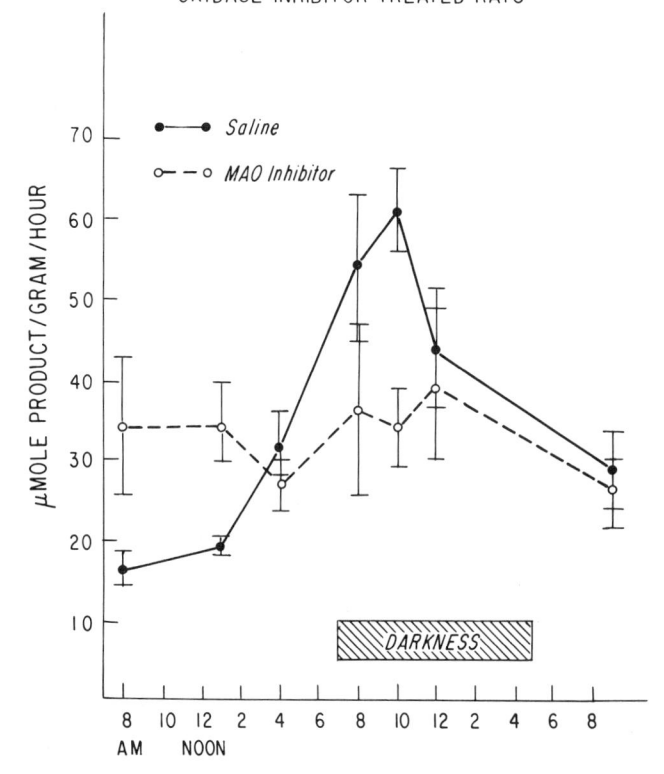

Figure 5. Effect of L-DOPA on the diurnal
rhythm of tyrosine transaminase activity.
Each group of rats was treated with a suspension of L-DOPA, 100 mg/kg given intraperitoneally in a volume of 1 ml 4 hr before death. Controls were injected with
saline at appropriate times. Vertical bars
indicate standard errors of the mean.
●——●, saline; O---O, L-DOPA. (Axelrod and
Black, 1968c.)

Interaction of norepinephrine and the tyrosine
transaminase system. Norepinephrine may alter tyrosine transaminase activity by direct interaction with
the enzyme system, or may operate indirectly through a

number of mediators. Since certain amines and amino acids are known to bind pyridoxal-5-phosphate, the cofactor required for tyrosine transaminase activity, the effect of norepinephrine upon enzyme catalysis was examined *in vitro*. In these experiments norepinephrine was incubated with the dialyzed, supernatant fraction liver enzyme and decreasing amounts of the pyridoxal-5'-phosphate cofactor. Up to 95% inhibition of tyrosine transaminase activity resulted (Black and Axelrod, in press). Norepinephrine, over the range of 10^{-5} M to 10^{-4} M competitively inhibited tyrosine transaminase (Fig. 6) (Black and Axelrod, in press).

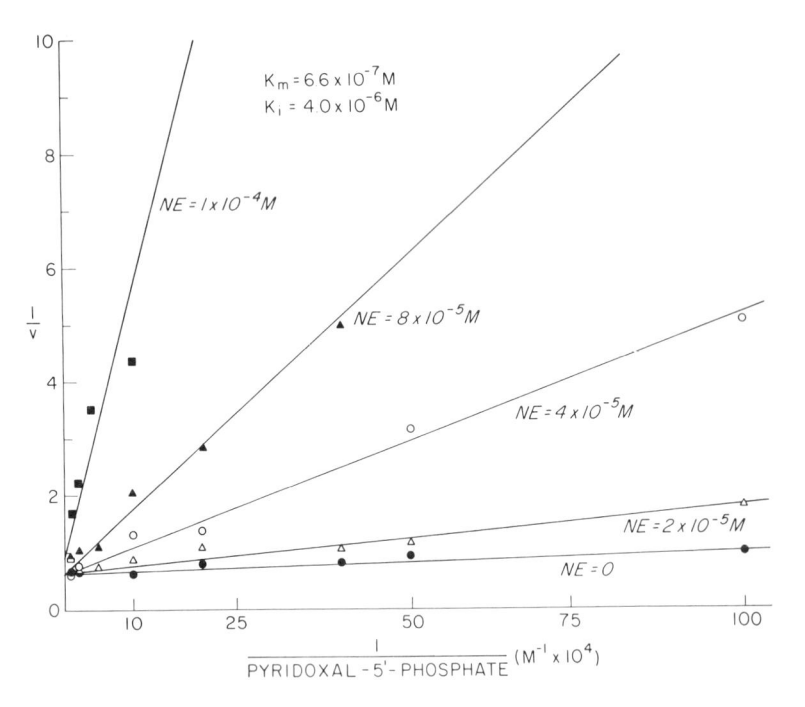

Figure 6. Effect of norepinephrine on tyrosine transaminase activity. The data are plotted by the method of Lineweaver and Burk. Assay was performed as described after 1 hr preincubation of tyrosine, enzyme preparation and indicated concentrations of norepinephrine and pyridoxal-5'-phosphate (Black and Axelrod, in press).

Maximal inhibition occurred only after preincubation of norepinephrine with the assay mixture (Fig. 7)

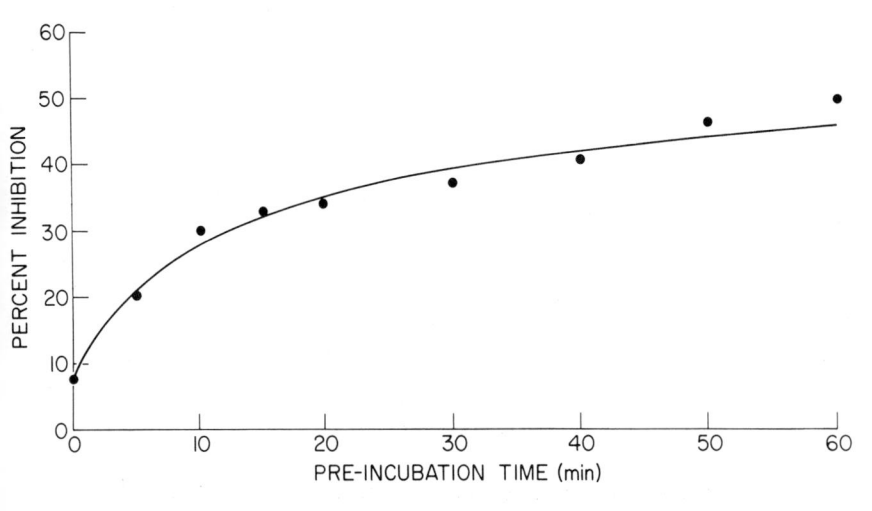

Figure 7. Effect of preincubation time on enzyme inhibition by norepinephrine. L-tyrosine, 7×10^{-3} M, 10 μl of dialyzed enzyme preparation, and 8×10^{-5} M norepinephrine were preincubated in 0.1 M phosphate buffer, pH 7.6, for varying lengths of time. Assay was begun by addition of α-ketoglutarate. Each sample was compared with its control preincubation for the same period in the absence of norepinephrine. (Black and Axelrod, in press).

(Black and Axelrod, in press).

To determine the site of action of norepinephrine, the amine was incubated with enzyme alone, pyridoxal-5'-phosphate alone, or both, and the assay was begun by addition of an aliquot of each to the reaction mixture. Full inhibition (90%) was obtained with preincubation of norepinephrine and pyridoxal-5'-phosphate alone. Preincubation of norepinephrine with enzyme alone afforded no greater inhibition (5%) than that seen with addition of norepinephrine at zero time.

The interaction of norepinephrine with pyridoxal-5'-phosphate was examined spectrophotometrically under conditions approximating those of the tyrosine transaminase assay. Incubation of norepinephrine with pyridoxal-5'-phosphate resulted in increased absorbance at 325 mµ and a decrease at 388 mµ, indicating the formation of a complex (Fig. 8) (Black and Axelrod, in

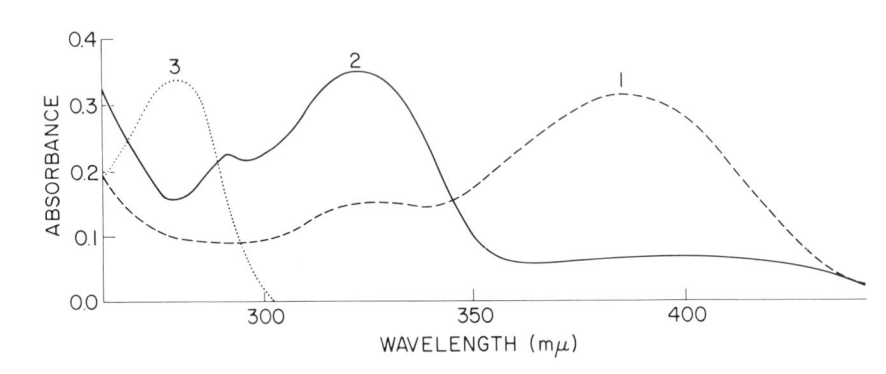

Figure 8. Spectrophotometric evidence of complex formation between norepinephrine and pyridoxal-5'-phosphate. Norepinephrine, 1 X 10^{-4} M, and pyridoxal-5'-phosphate, 1 X 10^{-4} M, were incubated either separately, or together in 0.1 M phosphate buffer, pH 7.6, for one hr at 37°C. Absorbance was determined at room temperature. Curve 1, spectrum of pyridoxal-5'-phosphate alone read against buffer blank. Curve 2, spectrum of norepinephrine and pyridoxal-5'-phosphate in combination read against norepinephrine in buffer. Curve 3, spectrum of norepinephrine alone read against buffer blank. (Black and Axelrod, in press).

press). This formation increased in a hyperbolic manner with time and the curve described was virtually superimposable on that seen with the plot of preincubation time against enzyme inhibition (Fig. 9).

For enzyme inhibition to proceed by the mechanism of norepinephrine, pyridoxal-5'-phosphate combination, the dissociation constant should favor complex formation. To establish this point, pyridoxal-5'-phosphate was incubated with varying concentrations of norepi-

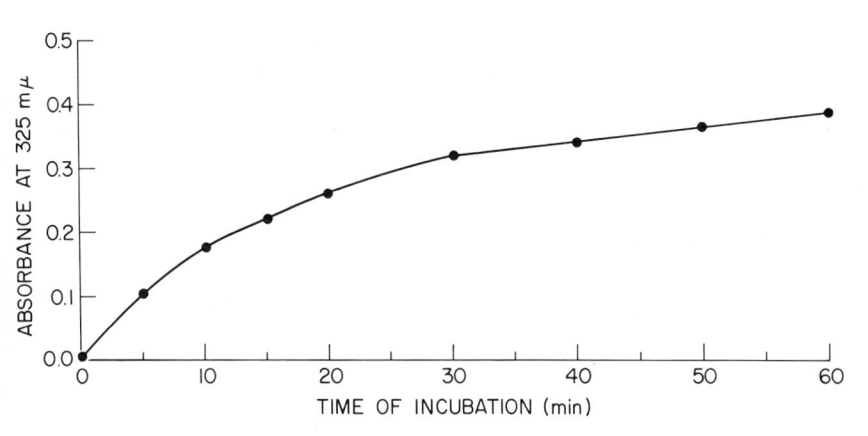

Figure 9. Characteristics of complex formation. Samples were treated as in Fig. 8. Absorbance at 325 mμ was read against appropriate controls containing pyridoxal-5'-phosphate alone. (Black and Axelrod, in press.)

nephrine. The presence of an isosbestic point at 345 mμ indicated that the reaction can be treated in terms of a single equilibrium between pyridoxal-5'-phosphate and the product of its reaction with norepinephrine (Fig. 10) (Black and Axelrod, in press). The sharp inflection point in the curve of absorbance at 325 mμ at a norepinephrine concentration of 6 X 10^{-5} M suggests two pyridoxal-5'-phosphates bound per norepinephrine (Fig. 11). On this basis a dissociation constant of 1.4 X 10^{-9} was calculated and is clearly consistent with the proposed model for the mechanism of inhibition (Black and Axelrod, in press).

Additional experiments indicated that the specificity of enzyme inhibition by structurally-related compounds agreed well with their ability to form complexes with pyridoxal-5'-phosphate. Primary catecholamines caused maximal enzyme inhibition (Black and Axelrod, in press). Inhibitory activity was markedly reduced by substitution on, or removal of, the m-hydroxyl group, but was independent of the catechol position. Substitution on the primary amine markedly decreased inhibition. β-Hydroxylation did not alter inhibition

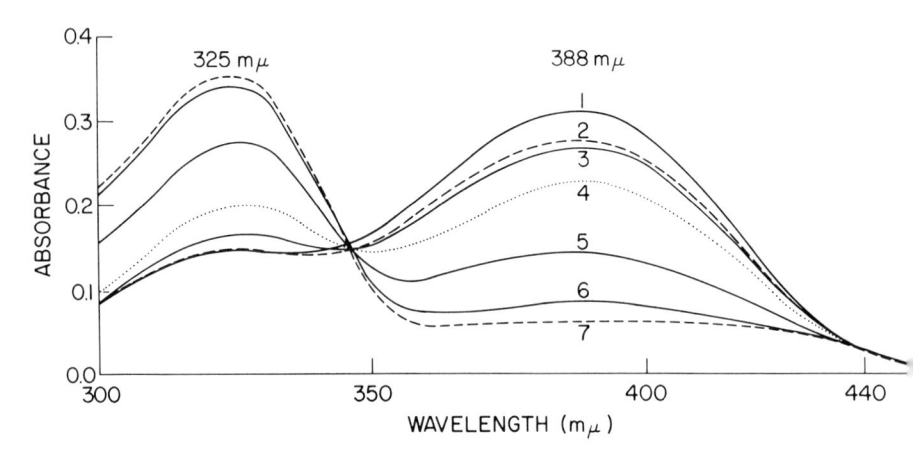

Figure 10. Effect of increasing concentrations of norepinephrine on complex formation. Pyridoxal-5'-phosphate, at 0.1 mM, was incubated as in Fig. 8, with the following concentrations of norepinephrine: 1, none; 2, 5 X 10^{-6} M; 3, 10^{-5} M; 4, 2 X 10^{-5} M; 5, 4 X 10^{-5} M; 6, 6 X 10^{-5} M; 7, 10^{-4} M. The isosbestic point occurs at 345 mμ (Black and Axelrod, in press).

and there was no evidence of stereospecificity. These structure-activity relationships paralleled those seen with complex formation (Black and Axelrod, in press).

The proposed *in vitro* model was tested *in vivo*. Pyridoxine hydrochloride induces tyrosine transaminase (Greengard and Gordon, 1963). If norepinephrine operates *in vivo* by competing with the enzyme for pyridoxal-5'-phosphate, norepinephrine should prevent the pyridoxine induction. Increasing doses of the catecholamine progressively reduced, and ultimately abolished the induction of tyrosine transaminase by pyridoxine (Fig. 12) (Black and Axelrod, in press).

Discussion

The above observations suggest a mode of action of norepinephrine *in vivo*. Enzyme molecules may accumulate secondary to either increased synthesis and/or decreased degradation of enzyme protein. Binding of

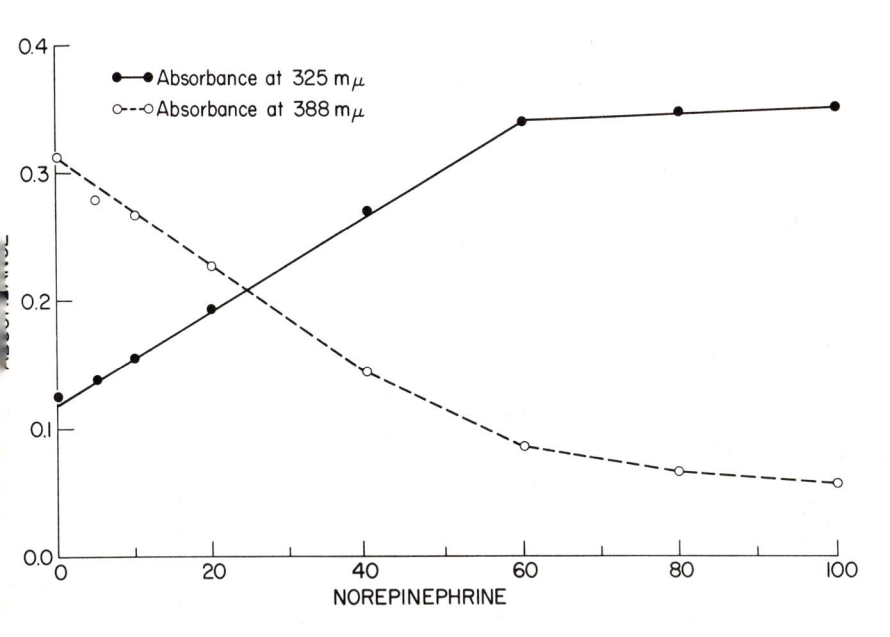

Figure 11. Stoichiometry of norepinephrine, pyridoxal-5'-phosphate interaction. The conditions were as in Fig. 10. At 325 mµ, absorbance was determined against pyridoxal-5'-phosphate in buffer. At 388 mµ, absorbance was determined against appropriate concentrations of norepinephrine in buffer. The units of norepinephrine concentrations are M X 10^{-6}. (Black and Axelrod, in press).

apo-tyrosine transaminase by pyridoxal-5'-phosphate may remove enzyme from equilibrium with its own turnover cycle. Norepinephrine, by depriving tyrosine transaminase of cofactor, may reintroduce enzyme into equilibrium with its metabolic cycle and thereby decrease synthesis or increase degradation. Such a formulation is represented schematically:[1]

[1]Symbols employed:
NE = Norepinephrine
Enzyme = Apoenzyme
PLP = Pyridoxal-5'-phosphate

318   Norepinephrine and Circadian Rhythm

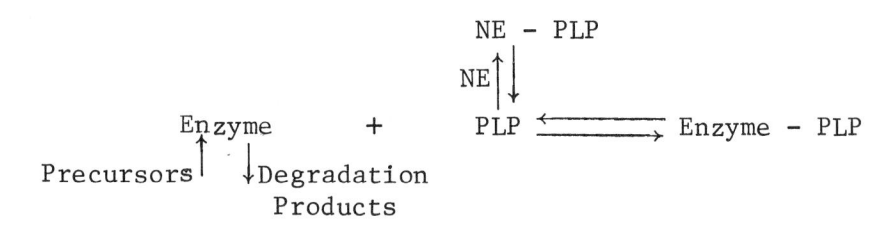

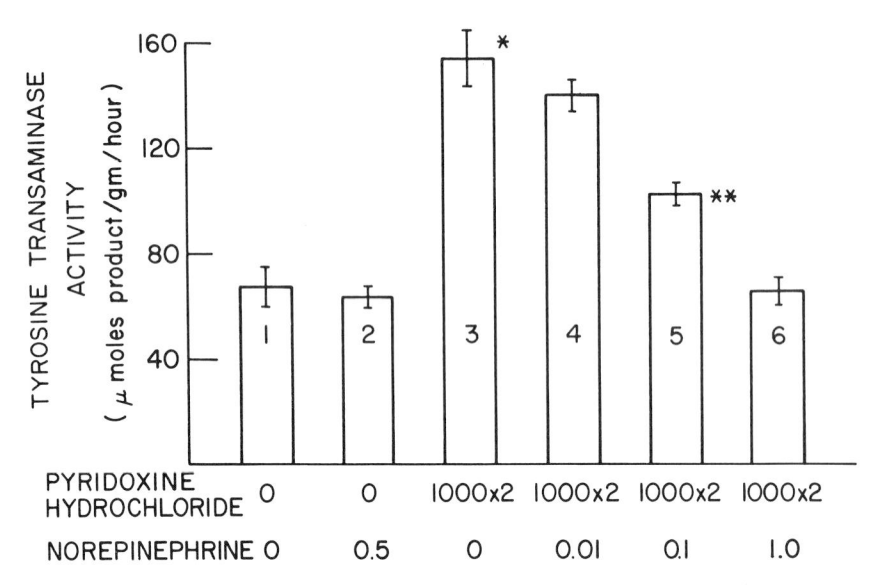

$$\text{Precursors} \uparrow \text{Enzyme} \downarrow_{\text{Degradation Products}} + \quad \text{PLP} \underset{\longrightarrow}{\overset{\longleftarrow}{\rightleftharpoons}} \text{Enzyme - PLP} \quad \overset{\text{NE - PLP}}{\underset{\text{NE}\uparrow\downarrow}{}}$$

Figure 12. Effect of norepinephrine on enzyme induction by pyridoxine hydrochloride. Groups of six to eight adrenalectomized rats were treated with pyridoxine hydrochloride, 1000 mg/kg, intraperitoneally at 8 A.M. and 10 A.M., and/or with norepinephrine subcutaneously in doses ranging from 0.01 to 1.0 mg/kg. Controls were treated with saline at appropriate times. Rats were killed at 12 noon. Results are expressed as mean ± standard errors of the mean (vertical bars).

*Differs from groups 1, 2, 5 and 6 at p < .001.

**Differs from all except group 4 at p < .01.

As indicated, enzyme concentration is dependent upon both norepinephrine and pyridoxal-5'-phosphate concentrations and control may be exerted through small and physiological variations in either component (Fig. 6).

The formulation presented suggests a mechanism by which norepinephrine may contribute to the generation of the tyrosine transaminase activity rhythm. Norepinephrine may prevent tyrosine transaminase accumulation by binding cofactor, while daily variations in the neural catecholamine release may regulate the enzyme cycle. Such a model is consistent with observations that the norepinephrine concentration of various areas of the brain varies daily (Friedman and Walker, 1968; Reis, et al., 1968). In turn the oscillation in neural norepinephrine may be determined by daily variations in the activity of rate-limiting steps in norepinephrine synthesis and/or activation.

REFERENCES

Aschoff, J. 1951. *Naturwissenschaften* 38: 506.
Axelrod, J. and I.B. Black. 1968c. *Nature* 220: 161.
Black, I.B. and J. Axelrod, 1968a. *Proc. Natl. Acad. Sci. U.S.* 61: 1287.
Black, I.B. and J. Axelrod. 1968b. *Proc. Natl. Acad. Sci. U.S.* 59: 1231.
Black, I.B. and J. Axelrod. 1969. *Federation Proc.* 28: 729.
Black, I.B. and J. Axelrod. In press. *J. Biol. Chem.*
Black, I.B. and J. Axelrod. Unpublished observations.
Civen, M., R. Ulrich, B.M. Trimmer and C.B. Brown. 1967. *Science* 157: 1594.
Friedman, A.H. and C.A. Walker. 1968. *J. Physiol.* 197: 77.
Granner, D.K., S. Hayashi, E.B. Thompson and G. Tomkins. 1968. *J. Molec. Biol.* 35: 291.
Greengard, O. and M. Gordon. 1963. *J. Biol. Chem.* 238: 3708.
Greengard, O., M.A. Smith and G.J. Acs. 1963. *J. Biol. Chem.* 238: 1548.
Greengard, O. and G.T. Baker. 1966. *Science* 154: 1461.
Holten, D. and F.T. Kenney. 1967. *J. Biol. Chem.* 242:

4372.
Kaempfer, R.O.R., M. Meselson and H.J. Raskas. 1968. *J. Molec. Biol.* 31: 277.
Kenney, F.T. 1967a. *J. Biol. Chem.* 242: 4367.
Kenney, F.T. 1967b. *Science* 156: 525.
Lin, E.C.C. and W.E. Knox. 1957. *Biochim. Biophys. Acta* 26: 85.
Ottolenghi, C. and R. Cavagna. 1968. *Endocrinology* 83: 924.
Potter, V.R., R.A. Gebert, H.C. Pitot, C. Peraino, C. Lamar, Jr., S. Lesher and H.P. Morris. 1966. *Cancer Res.* 26: 1547.
Potter, V.R., E.F. Baril, M. Watanabe and E.D. Whittle. 1968. *Federation Proc.* 27: 1238.
Reis, D.J., M. Weinbren and A. Corvelli. 1968. *J. Pharmacol. Exptl. Therap.* 164: 135.
Rosen, F. and R.J. Milholland. 1963. *J. Biol. Chem.* 238: 3730.
Schepartz, B. 1951. *J. Biol. Chem.* 193: 293.
Shambaugh, G.E., III, D.A. Warner and W.R. Beisel. 1967. *Endocrinology* 81: 811.
Wurtman, R.J. and J. Axelrod. 1967. *Proc. Natl. Acad. Sci. U.S.* 57: 1594.

Serotonin and Sleep*

M. Jouvet

Department of Experimental Medicine
School of Medicine
Lyons, France

Neuropharmacology and biochemistry are more and more involved in the study of sleep mechanisms. These new fruitful approaches are based upon three main findings which have been acquired with neurophysiological techniques in the past ten years:

1. Sleep is no longer considered as a unique phenomenon opposed to waking. On the contrary, in birds and mammals, behavioral sleep is the result of the succession of two different states, the polygraphic aspects of which are shown in Fig. 1: Slow Wave Sleep (SWS) and Paradoxical

*This work was supported by Direction des Recherches et Moyens d'essais (Grant 68-0039), the Institut National de la Santé et de la Recherche Médicale and the European Office of Aerospace Research (Grant EOAR 62.67).

322 Serotonin and Sleep

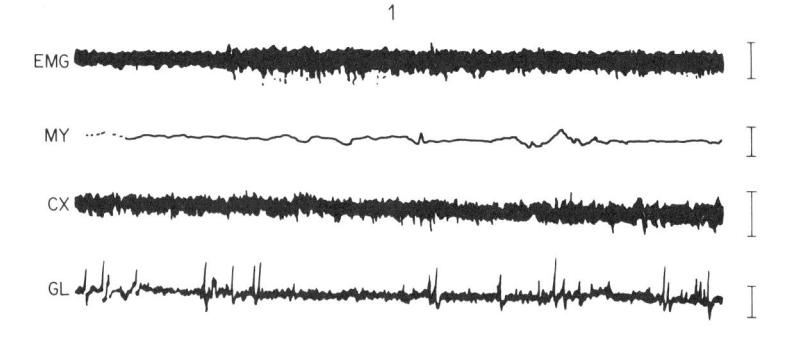

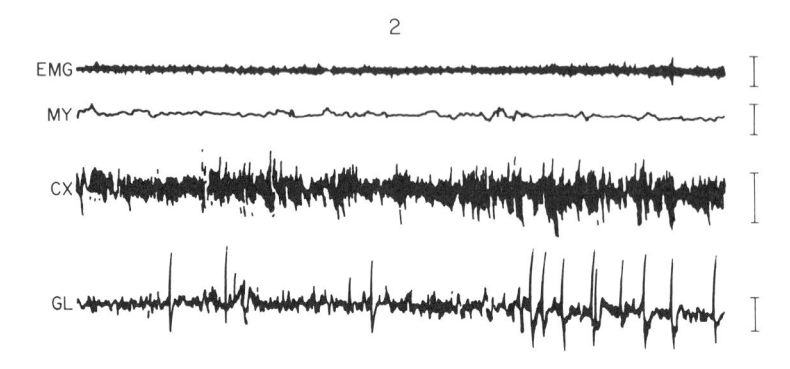

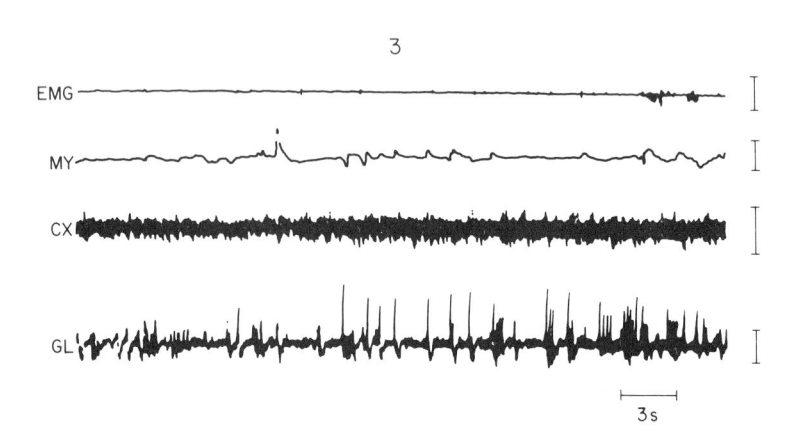

Sleep (PS) (or REM sleep). These two states appear quite regularly and can be quantitatively measured. In brief, a cat spends about 50% of its time sleeping (75% in SWS, 25% in PS occurring in phases of 6 min duration separated by intervals of about 25 min).

2. The neural mechanisms which are responsible for both SWS and PS are mainly, if not exclusively, located in the lower brain stem. This can be illustrated by two experiments:

a) A total intercollicular transection of the brain stem (cerveau isolé) (Bremer, 1935) is followed by a synchronized EEG analogous to SWS. This is explained by the fact that this transection is situated in front of the waking system located in the mesencephalic reticular formation (reticular activating system) (Moruzzi and Magoun, 1949). This transection prevents the activating influences from reaching the cortex. However, a section situated immediately caudal to the waking system in the middle of the Pons (Medio Pontine Pretrigeminal preparation) is followed on the contrary by a significant increase in EEG arousal (Batini, *et al.*, 1959). The probable

---

Figure 1. (Opposite page.) Polygraphic recording of the adult cat, showing waking and the two states of sleep. 1) Waking: muscular activity of the neck (EMG), eye movements (MY), fast low-voltage activity of the occipital cortex (CX), and sharp waves in the lateral geniculate (GL) which accompany eye movements. 2) Slow Wave Sleep: decreased muscular activity of the neck, absence of eye movements, high-voltage slow waves on the cortex, and high-voltage geniculate waves which directly precede paradoxical sleep by 1 or 2 min. 3) Paradoxical Sleep: total disappearance of EMG activity, together with rapid eye movements, fast low-voltage cortical activity, and clusters of high-voltage lateral geniculate waves. Each line represents 1 min of recording; amplitude calibration, 50 microvolts.

explanation is that the medio pontine transection suppresses ascending deactivating influences originating in the lower brain stem, caudal to the transection. Thus, somewhere in the lower brain stem (caudal Pons and/or Medulla) there should be some neurons which are responsible for synchronizing the EEG and possibly for inducing SWS.

b) The evidence for the structures responsible for PS in the lower brain stem are also well established: a chronic Pontile cat (the entire brain of which in front of the Pons has been removed) still shows clearly the periodical behavioral and central manifestations of PS. On the other hand, a limited lesion situated in the dorso lateral part of the Pontile tegmentum may selectively suppress PS (see references in Jouvet, 1967).

Thus there should be some group of neurons located in the Pontine tegmentum which are responsible for the periodical occurrence of PS.

3. All these findings indicate also that the states of sleep are dependent upon *active* mechanisms since limited lesions of the brain are able to suppress sleep.

Thus in order to explain the mechanisms of both SWS and PS, one should delimitate some groups of cell bodies in the lower brain stem, which actively and periodically dampen the waking system. It is also necessary to look for some peculiar nerve cell, possibly with secretory function. Indeed, it is impossible to explain many aspects of sleep with only the short time constant of the synaptic potential. The milliseconds which is the unit of time of electrophysiologists is not a unit of time convenient for explaining the circadian periodicity of sleep. It appears fruitful therefore, to explore the domain of the "wet" neurophysiology in order to explain sleep.

The first entry into this domain was through the

door of neuropharmacology.

Neuropharmacological Approach to Sleep

Thanks to the use of quantitative recordings of the states of sleep in chronically recorded animals, it became evident that most drugs which act upon brain monoamines also interfere with sleep. Among many drugs which were used, I will only summarize the effects of reserpine and monoamine oxidase (MAO) inhibitors.

Drugs producing a decrease in brain monoamines. Reserpine (0.5 mg/kg in the cat) suppresses SWS for 6-8 hr and PS for one day. A secondary injection of 5-hydroxytryptophan (5-HTP), which is believed to restore a normal level of brain 5-HT, results in the immediate reappearance of SWS, whereas the injection of DOPA following reserpine leads to the reappearance of PS. Thus it was deduced that serotonin may be involved in SWS whereas PS may necessitate catecholaminergic mechanisms (Matsumoto and Jouvet, 1964).

Drugs producing an increase in brain monoamines. MAO inhibitors which act upon brain monoamines by inhibiting their catabolism and thus by increasing their level in the brain were shown to act dramatically upon sleep states. Most of the MAO inhibitors utilized (Nialamide, Pheniprazine, Iproniazide) have a very selective suppressive effect upon PS in the cat. This suppressive effect is so intense that it is even obtained when the "need" for PS is greatly enhanced following paradoxical sleep deprivation (Jouvet, et al., 1965) (Fig. 2). This result suggests that monoamine oxidase could be involved in the transition from SWS to PS. However, these facts had only a limited significance in view of the complexity of the biochemical mechanisms of the brain. Most of these drugs were acting upon both indolamines and catecholamines, and therefore, might interfere with the cyclic alteration of the brain's electrical activity.
More recently, some drugs acting selectively upon 5-HT metabolism have been discovered. The most interesting ones are those which inhibit the synthesis

326 Serotonin and Sleep

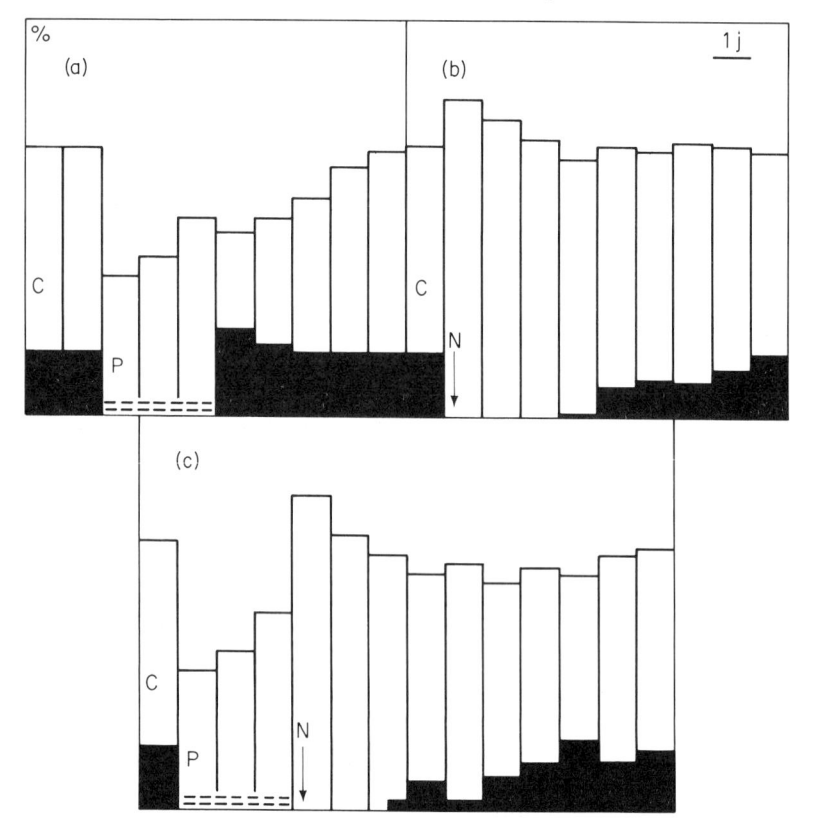

Figure 2. *A*. When paradoxical sleep is instrumentally suppressed (by placing the cat upon a small support surrounded with water) a rebound of PS occurs during recovery sleep. Ordinates, daily percentage of SWS (white) or PS (black); abscissae, time in days; P, instrumental deprivation of PS; C, control recording. *B*. A single injection of 10 mg/kg of Nialamide (N) selectively suppresses PS for 4 days. There is no immediate rebound thereafter. *C*. If Nialamide is injected (N) at the beginning of recovery sleep after a previous instrumental deprivation of PS (p), it suppresses PS for 3 days.

of 5-HT. It is thus possible to alter selectively the metabolism of one monoamine in the study of the states of sleep.

Pharmacologically Induced Decrease of Cerebral Serotonin

It has been shown that p-chlorophenylalanine (PCPA) selectively decreases the level of 5-HT in the brain without changing the level of catecholamines (Koe and Weissman, 1967).

As shown in Fig. 3, slices of cat cerebral cortex or brain stem were incubated *in vitro* with ^{3}H-tryptophan and D,L-^{3}H-5-HTP and the amount of ^{3}H-5-HT was measured subsequently. There is an 80% decrease of ^{3}H-5-HT synthesized from ^{3}H-tryptophan (as compared with control animal) in cats pretreated with PCPA. On the other hand, the synthesis of ^{3}H-5-HT from ^{3}H-5-HTP is normal (Pujol, *et al.*, 1969). Thus it is most likely that PCPA decreases brain serotonin in the cat brain by inhibiting tryptophan hydroxylase, as was previously shown for rat brain and liver by Koe and Weissman (1967). The action of PCPA upon the sleep states of the cat and the rat has been extensively studied (Delorme, *et al.*, 1966; Mouret, *et al.*, 1968; Koella, *et al.*, 1968; Weitzman, *et al.*, 1968). After a single injection of 400 mg/kg of PCPA in the cat, no apparent variation of behavior or of EEG recordings is observed during the first 24 hr. This fact demonstrates that the drug in itself has no direct pharmacological action upon the brain. Following this period an abrupt decrease of both states of sleep occurs and after about 30 hr an almost total insomnia appears as illustrated by a permanent and quiet waking behavior, mild mydriasis, and a permanent low-voltage fast cortical activity. The recovery of sleep begins after the 40th hr and is accompanied by the appearance of permanent phasic waves in the lateral geniculate body and occipital cortex (similar to the phasic activity which is observed in the lateral geniculate during SWS immediately preceding PS or during PS). Very discrete episodes of PS may appear, either following short episodes of SWS or even directly following waking. SWS episodes of longer duration gradually reappear at shorter intervals. Normal qualitative and quantitative patterns of sleep are resumed after about 200 hr.

A significant correlation has been found to exist

328   Serotonin and Sleep

between the decrease of SWS and the decrease of cerebral serotonin caused by PCPA administration (Fig. 4) (Mouret, et al., 1968).

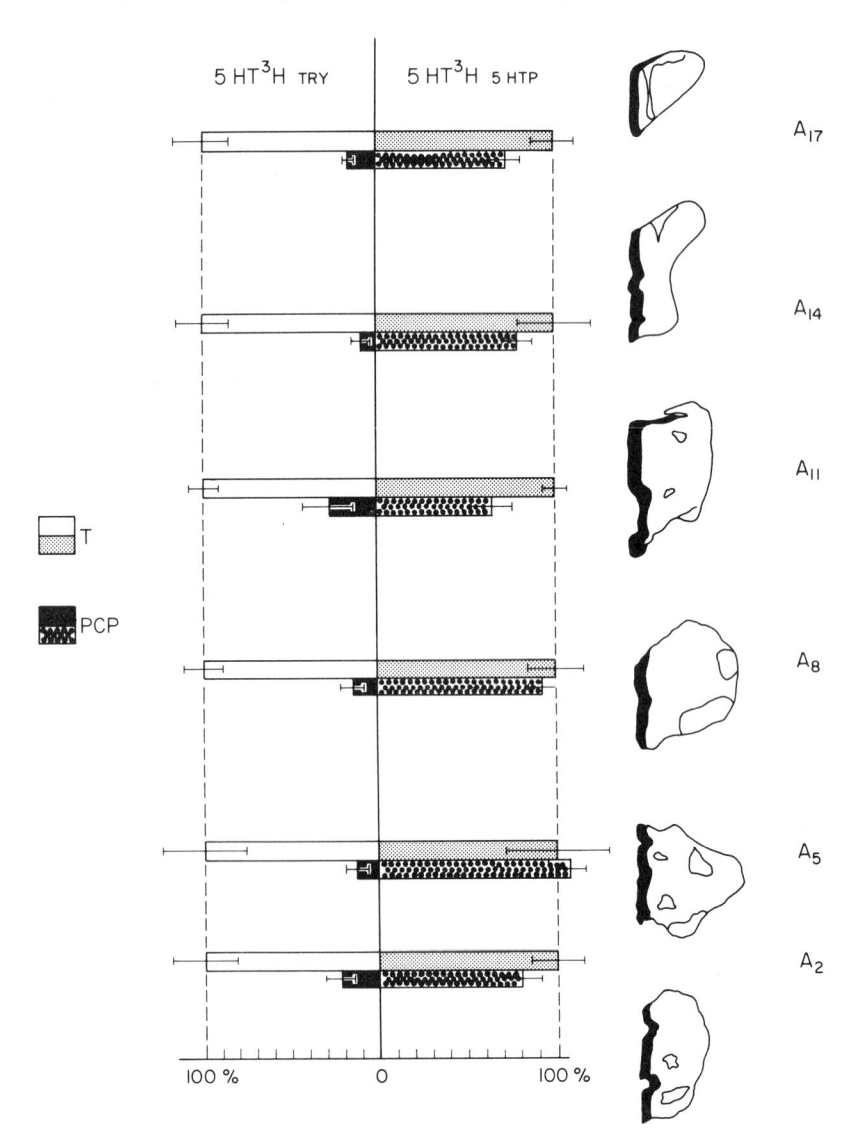

Since PCPA inhibits only the first step of the synthesis of 5-HT (at the level of tryptophan hydroxylase), it is possible to by-pass its blocking action and thus re-establish the level of 5-HT by injecting the direct precursor of 5-HT. This has been performed by the injection of 5-HTP (which readily crosses the blood brain barrier) following PCPA: with this procedure, it is possible to manipulate the state of sleep of the animal. A single injection (IV or IP) of a very small dose of 5-HTP (2-5 mg/kg) performed when the insomnia has reached its maximum (30 hr following PCPA) is able to restore a quantitatively and qualitatively normal pattern of both states of sleep. This restoration may last for a period of 6-10 hr after which there is a rapid return to insomnia. Other experiments have shown that whereas a cat receiving a multiple dose of PCPA presents a severe and long-lasting insomnia, an animal receiving balanced daily doses of 5-HTP with the same amount of PCPA may present normal or even increased sleep during several days (Fig. 5). These experiments show rather conclusively that sleep mechanisms can be manipulated by interfering *only* with the synthesis of 5-HT.

This study (which is in agreement with the previous findings that 5-HTP is able to restore SWS when injected after reserpine) led to the hypothesis that SWS requires the presence of 5-HT at the terminals of serotonin-containing neurons. In view of other pharmacological data demonstrating that PS is eliminated for a long period of time by MAO inhibitors, it is suggested that a deaminated metabolite of 5-HT may be responsible for the triggering of PS. Among those demaminated metabolites, 5-hydroxyindolacetaldehyde (5-HIAA) could possibly by involved since it has been shown that this product may have some central hypnogenic effect in the rat (Sabelli, *et al.*, 1969).

However, these neuropharmacological results, alone,

Figure 3. (Opposite page.) On the left-hand side, the amount (in percentage) of ^{3}H-5-HT synthesized by brain stem slices incubated in Krebs-Ringer, from ^{3}H-tryptophan in normal (white) and PCPA pretreated cats (black) (500 mg/kg 40 hr before the sacrifice). The absolute values are 83.6 ± 9.9 mµC/g for the control (T) and 14.8 ± 2.8 for p-chlorophenylalanine treated cats (PCP). On the right-hand side, ^{3}H-5-HT synthesized from ^{3}H-5-HTP. Absolute values are respectively 225.9 ± 50 mµC/g and 236 ± 36 mµC/g. On the far right, representation of the structures incubated according to Horsley-Clarke coordinates.

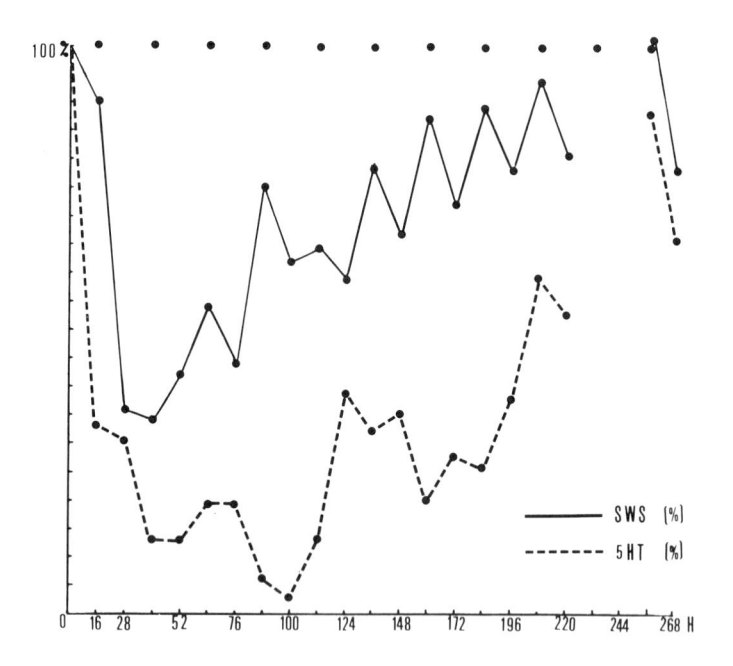

Figure 4. Effects of p-chlorophenylalanine on sleep and brain concentrations of serotonin in the rat. After intraperitoneal injection (500 mg/kg), there is a decrease in the amount of Slow Wave Sleep (SWS) and in serotonin (5-HT) concentration, followed by a slow return to normal values after 268 hr. (Solid line), percentage of Slow Wave Sleep relative to the amount for a control rat; each point represents the mean percentage per 12-hr period (black dots at top indicate night hours, 7 P.M. to 7 A.M.). (Dashed line), percentage of serotonin (relative to the concentration for a control rat); for the purpose of this analysis, two animals were killed every 12 hr. (From Mouret, et al., 1968).

cannot be accepted as a conclusive evidence for the decisive role of 5-HT in the triggering of sleep. It must be admitted that PCPA alters not only brain 5-HT but also the serotonin level in the total body. It is therefore very difficult to eliminate a possible pe-

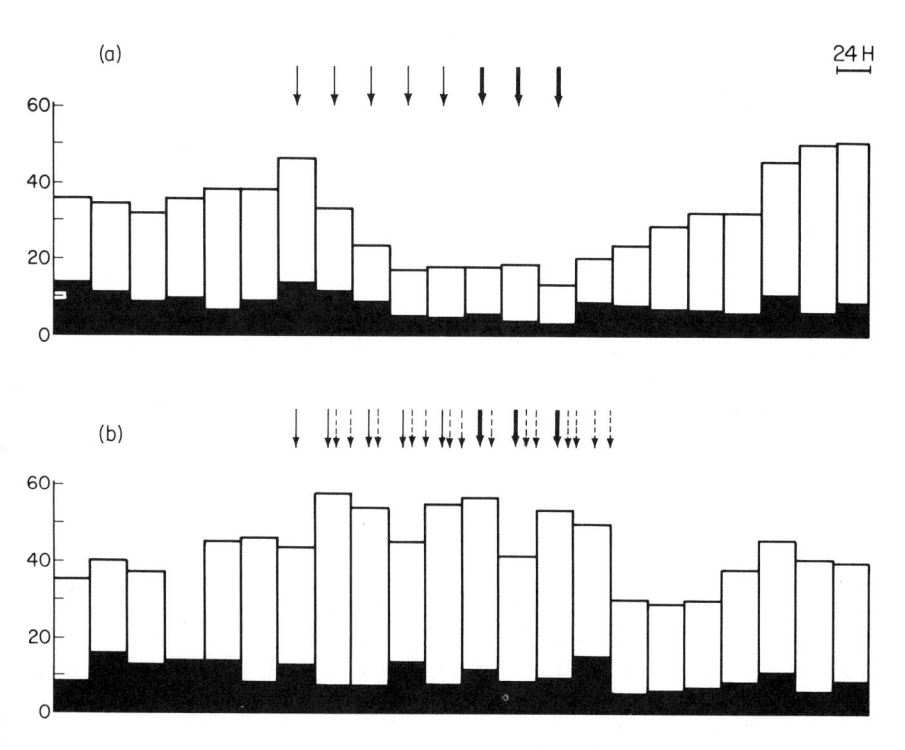

Figure 5. *A*. Repeated daily injections (100-150 mg/kg) of PCPA (black arrows) decrease both states of sleep. *B*. Another animal receiving the same daily amount of PCPA (black arrows) together with appropriate dose of 5-HTP (5 to 20 mg/kg) (dotted arrows) present on the contrary an increase of sleep. Ordinates: percentage of SWS (white) and PS (black) in a 24-hr day. Abscissa: days.

ripheral factor in the very significant decrease of sleep which follows PCPA.

Thus, only a method permitting the selective decrease of brain 5-HT should give unequivocal results. Fortunately, new techniques became available recently which permitted us to close the gap between neurophysiological and neuropharmacological data. It became possible, with the help of biochemistry and histofluorescence to map out serotoninergic neurons and thus to reconcile the results obtained with neurophysiology (showing that neurons located in the lower brain stem are responsible for sleep) and neuropharmacology (showing that 5-HT is involved in sleep).

Histochemical Anatomy of the Lower Brain Stem

The histochemical method of Glenner has demonstrated that MAO could be localized in some restricted regions of the brain stem, particularly in the dorso lateral part of the Pontine tegmentum (Nucleus locus coeruleus and sub-coeruleus) (Hashimoto, *et al.*, 1962) (Fig. 7B).

Thanks to the histofluorescence technique of Hillarp, Falk, *et al.* (1962) and of Dahlström and Fuxe (1964), it has become possible to map out precisely the location of nerve cells containing monoamines in the rat. The 5-HT containing cell bodies (which show a bright yellow fluorescence) are located mostly on the midline, in the Raphe system, from the Nucleus Raphe obscurus in the caudal part of the medulla to the Nucleus Raphe linearis at the junction between the Pons and the mesencephalon.

These neurons have about the same distribution in the cat (Pin, *et al.*, 1968) as shown in Fig. 6. This localization of 5-HT containing cell bodies in a concentrated group of cells of the Raphe system makes possible their destruction by stereotaxic technique.

In the Pons, the norepinephrine (NE) containing neurons are located mostly in the dorso lateral part of the Pontine tegmentum and are very closely correlated with the localization of MAO activity in the cat (regions of the N. locus coeruleus, sub-coeruleus, parabrachialis medialis and lateralis). The destruction of these neurons selectively suppresses paradoxical sleep in the cat (Roussel, *et al.*, 1967) (Fig. 7B). The possible mechanisms triggering this state of sleep are discussed in detail elsewhere (Jouvet, 1969).

Figure 6. (Opposite page.) Topography of 5-HT containing cell bodies (black dots) in the brain stem of the cat (Horsley-Clarke coordinates). (From Pin, *et al.*, 1969).

The cell bodies of the 5-HT and NE-containing neurons were shown to send terminals in widespread regions of the brain and spinal cord (Dahlström and Fuxe, 1964;

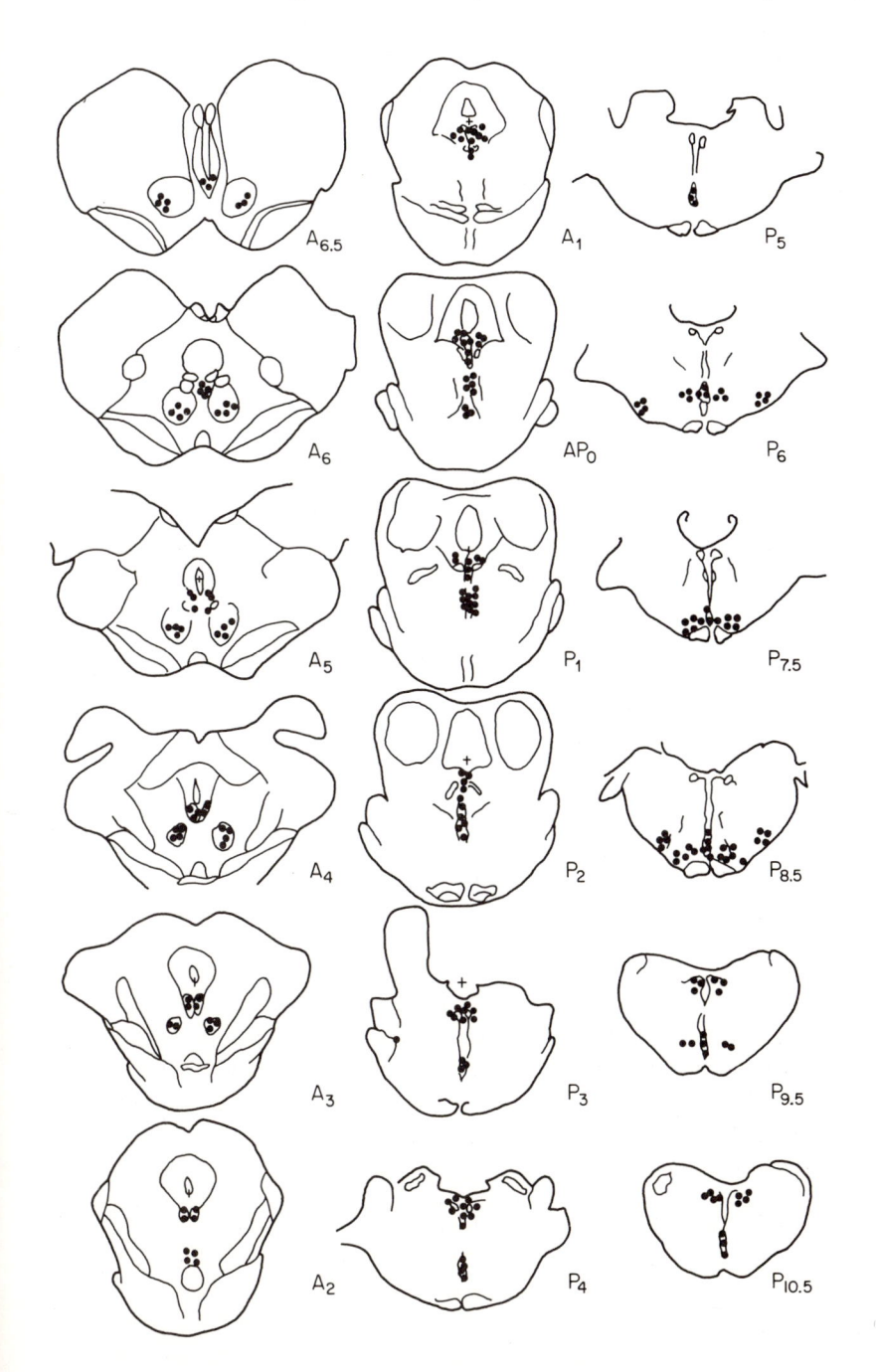

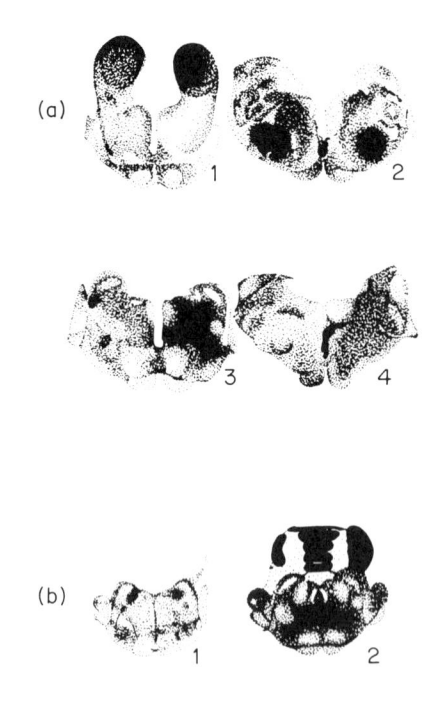

Figure 7. *A*. Subtotal lesion of the raphe system in the cat. Rostral section at the level of (1) the raphe centralis superior and dorsalis nuclei, (2) the Raphe pontis, and (3) and 4) the Raphe magnus and pallidus. *B*. Rostral sections of the pons: (1) Glenner coloration shows the monoamine oxidase (dark areas in the nucleus locus coeruleus, (2) bilateral destruction of the nucleus locus coeruleus which selectively suppresses paradoxical sleep.

Heller and Moore, 1965). Moreover, it was shown that the sectioning of the axon would suppress specific fluorescence of the corresponding terminals after a delay of about 8-10 days. This finding permitted the destruction of specific groups of NE-containing neurons by stereotaxically oriented coagulation and the correlation of such destruction with biochemical analysis, allowing a suitable time between destruction of the nerve cells and sacrifice of the animal. Thus, it was possible to act selectively upon the 5-HT level of selected regions of the brain. Since biochemical analysis could give the quantitative results of such a procedure, since topographical methods could give an arbitrary measurement of the extent of the lesion, and since quantitative determination of sleep states was available, it was possible to correlate these three variables.

Selective Decrease of Brain Serotonin
after Destruction of the Raphe System
(Jouvet, et al., 1966; Jouvet, 1969)

The destruction of the serotonin-containing neurons of the Raphe system was performed stereotaxically in chronically implanted cats (Fig. 7). Following the operation, the aninals were continuously recorded for a period of 10-13 days (this being the critical duration for the voiding of the serotonergic terminals). On the 13th day, the cats were sacrificed [always at the same hour (11 A.M.) in order to avoid possible circadian variation] for the histological evaluation of the damaged area of the brain and for the biochemical analysis of the intact regions of the brain. By this method, the following information was obtained: a valid quantification of the sleep states (obtained by the mean percentages of SWS and PS for 10-13 days recording), a measurement of the volume of the lesion by topographical analysis (represented by the percentage of the total Raphe system destroyed), and an analysis of the monoamine levels in the brain (represented by the percentage of 5-HT and NE in the operated cats compared to the levels found in normal cats sacrificed under the same conditions).

Following a subtotal (80-90%) coagulation of the Raphe system, a state of permanent insomnia is observed during the first 3-4 days (Fig. 8): the animals have permanent running movements and a large mydriasis. They can react immediately to a moving stimulus with vertical eye movements (lateral eye movements are suppressed by the midline lesion). Their cortical activity is permanently fast and exhibits the well known pattern of the arousal reaction. In the period that

336   Serotonin and Sleep

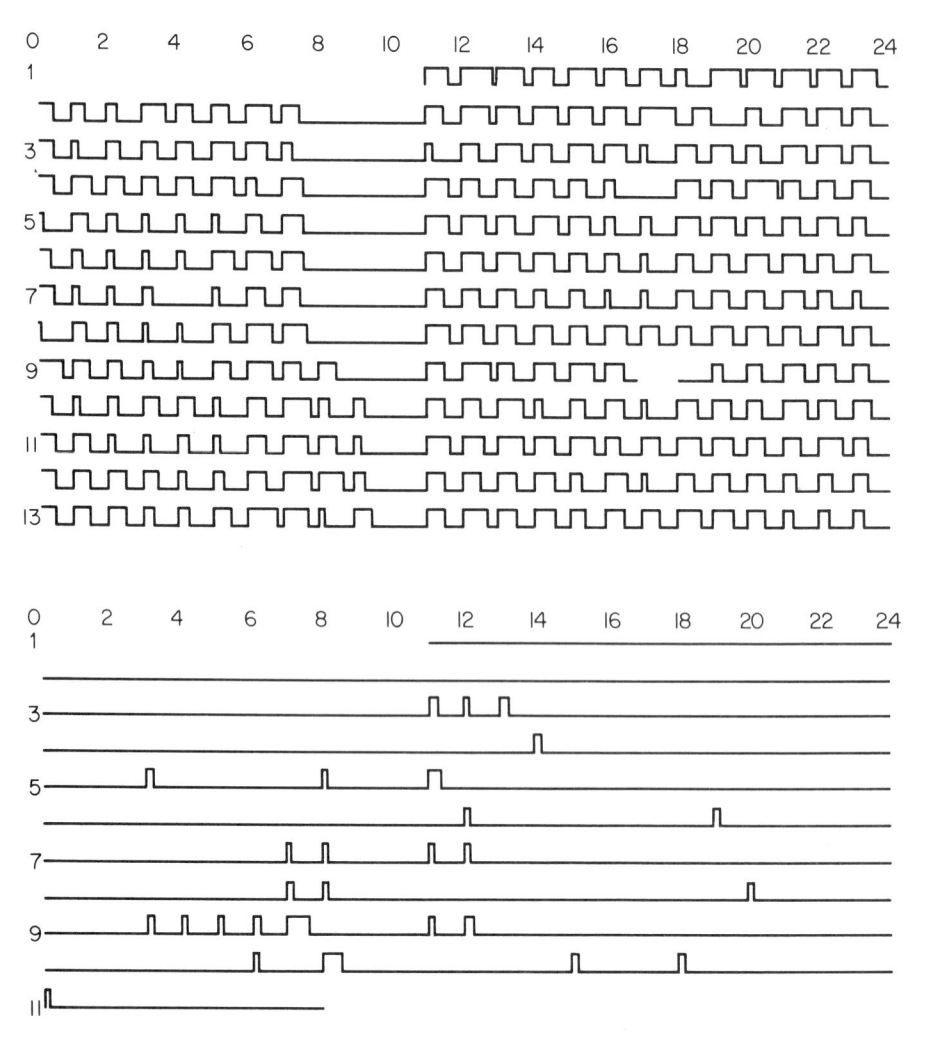

Figure 8. Upper part of figure: sham operated cat recorded during 13 days—the amount of sleep (white rectangle) is 57% of the recording. Lower part of figure: insomnia following a subtotal destruction of the Raphe system (percentage of total sleep (SWS) during 11 days: 3.5%). Ordinates: sleep. Abscissae: time in hr, each line represents one day of recording.

follows, the percentage of SWS does not exceed 10% of the recording time, and PS is never observed. The injection of 5-10 mg/kg of 5-HTP (which is able to restore sleep in an insomniac cat pretreated with p-chlorophenylalanine) has no effect upon the waking EEG and behavior of these cats. Partial lesions of the Raphe system (rostral or caudal regions) result in an insomnia which is less pronounced (due to a gradual recuperation after the first two days). In these preparations, PS is found to occur in correlation with the daily percentage of SWS (only when this value surpasses 15% is any PS observed). Destruction involving less than 15% of the Raphe system did not provoke a significant change in the states of sleep. It was thus demonstrated that a significant correlation existed between the volume of the Raphe system and thus the quantity of 5-HT containing cell bodies destroyed and the amount of sleep. The biochemical analysis of these insomniac preparations revealed a significant decrease in cerebral serotonin and 5-HIAA with no variation in NE (Table 1). It was therefore demonstrated that a significant correlation exists between the amount of destruction of the Raphe system, the intensity of the resulting insomnia, and the selective decrease of cerebral serotonin (Fig. 9).

In view of the neuropharmacological findings which demonstrated that the inhibition of the synthesis of serotonin (by PCPA) produces total insomnia and that this insomnia is immediately reversible by the administration of the precursor of 5-HT (5-HTP), the evidence in favor of a serotonergic hypnogenic mechanism is very convincing. Both neuropharmacological and neurophysiological data suggest that the Raphe system, which contains the large majority of the serotonin-containing neurons, is the important sleep inducing structure in the brain and that cerebral serotonin plays a determinant role in the active process underlying sleep.

TABLE 1

Comparison, for various groups of cats with brain lesions, of (i) the amount of sleep following surgery, expressed as a percentage of total recording time (10–13 days); (ii) the percentage of the Raphe system left intact; and (iii) the amounts of serotonin and norepinephrine in the brain rostral to the lesion, expressed as percentages of the amounts in the brains of normal control cats.

Group	No. in group	Amount of sleep (%) Slow-wave	Amount of sleep (%) Paradoxical	Percentage of Raphe intact	P*	Percentage of cerebral serotonin	P*	Percent of cerebral norepinephrine	P*
A	12	48.5 ± 7.5	9.5 ± 2.5	95.4 ± 12		90 ± 23		102 ± 17	
B	6	30 ± 2.7	5.5 ± 2.7	77.5 ± 19	.02	68 ± 18	.10	99 ± 22	NS
C	10	16.5 ± 2.2	1 ± 0.7	64 ± 8.5	.001	54 ± 22	.01	93 ± 17	NS
D	6	9 ± 1.5	0	35 ± 10	.001	29 ± 11	.001	92.5 ± 17	NS

The group divisions are based on the extent of the lesion and the amount of sleep following surgery: (A) cats with insignificant destruction of the Raphe system—an amount having no effect on sleep; (B and C) cats with major but less than total destruction of the Raphe system; (D) cats with almost total destruction of the Raphe system. The percentages are mean values for an entire group, plus standard deviation.

*Student's t-test values for P obtained by comparison with group A; NS, not significant. Each P column refers to the column that immediately precedes it.

From Jouvet (1969).

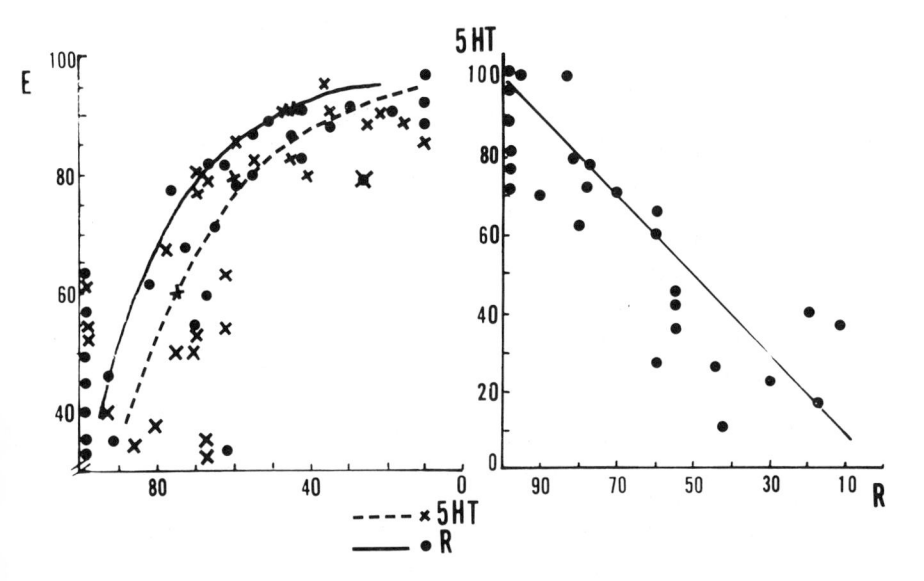

Figure 9. On the left: ordinates; waking (E) as percentage of recorded time (10-13 days) in operated cats. Abscissae; percentage of serotonin (5-HT) in the brain rostral to the destruction of the Raphe system. Dots: percentage of Raphe system left intact by the lesion (R). There is a significant correlation between the increase of waking (insomnia), the decrease of 5-HT, and the intensity of the destruction of the Raphe system. On the right: ordinates; percentage of serotonin in the brain rostral to the destruction of the Raphe (as compared with normal cats). Abscissae; percentage of Raphe system left intact by the lesion.

REFERENCES

Batini, C., F. Magni, M. Palestini, G.F. Rossi and A. Zanchetti. 1959. *Arch. Ital. Biol.* 97: 13.
Bremer, F. 1936. *C.R. Soc. Biol.* 122: 646.
Dahlström, A. and K. Fuxe. 1964. *Acta Physiol. Scand. Suppl.* 232: 1.

Delorme, F., L. Froment and M. Jouvet. 1966. *C.R. Soc. Biol.* 160: 2347.
Falck, B., N.A. Hillarp, G. Thieme and A. Torp. 1962. *J. Histochem. Cytochem.* 10: 348.
Hashimoto, P.H., T. Maeda, K. Torii and N. Shimizu. 1962. *Med. J. Osaka Univ.* 12: 425.
Heller, A. and R.Y. Moore. 1965. *J. Pharmacol. Exper. Therap.* 150: 1.
Jouvet, M. 1967. *Physiol. Rev.* 47: 117.
Jouvet, M. 1969. *Science* 163: 32.
Jouvet, M., P. Vimont and J.F. Delorme. 1965. *C.R. Soc. Biol.* 159: 1595.
Jouvet, M., P. Bobillier, J.F. Pujol and J. Renault. 1966. *C.R. Soc. Biol.* 160: 2343.
Koe, K.B. and A. Weissman. 1967. *J. Pharmacol. Exp. Therap.* 154: 499.
Koella, W.P., A. Feldstein and J.S. Czicman. 1968. *EEG Clin. Neurophys.* 25: 481.
Matsumoto, J. and M. Jouvet. 1964. *C.R. Soc. Biol.* 158: 2135.
Moruzzi, G. and H.W. Magoun. 1949. *EEG Clin. Neurophys.* 1: 445.
Mouret, J., P. Bobillier and M. Jouvet. 1968. *Eur. J. Pharmacol.* 5: 17.
Pin, C., B. Jones and M. Jouvet. 1968. *C.R. Soc. Biol.* 162: 2136.
Pujol, J.F., P. Bobillier, A. Buguet, B. Jones and M. Jouvet. 1969. *C.R. Soc. Biol.* 268: 100.
Roussel, B., A. Buguet, P. Bobillier and M. Jouvet. 1967. *C.R. Soc. Biol.* 161: 2537.
Sabelli, H.C., W.J. Giardina, S.G.A. Alivisatos, P.K. Seth and F. Ungar. 1969. *Nature* 223, No. 5201: 73.
Weitzman, E.D., M. Rapport, P. McGregor and J. Jacoby. 1968. *Science* 160: 1361.

Author Index

Abe, K., 20, 21, 30
Abraham, K. A., 290
Ackerman, D., 75
Ahlfours, C. E., 135
Ahlquist, R. P., 9, 10, 14
Ahlquist, W. B., 190
Ahlström, C. G., 231
Aiello, E. L., 131, 132
Aketa, K., 133
Akgün, S., 197
Alexander, R. S., 243
Alonso, D., 31
Altura, B. M., 240, 241, 248
Andem, M. R., 248
Andén, N.-E., 88
Anderson, J., 194
Appleman, M. M., 154, 162
Ariens, E. J., 4
Armenian, A. R., 270
Armstrong, D., 78
Arndt, H. J., 181
Arnold, A., 27
Aschoff, J., 301
Ascoli, F., 295
Atkinson, D. E., 147
Aulich, A., 11
Aurbach, G. D., 40, 41, 45, 47
Awapara, J., 254
Axelrod, J., 209, 302, 303, 304, 305, 306, 307, 309, 310, 311, 312, 313, 314, 315, 316, 317

Baba, W. I., 29
Bachrach, W. H., 213
Baggett, B., 47
Baguet, F., 132
Bain, J. A., 269
Baird, C. E., 38, 39, 40, 50
Baker, G. T., 302

Ball, E. G., 183, 184, 186, 190
Ballard, P. L., 290
Balzer, H., 115
Bar, H., 43, 46
Bar, U., 143
Barber, A. A., 96
Barchas, J., 199
Barcroft, H., 241
Bargmann, W., 193
Barrett, R. J., 183
Barros, C., 289
Barter, R., 210
Batini, C., 323
Baudhuin, P. M., 103
Baxter, C. F., 273
Beaven, M. A., 209, 210, 213, 214, 216, 217, 218, 223
Becker, F. E., 47
Belocopitow, E., 163
Bennett, C. M., 45
Berliner, R. W., 45
Bernard, Claude, 209
Berridge, M. J., 29
Bhide, N. K., 21
Bhoola, K. D., 83
Bieck, P., 199
Bilodeau, F., 258
Birnbaumer, L., 184, 188
Bishop, J. S., 163, 171, 175
Bitensky, M. W., 47
Bizzi, A., 193
Björntorp, P., 197
Black, I. B., 302, 303, 304, 305, 306, 307, 309, 310, 311, 312, 313, 314, 315, 316, 317
Blanchaer, M. C., 143
Blatt, L. M., 172
Blecher, M., 183, 188, 194
Bleicher, S. J., 197

Author Index

Blum, J. J., 97, 98, 99, 100, 101, 102, 104, 105, 109, 110, 116, 270
Booker, B., 62
Bose, B. C., 269
Bowman, W. C., 142, 143, 145, 156
Boyer, J., 197
Boyle, S. M., 285
Bradley, P. B., 54, 56
Brana, H., 102
Bray, G. A., 194
Breckenridge, B. McL., 143, 156
Bremer, F., 323
Brodie, B. B., 46, 63, 66, 67, 191, 209, 212, 218
Brown, E., 38, 39
Brownstein, M. J., 57
Bueding, E., 40
Bülbring, E., 211
Buniatian, H. Ch., 258, 270
Burgen, A. S. V., 210
Burn, J. H., 223
Burns, T. W., 23, 24
Busch, H., 293
Butcher, R. W., 8, 11, 12, 23, 38, 39, 40, 45, 50, 56, 101, 183, 184, 197, 198, 200
Buznikov, G. A., 133

Cahill, G. F., 182, 194
Caldarera, C. M., 277, 281, 282, 289, 291
Cameron, L. E., 100
Cannon, W. B., 9, 21, 26
Carlson, L. A., 184
Carlsson, A., 209, 211
Castaneda, M., 37, 56
Cavagna, R., 302
Chalmers, T. M., 183
Changeux, J., 131
Chase, L. R., 40, 41, 45, 47
Cherkes, A., 182
Cheung, W. Y., 102
Chytil, F., 102
Civen, M., 303
Clark, A. J., 3, 5
Clausen, T., 188
Coggeshall, R. E., 76, 88
Cohen, P. S., 285
Cohen, S. S., 276, 291
Columbo, J. P., 134
Conrad, V., 248
Cook, D. E., 104
Cooke, I. M., 86-87
Cori, C. F., 124, 135
Cornblath, M., 148
Correll, J. T., 134
Correll, J. W., 191
Costa, E., 36, 41, 45, 47, 51, 59, 61, 64

Cottrell, G. A., 81, 82
Courvoisier, S., 54
Craig, J. W., 40, 153, 163, 168
Cryer, P. E., 46, 50
Curtis, D. R., 254, 255

Dahl, E., 81, 83, 86
Dahlström, A., 88
Dale, H. H., 8, 9, 223
Danforth, W. H., 154, 155, 163
Darnell, J. E., Jr., 291
Davies, J. I., 50
Davies, R. E., 132
Davis, J. S., 40
Davoren, P. R., 7, 39, 41, 46
Davtian, M. A., 258
Dawson, R. M. C., 256
De Duve, C., 103
De Renzo, E. C., 198, 199
De Wulf, H., 163, 171
Dean, P. M., 31
Debons, A. F., 63
DeCaro, L. G., 190
DeFeudia, F. V., 256
DeLange, R. J., 152
Delorme, F., 327
Denton, R. M., 188
DeRobertis, E., 7, 41
Derry, D. M., 193
Dewey, V. C., 96, 97, 116
Deykin, D., 63
Di Girolamo, M., 190
Dion, A. S., 288
Dobbs, J. W., 14, 28
Dorrington, J. H., 47
Dousa, T., 41, 47
Dragstedt, Carl, 223
Drummond, G. I., 149, 153
Duarte, C. G., 45
Dykstra, W. G., Jr., 277

Ebadi, M. S. R., 270
Edery, H., 249
Ehinger, B., 85
Ehrlich, P., 3
Elliott, A. M., 96
Elliott, K. A. C., 254, 255, 257, 259, 260, 261, 262, 263, 264, 266, 267, 269
Ellis, S., 142, 143, 145
Elofsson, R., 86
Emmelin, N. G., 210
Engel, F. L., 182, 190
Engelman, K., 211
Entman, M. L., 7, 46
Epstein, S. E., 13, 47, 100

Author Index

Erjavec, F., 213, 218
Erspamer, V., 77, 78, 80, 136, 210
v. Euler, U. S., 82
Everson-Pearse, A. G., 210

Falck, B., 76, 83, 86, 88, 209
Fausto, N., 282
Feldberg, Wilhelm, 223
Felt, B., 63
Fleming, W. W., 61
Folkow, B., 239, 243
Ford, D., 255
Fox, C. F., 290
Fredrick, J. F., 110
Frerichs, H., 183
Friedman, A. H., 319
Friedman, D. L., 162
Frontali, N., 87
Fuchs, E., 290
Fujii, S., 196, 197
Fujioka, M., 277
Fujita, S., 269
Furchgott, R. E., 4
Furman, R. H., 197
Fuxe, K., 82, 88, 332, 333

Gaddum, J. H., 3, 223
Gaitonde, M. K., 260, 263, 265
Gale, E. F., 215
Garattini, S., 210
Gelder, N. M. van, 255, 267, 269
Genghof, D. S., 96
Gerschenfeld, H. M., 79
Gillis, J. M., 132
Gilman, A. G., 40, 45
Girardier, L., 195
Giudice, G., 289
Goldberg, N. D., 40, 163, 171
Goldstein, J., 288
Goldstone, M., 81
Goodman, H. M., 8, 31, 194, 200
Gordon, M., 316
Gordon, R. S., Jr., 182
Gorin, E., 187
Gosselin, R. E., 131, 132
Gottesfeld, Z., 260, 261, 262
Graham, L. T., 255
Grahame-Smith, D. G., 38, 44
Grahn, B., 235
Gray, J., 132
Greenberg, M. J., 82
Greengard, O., 302, 316
Greengard, P., 7
Greenough, W. B., 47
Gruener, R., 155

Gupta, I., 21
Gustafsson, B., 218, 224
Gyermek, L., 54

Haddy, F. J., 211
Hadzi, J., 95
Hagen, J. H., 184, 190
Hagen, P., 134
Håkanson, R., 211
Haley, T. J., 248
Halperin, M. L., 188
Handler, J. S., 20
Hardman, H. F., 13
Harris, J. B., 31
Hashimoto, P. H., 332
Haugaard, N., 36, 100
Hausberger, F. X., 191
Havel, R. J., 191
Haverback, B. J., 213
Hayaishi, O., 7, 37
Hayashi, T., 255
Haynes, R. C., 38, 44
Hechter, O., 43, 46
Heller, A., 57, 334
Helmreich, E., 100, 147, 158
Henderson, N., 260
Henion, W. F., 15
Herbst, E. J., 277, 288
Herrera, M. G., 188
Hers, H. C., 163, 171
Herz, R., 157
Hess, M. E., 36, 100
Hickenbottom, J. P., 150
Hillarp, G., 83, 86, 88, 209
Hilton, S. M., 245
Himms-Hagen, J., 190, 195
Hirata, M., 7, 37
Ho, S. J., 196
Hobbiger, F., 255
Hoffman, G. F., 211
Hogg, J. F., 96, 103, 115
Holten, D., 302
Holz, C. G., 96
Honda, F., 56
Horakova, Z., 214, 218
Hornbrook, K. R., 171
Hunt, A. E., 104
Huston, R. B., 152

Ic'en, R., 277, 285
Imamura, H., 56
Isaacs, L., 218
Iverson, L. L., 209
Ivy, A. C., 213
Iwata, H., 100

Jacobsen, S., 210, 213
Jaeger, C. P., 86
Janakidevi, K., 97, 116
Jänne, J., 235, 276, 277, 278, 279, 280, 281, 282, 283, 285, 287, 288, 289, 292, 293
Janssen, P. A. J., 54
Jaques, R., 82
Jasper, H. H., 254
Jeanrenaud, B., 188, 189, 190, 200
Jefferson, L. S., 39, 56, 171, 175
Johnson, H. L., 212, 213, 214
Johnston, Marian, 232, 235
Jouvet, M., 324, 325, 332, 335, 338
Jungas, R. L., 46, 56, 163, 184, 186, 190, 196, 200

Kaempfer, R. O. R., 302
Kahlson, G., 213, 218, 224, 225, 226, 227, 228, 229, 231, 234, 235, 247
Kahn, V., 104, 110
Kakiuchi, A., 31
Kakiuchi, S., 38, 39, 44, 50, 54
Kaneko, T., 40
Kappers, J. A., 59
Katz, J., 188, 189
Katzen, R., 45, 47
Kay, R. H., 133
Keele, G. A., 78
Kenney, F. T., 302
Kerkut, G. A., 81, 82, 88
Kidder, G. W., 96, 97, 116
Kidman, A. D., 36, 52, 61
Killam, K. F., 269
Kim, K. S., 211
King, 81
Kipnis, D. M., 20, 38, 39
Klainer, L. M., 44, 46, 184
Knox, W. E., 302
Kobayashi, B., 134
Koe, K. B., 327
Koella, W. P., 327
Kono, T., 25
Korn, E. D., 197
Kornberg, H. L., 103, 105
Kostyo, J. L., 279
Krakow, J. S., 290
Krause, E.-G., 148
Krebs, E. G., 146, 147, 152
Kreutner, W., 163, 171
Krishna, G., 44, 46, 51, 64
Krnjevic, K., 254, 256
Kukovetz, W. R., 13
Kuntzman, R., 193
Kuo, J. F., 198, 199

Lajtha, A., 265
Lambert, R., 215

Lands, A. M., 10
Langer, S. Z., 60–61
Langley, J. N., 3
Langley, P., 23, 24
LaRaia, P. J., 39
Larner, J., 153, 161, 162, 163, 171, 179
LeBoeuf, B., 194
Leboy, P. S., 294
Lentz, T. L., 84
Leonard, S. L., 281
Levey, G. S., 13, 47, 100
Levi-Montalcini, R., 62
Levine, R. A., 13, 134
Levine, R. J., 218, 235
Levy, B., 14
Levy, M. R., 104, 115
Lewis, G. P., 240, 249, 250
Lewis, T., 223, 246
Li, C. H., 183
Lilja, B., 233
Lin, E. C. C., 302
Lincová, D., 191
Lindberg, O., 133
Lindberg, S., 227
Little, J. M., 29
Lopez, E., 43
Loveland, R. E., 79
Lovell, R. A., 254, 269
Lovenberg, W., 211
Low, A., 181
Lundholm, L., 40
Lynn, W. S., 182
Lyon, J. B., Jr., 155

MacAuliff, J. P., 27
McElroy, W. D., 133
McGeer, E. G., 57
McGeer, P. L., 57
MacGregor, R. R., 291
Machado, A. B. M., 62
Machiyama, Y., 257
Mackay, D., 230
McKeon, W. B., 54
McLennan, H., 82
Magoun, H. W., 323
Mahler, H. R., 291
Mahler, R., 184
Maickel, R. P., 191
Makman, R. S., 39
Malamud, D., 47
Malhortra, C. L., 269
Manners, D. J., 103
Mans, R. J., 290
Mansour, J. M., 128
Mansour, T. E., 29, 39, 113, 121, 122, 123, 124, 125, 126, 127, 128, 129, 130, 135
Marcucci, F., 269

Author Index

Marsh, J. M., 39, 45
Marshall, F. D., 269
Martinson, J., 210
Matsumoto, J., 325
Matthies, H., 269
Mayer, S. E., 13, 26, 140, 142, 144, 145, 146, 148, 149, 150, 151, 153, 154, 155, 163
Maynard, D. M., 87
Meester, W. D., 13
Mersmann, H. J., 171
Mettrick, D. E., 82
Miles, A. A., 249
Milholland, R. J., 302
Mirsky, I. A., 134
Mitchell, J. F., 256
Miyamoto, E., 41, 115
Mongar, J. L., 212
Monod, J., 119, 131
Moore, K. E., 131, 132
Moore, R. Y., 334
Moorhead, M., 78, 80, 87
Moran, N. C., 10, 14, 144
Morgan, H. E., 147, 148
Moruzzi, G., 323
Mosinger, B., 196, 197, 200, 201
Moulton, B. C., 281
Mouret, J., 327, 328, 330
Muller, M., 103
Murad, F., 11, 13, 29, 39, 45, 46, 47, 49, 100
Mussini, E., 269
Myhrberg, H. E., 81, 84
Myles, M. S., 257

Namm, D. H., 13, 151, 152
Nauss, K. M., 132
Newenschwander-Lemmer, N., 191
Nordenfelt, L., 210
Nott, M. W., 142, 143, 156

Ohlin, P., 210
Okuda, H., 195
Orö, L., 184, 191
Örstrom, A., 133
Otsuka, M., 256
Ottolenghi, C., 302
Owman, C., 76, 88, 209
Oye, I., 39, 41, 42, 46, 48
Ozawa, E., 152

Page, I. H., 211
Palm, D., 269
Palmer, E. C., 29, 31
Park, C. R., 25, 195, 198
Pastan, I., 45, 47, 115
Patel, N. G., 29

Paton, W. D. M., 212
Pegg, A. E., 279, 281
Perlman, R. L., 115
Permeggiani, A., 147
Petersen, E. E., 290
Pin, C., 332
Piras, R., 153, 154
Pitts, R. F., 45
Pöch, G., 13
Pohl, S. L., 45, 46
Popielski, L., 208
Popov, N., 269
Porter, J., 155
Posner, J. B., 40, 152
Potter, V. R., 303
Powell, C. E., 10
Pujol, J. R., 327

Questel, J. H., 254

Rabinowitz, M., 7, 41
Raina, A., 235, 276, 277, 278, 279, 280, 283, 285, 287, 288, 289, 291, 292, 293
Rall, T. W., 7, 11, 13, 31, 38, 39, 40, 42, 44, 45, 50, 54, 120, 125, 145
Randle, P. J., 188
Raper, C., 145
Reddy, W. J., 42
Reeves, R. B., 149
Reid, G., 134
Reilly, M. A., 247, 248
Reinert, M., 82
Reis, D. J., 319
Reite, O. B., 212
Reiter, R. J., 64
Renold, A. E., 181, 182, 188
Reshef, L., 182
Richter, D., 256, 265
Riley, J. F., 83, 211
Rizack, M. A., 185, 196, 197, 198
Roberts, E., 254
Roberts, R. B., 265
Robinson, G. A., 7, 10, 11, 13, 14, 19, 20, 21, 22, 26, 28, 36, 39, 48, 120, 126, 127, 148, 149
Rocha e Silva, M., 83, 241
Rodbell, M., 41, 42, 46, 184, 188, 190, 194
Rosell-Perez, M., 162
Rosen, F., 302
Rosen, O. M., 7, 37, 42, 56
Rosen, S. M., 7, 37, 42, 56
Rosenblueth, A., 9, 21, 26
Rosengren, E., 225, 229, 235, 247
Rosenthaler, J., 215, 217
Roussel, B., 332
Rubinstein, D., 196

346 Author Index

Rude, Sonia, 76, 81, 84, 86, 88, 89, 91
Rudman, D., 183, 190, 197
Russell, D. H., 235, 278, 283
Rychlik, I., 41, 47
Ryley, J. F., 96, 103, 104

Sabelli, H. C., 329
Sagan, L., 95
Saha, J., 195
Salganicoff, L., 102
Salvador, R. A., 205
Salzman, N. P., 55
Sandberg, N., 228
Sanger, F., 162
Sayoc, E. F., 29
Scaltrini, G. C., 134
Schachter, M., 82, 250
Schayer, R. W., 212, 223, 247, 248, 249
Scheparta, B., 302
Schild, H. O., 212
Schlender, K. K., 163
Schlesinger, K., 269
Schonheyder, F., 197
Schueler, F. W., 4, 5
Schur, H., 181
Schwartz, I. L., 63
Schwartz, J., 186
Scow, R. O., 187
Sedden, C. B., 86
Segal, H. L., 171
Seiler, N., 283
Seliger, H. H., 133
Senft, G., 8
Severs, W. B., 210, 214, 218
Shafrir, E., 187
Shambaugh, G. E., 303
Shapiro, B., 181, 182
Shore, P. A., 211, 213, 218
Shulman, N. R., 47
Siggins, G. R., 41, 55
Singh, S. I., 269
Sirek, A., 134
Slater, I. H., 10
Smith, E. R., 208
Snyder, S. H., 235, 278, 283
Sobel, B. E., 46
Söling, H. D., 188
Srinivasan, V., 256
S.-Rózsa, K., 132
Stadtman, E. R., 135
Staneloni, R., 153, 154
Stefani, F., 79
Steinberg, D., 184, 186, 188, 198
Steiner, D. F., 163
Stock, K., 186
Stone, D. B., 121, 122, 128, 129, 130

Stout, E. R., 290
Strand, O., 186
Strasberg, P., 255, 259
Streeto, J. M., 42
Stromblad, R., 210
Stubbs, S. St.G., 143
Sutherland, E. W., 6, 7, 8, 11, 16, 26, 35, 36, 37, 39, 41, 42, 43, 45, 46, 48, 101, 120, 125, 126, 145, 183, 200
Suyter, M., 182
Svensson, S. E., 233
Sweeney, D., 81, 82, 86
Szego, C. M., 40

Tabor, C. W., 276
Tabor, H., 276
Takeda, Y., 276
Tallan, H. H., 254, 269
Tanner, M. J. A., 276
Tata, J. R., 279
Taunton, O. D., 42, 46
Taylor, A. N., 61
Telford, J. M., 82
Tewari, S., 273
Threlfall, G., 281, 285
Touabi, M., 188
Trendelenburg, U., 60-61
Turtle, J. R., 20, 38, 39
Twarog, Betty M., 79, 132
Tyler, A., 37, 56

Valzelli, L., 210
Van Orden, L. S., III, 76, 88
Vaughan, M., 13, 46, 47, 63, 183, 186, 187, 188, 189, 194, 197, 198, 199, 200, 201
Villalobos, J. G., 282
Villar-Palasi, C., 161, 169, 170
Vincent, N., 143
Volqvartz, K., 197
Vrba, R., 258, 263, 265
Vugman, I., 83

Walaas, E., 193
Walaas, O., 193
Walker, C. A., 319
Walker, R. J., 82
Walsh, D. A., 152
Warburg, O., 270
Watkins, J. C., 254
Waton, N. G., 214
Watson, J. F., 45
Weber, A., 100, 157
Webster, M. E., 249
Wegner, J. I., 169, 170

Author Index

Weil-Malherber, H., 254, 258
Weiss, B., 36, 41, 42, 43, 45, 47, 48, 49–50, 51, 52, 57, 59, 61, 64, 191
Weissbach, H., 214
Weissman, A., 327
Weitzel, G., 134
Weitzman, E. D., 327
Welsh, J. H., 78, 80, 81, 84, 87, 136
Wenke, M., 190, 191, 192
Werman, R., 255
Wertheimer, E., 181
West, G. B., 83
West, T. C., 13
Westermann, E., 186
Wexler, J., 98
Wheatley, A. H. M., 254
Whitby, L. G., 209
White, J. E., 182, 190
Whittaker, R. H., 95
Widnell, C. C., 279
Wiedeman, M. P., 241
Wieland, O., 182
Wilkenfeld, B. E., 14
Williams, B. J., 142, 148, 153, 154, 163
Williams, Lois D., 84
Williams, R. H., 42
Williams-Ashman, H. G., 279, 281, 290

Williamson, J. R., 142, 148
Wilson, I. B., 4
Wilson, J. E., 156
Wilson, R. W., 61
Winegrad, A. I., 181
Wingo, W. J., 100
Wirsen, C., 192
Wolfe, S. M., 47
Wollenberger, A., 148
Wood, J. D., 257, 269
Wu, C., 96
Wurtman, R. J., 57, 64, 303
Wyman, J., 131

Yanagi, I., 195, 196, 197
Yessaian, N. H., 270
Yip, Agens, T., 162
Yockey, W. C., 269
Yoshino, Y., 255, 258, 263, 264, 265, 266, 267

Zieve, P. D., 47
Zimmerman, K. W., 208
Zs.-Nagy, I., 132
Zweifach, B. W., 239, 240, 241, 248

Subject Index

Absorption, 259–262
Acetone powders, fractionation of, 110
N-Acetyl aspartic acid, 257
Acetylcholine, 13, 29, 143–144, 253, 259, 260, 262, 271
  microcirculation effects:
    pharmacological, 240
    physiological, 240–241
Acetylcholinesterase, 4
Acids:
  amino, 96, 304
    in brain, 253–273
    glucose carbon entry, 263–266
    glucose metabolism relation, 256–258
    neurophysiological aspects, 254–255
    release of, 266–267
  γ-aminobutyric (GABA), 253–273
    absorption, 259–262
    binding, 259–262
    in brain, 253–273
    convulsions and, 267–269
    neurophysiological aspects, 254–255
    occlusion, 259–262
  γ-amino-β-hydroxybutyric (GABOB), 255–256
  α-aminoethanesulfonic, 256
  aspartic, 257, 265, 268
    N-acetyl, 257
  deoxyribonucleic (DNA), 277
  folic, 96, 281
  free fatty (FFA), 182, 184, 185, 188, 190, 191, 194, 198–199
  glutamic, 256, 265, 268, 270
  L-glutamic, 253–254
  lactic, 122, 124, 125
  lysergic diethylamide (LSD), 121, 127
  nicotinic, 38
  nucleic, 279, 282
  ^{14}C-orotic, 285, 288
  polyamine, 279

Acids (*cont.*):
  ribonucleic (RNA), 276–297
    accumulation, 277–295
    correlation, 277–284
    tenuazonic, 248
    tricarboxylic cycle, 103
ACTH, *see* Adrenocorticotropic hormone (ACTH)
Actinomycin D, 283, 289, 292, 296
Adenine derivatives, 50
Adenosine, 50
Adenosine diphosphate, 50, 51, 52
3′-Adenosine monophosphate, 51, 52
5′-Adenosine monophosphate, 50, 51, 52
Adenosine 3′, 5′-phosphate, *see* Cyclic AMP
Adenosine triphosphate, 36
S-Adenosylmethionine decarboxylase, 279, 281
Adenyl cyclase, 6–7, 27–28, 163
  activation of, 36, 43–48, 65–66
  in adipose tissue, 37, 38, 46, 50
  in adrenal glands, 37, 38, 42, 46
  basal, 53, 64–65
  in brain, 37–41, 46
  circadian variation, 57–59
  distribution of, 37–41
    gross, 37–41
    species, 37
    subcellular location, 41
  estrus cycle in, 65
  factors influencing activity of, 42–67
    adrenergic blocking agents, 49–50
    chronic effects, 56
    endocrine factors, 63–66
    environmental lighting, 57–59
    fasting, 66–67
    hormone activation, 43–45, 48
    inhibitors of, 48–49
    insulin, 56
    ionic environment, 42–43

350　Subject Index

Adenyl cyclase (*cont.*):
　neurohormone activation, 43–45
　neuronal activity, 59–63
　ontogenetic development, 56–57
　phenothiazines, 52–56
　prostaglandin E_1, 50
　purine derivatives, 50–52
　sodium fluoride activation, 45–48, 65–66
　molecule, 48
　neurotransmitters, 64
　pineal, 49, 52–55, 64
　prevalence of, 37
Adenyl 3′, 5′-cyclase, activation by serotonin, 125–127
Adenylates, 154
Adipocytes, 23, 25
Adipose tissue, 11, 42, 45, 181–206
　adenyl cyclase in, 37, 38, 46, 50
　calcium role, 196–197
　cyclic AMP role, 44, 193–198, 200
　epinephrine role, 187–190, 193–197, 200
　fat mobilization, 182–187
　innervation, 191–193
　insulin inhibitory action, 56
　norepinephrine role, 187
　receptors, 190–191
　serotonin role, 198–200
Adrenal glands:
　adenyl cyclase in, 37, 38, 42, 46
　dopamine in, 81
　norepinephrine in, 82
Adrenalectomy, 66
Adrenergic amines, muscle effects, 139–157
Adrenergic *beta* receptors, 16
Adrenergic blocking agents, 9–10
　in adenyl cyclase activity, 49–50
　alpha, 10–11, 20–21, 23, 25–28, 49
　　effects on *Tetrahymena pyriformis*, 98–100
　beta, 10–20, 21, 23, 25–28, 48, 49
　　effects on *Tetrahymena pyriformis*, 98–100
Adrenergic drugs:
　effects on glycogen phosphorylase, 104–109
Adrenocorticotropic hormone (ACTH), 7, 11, 12, 38, 43, 44
Alanine, 258, 265, 268
β-Alanine, 256
Aliphatic polyamines, 275
Allosteric regulation, 120
Alloxan, 174
Alpha adrenergic blocking agents, 10–11, 20–21, 23, 25–28, 49
　effects on *Tetrahymena pyriformis*, 98–100
Alpha receptors, 9–28
American Heart Association, 3

Amines:
　adrenergic, 139–157
　in carbohydrate metabolism, 121–124
Amino acids, 96, 304
　in brain, 253–273
　glucose carbon entry, 263–266
　glucose metabolism relation, 256–258
　neurophysiological aspects, 254–255
　release of, 266–267
γ-Aminobutyric acid (GABA), 253–273
　absorption, 259–262
　binding, 259–262
　convulsions and, 267–269
　neurophysiological aspects, 254–255
　occlusion, 259–262
γ-Aminobutyryl choline, 256
γ-Aminobutyryl histidine, 256
α-Aminoethanesulfonic acid, 256
γ-Amino-β-hydroxybutyric acid (GABOB), 255–256
Aminotransferase, tyrosine, 280
Ammonia, 258
Ammonium sulfate, 292
　fractionation, 110
Anaerobic glucolysis, 265
　effect of cyclic 3′, 5′-AMP on, 127–128
Annelids:
　dopamine in, 81
　histochemical fluorescence studies, 84–86
　5-hydroxytryptamine in, 79, 80
　norepinephrine in, 81
Anoxia, 148, 150–151
Anoxic heart, 149
Antihistamines, 248–249
Apoenzyme, 258
Argentaffin cells, 210–211
Arterenols:
　butyl, 190
　isopropyl, 190
Arthropods:
　dopamine in, 81
　histochemical fluorescence studies, 86–87
　5-hydroxytryptamine in, 79, 80
　norepinephrine in, 81
Ascaris lumbricoides, 127
Asparagine, 268
Aspartic acid, 257, 265, 268
　N-acetyl, 257

Barium, 42
Basal adenyl cyclase, 52–53, 64–65
Beta adrenergic blocking agents, 10–20, 21, 23, 25–28, 48, 49
　effects on *Tetrahymena pyriformis*, 98–100
Beta cells, pancreatic, 20
Binding, 259–262
Birds, 5-hydroxytryptamine in, 80

Subject Index

Bivalves:
  histochemical fluorescence studies, 86
  5-hydroxytryptamine in, 79, 80
Blood pressure, epinephrine and, 8
Bradykinin, 249–250
  microcirculation effects:
    pharmacological, 240
    physiological, 241
Brain:
  adenyl cyclase in, 37–41, 46
  amino acids in, 253–273
    glucose carbon entry, 263–266
    glucose metabolism relation, 256–258
    neurophysiological aspects, 254–255
    release of, 266–267
  γ-aminobutyric acid (GABA) in, 253–273
    absorption, 259–262
    binding, 259–262
    convulsions and, 267–269
    neurophysiological aspects, 254–255
    occlusion, 259–262
  cyclic AMP in, 44
  drugs producing decreases in monoamines, 325
  drugs producing increases in monoamines, 325–326
  5-hydroxytryptamine in, 79
  selective decrease of serotonin, 335–339
Brain stem, 323, 327
  histochemical anatomy of, 332–335
Brocresine, 235
4-Bromo-3-hydroxybenzyloxyamine, 235
Bufo americanus, 5-hydroxytryptamine in, 80
Butyl arterenols, 190

Caffeine, 8, 154–155, 156–157
  effects on *Tetrahymena pyriformis*, 101–102, 103
Calcium, 42, 43
  adipose tissue role, 196–197
Calcium ion, 152–153, 260
Calliactis parasitica, 5-hydroxytryptamine in, 78
Carbohydrate metabolism:
  amines in, 121–124
  epinephrine in, 134–135
  serotonin in, 134–135
Carbon, glucose entry, 263–266
Cardiac muscles, 143–145, 146
  glycogen metabolism, 148–152
Catalase, 104
Catecholamines (CA), 7, 8–28, 36, 39, 46, 47, 259, 260, 269–270
  in adenyl cyclase activity, 53–55
  distribution of, 75–77, 80–82
  environmental lighting, 58

Catecholamines (*cont.*):
  exocrine organ role, 208–210
  inotropic response to, 13–14
  lipolysis stimulation, 23–26
  microcirculation effects:
    pharmacological, 240
    physiological, 240–241
  order of potency of, 14
  in uterine relaxation, 14–18
Catron, 306, 309, 310
Cells:
  argentaffin, 210–211
  beta, pancreatic, 20
  chromaffin, 88–90
  enterochromaffin, 77, 78
  eukaryotic, 95
  fat, 23–25
  mast, 212
  myoepithelial, 208
  procaryotic, 113
  Renshaw, 254–255
  Retzius, 88–91
  secretory, 207–208
Centrioles, 95
Cephalopods, 5-hydroxytryptamine in, 80
Cerebellar tissue, 5-hydroxytryptamine in, 79
Chloramphenicol, 287
Chlorisondamine, 309
p-Chlorophenylalanine 327–332
Chloroplasts, 95
Chlorpromazine:
  in adenyl cyclase activity, 52–56
  potency of, 54–55
Choline, γ-aminobutyryl, 256
Choline esters, 29
Cholinesterase, 260
Chorionic gonadotropin, 47
Chromaffin cells, 88–90
Chromatography, 90
Chronotropic effects of epinephrine, 9
Ciliates, 95–118
  adrenergic drugs effects, 104–109
  alpha-adrenergic blocking agents effects, 98–100
  beta-adrenergic blocking agents effects, 98–100
  gluconeogenesis in, 103–104
  glycogen content, 96, 97–98, 101–102
  glycogen phosphorylase, 104–109
    properties of, 109–113
  glycogen synthetase, 104–109
  peroxisomes in, 103–104
  phosphodiesterase activity, 101, 102–103
  reserpine effects, 97–98
  theophylline effects, 99, 101–102, 103
  triiodothyronine effects, 100–101
Circadian rhythm, 301–320

Citrulline, 258
Clostridium welchii, 215
Cockroach, histochemical fluorescence studies, 87
Coelenterates:
 histochemical fluorescence studies, 83–84
 5-hydroxytryptamine in, 78
Coelenteric tissue, 5-hydroxytryptamine in, 78
Coenzymes, 258
Collagen formation, 227–228
Convulsions, 258, 265, 267–269
Crabs, histochemical fluorescence studies, 86–87
Crayfish, histochemical fluorescence studies, 86
Crithidia fasciculata, 116
Crustaceans:
 dopamine in, 81, 82
 5-hydroxytryptamine in, 78, 80
 norepinephrine in, 81, 82
 planktonic, serotonin effects, 133–134
Cyclic AMP (adenosine 3', 5'-phosphate), 6–8, 44, 125–126, 129, 130
 adipose tissue, 44, 193–198, 200
 in anaerobic glycolysis, 127-128
 biosynthesis of, 42
 in brain, 44
 changes in level of, 13–16, 20–31
 concentrations in tissue, 43–44
 cyclic 3', 5'-adenosine monophosphate, 35
 3', 5'-cyclic adenylate, 163, 165, 168, 169–170, 174–176, 179
 dibutyryl derivative of, 29
 exogenous, 13
 formation of, 6–8
 hormones in the level of, 6
 importance of, 36
 intracellular level of, 11
 lack of data about, 25–26
 mechanism of action, 146
 metabolism, 6–8
 pronethalol effect on level of, 12
 as a second messenger, 6
 in anaerobic glycolysis, 127-128
Cyclic 3', 5'-guanosine monophosphate, 35–36
3', 5' Cyclic nucleotides, 163
Cycloheximide, 248

Decarboxylase:
 S-adenosylmethionine, 279, 281
 dihydroxyphenylalanine, 270
 histidine, 244, 246–248
 ornithine (ODC), 278, 279, 281, 282, 283, 285, 286
Denervation, 59–63

Deoxyribonucleic acid (DNA), 277
Dibenzyline, 100, 116
Dichloroisoproterenol (DCI), 9–11, 49–50, 99–100, 107–109
Diglyceride, 187
Diglyceride lipase, 187
Dihydroergotamine, 20, 30
L-Dihydroxyphenylalanine (DOPA), 215, 306, 308, 309, 310, 311
Dihydroxyphenylalanine decarboxylase, 270
Dilator metabolites, 242–243
Direction des Recherches et Moyens d'essais, 321
DOPA, see L-Dihydroxyphenylalanine (DOPA)
Dopamine (DA), 270
 distribution of, 80–82
Drosophila melanogaster, 288
Dugesia tigrina:
 dopamine in, 81
 norepinephrine in, 81

Earthworm, histochemical fluorescence studies, 84
Echinoderms:
 dopamine in, 81
 5-hydroxytryptamine in, 79, 80
 norepinephrine in, 81
Embden-Meyerhof glycolytic pathway, 189
Embryogenesis, serotonin in, 133
Endocrine factors in adenyl cyclase activity, 63–66
Enteramine, 77
Enterochromaffin cells, 77, 78
Environmental lighting, 57–59
Enzymes, 6–8, 25, 36; *see also* names of enzymes
 activation, 49
 environmental lighting, 57–59
 glycolytic, serotonin effects, 124–125
 inhibition, 49
 in the metabolism of glycogen, 104–109
 peroxisomal, 105
 pineal, 43
 rate-limiting, 135
Epinephrine, 8, 12, 13, 14, 20, 21–24, 38, 39, 40, 46, 47, 48, 100, 115–116, 124, 125
 adipose tissue role, 187–190, 193–197, 200
 blood pressure and, 8
 in carbohydrate metabolism, 134–135
 chronotropic effects, 9
 distribution of, 81–82
 exocrine organs, 208, 210
 in glycogen metabolism, 140–143, 149–156
 glycogen synthetase effects, 163, 165–166, 168, 169
 hepatic glycogenolytic effect of, 11

Subject Index

Epinephrine (cont.):
  inotropic effects, 9
  insulin inhibition, 26
Ergotamine, 11
Ergotoxine, 8, 9
Erythrocytes:
  adenyl cyclase in, 37, 39, 46
  frog, 42
Escherichia coli, 215, 285
Esters, choline, 29
Estradiol, 40
Estrogen-progesterone balance, 26-27
Estrus cycle, 65-66
Eukaryotic cells, 95
European Office of Aerospace Research, 321
Exocrine organs, 207-222
  catecholamines, 208-210
  histamine, 208, 212-220
    bacterial origin in stomach, 214-218
  serotonin, 208, 210-212

Fasciola hepatica, 45, 113
Fasting in adenyl cyclase activity, 66-67
Fat mobilization, 182-187
FFA, *see* Free fatty acids (FFA)
Fishes, 5-hydroxytryptamine in, 80
Flatworms, histochemical fluorescence studies, 84
Florey's Factor I, 259
Fluoride, 42, 43
Foetal tissues, histamine formation in, 224-227
Folic acid, 96, 281
[14]C-Formate, 288-289
Fractionation
  of acetone powders, 110
  ammonium sulfate, 110
Free fatty acids (FFA), 182, 184, 185, 188, 190, 194, 198-199
Frog:
  erythrocyte, 42
  gastric secretion stimulation, 31
Frog skin:
  melanocyte-stimulating hormone antagonism, 30
  sodium transport across, 29
Fructose-1, 6-di-P, 124
Fructose-6-P, 124-125, 129, 130

GABA, *see* γ-Aminobutyric acid (GABA)
GABA shunt, 256-257
GABOB, *see* γ-Amino-β-hydroxybutyric acid (GABOB)
Ganglionectomy, 62
Gastric secretion, 31
Gastrin, 213

Gastrocnemius, 155
Gastrointestinal (GI) tract, 208-209, 211
Gastropods
  histochemical fluorescence studies, 86
  5-hydroxytryptamine in, 80
Glands:
  adrenal:
    adenyl cyclase in, 37, 38, 42, 46
    dopamine in, 82
    norepinephrine in, 82
  hypobranchial, 5-hydroxytryptamine in, 78
  mammary, 41, 233
  parotid, 37, 47
  pineal, 37, 41, 45
    adenyl cyclase activity, 37, 42, 47, 48, 50, 51-52, 59-63
    denervation, 59-63
    environmental lighting, 57-59
  salivary, 5-hydroxytryptamine in, 78
  thyroid, 45
    adenyl cyclase in, 37, 40, 47
Glucagon, 7, 11, 12, 13, 19, 38, 39, 45, 47
Glucocorticoids, 163, 171, 302, 304
Glucolysis, anaerobic, 265
  effect of cyclic 3', 5'-AMP on, 127-128
Gluconeogenesis:
  enzymes in, 115
  in *Tetrahymena pyriformis*, 103-104
Glucose, 104, 107-109, 112-113, 124-125, 134, 188, 194
  amino acid relation to metabolism of, 256-258
  in carbohydrate metabolism, 121-123
  carbon entry, 263-266
  glycogen synthetase effects, 163, 171, 172, 173, 176-177
Glucose-6-P, 124-125
Glucose-6-phosphate, 147, 154, 161-162, 189
  dependent, 162
  independent, 162
Glucose transport, 25
Glutamate, 256-257
Glutamic acid, 256, 265, 268, 270
L-Glutamic acid, 253-254
Glutamine, 257, 267, 268
Glycine, 255, 258, 264, 265, 270
Glycogen:
  in carbohydrate metabolism, 121-123
  liver, 96
  metabolism, 104-109, 139-159
    cardiac, 148-152
    epinephrine in, 140-143, 149-156
    norepinephrine in, 155-156
    skeletal muscle, 152-157
  in *Tetrahymena pyriformis*, 96, 97-98, 101-102

Glycogen phosphorylase:
 adrenergic drugs effects, in *Tetrahymena pyriformis*, 104–109
 properties of, 109–113
 theophylline, 104–109
Glycogen synthetase, 115, 161–180
 adrenergic drugs effects, 104–109
 epinephrine effects, 163, 165–166, 168, 169
 glucose effects, 163, 171, 172, 173, 176–177
 insulin effects, 161, 163, 165, 166, 171, 173, 175–179
 in *Tetrahymena pyriformis*, 104–109
 theophylline in, 104–109
Glycogenolysis, 134, 139–142, 144, 145, 148, 153
Glycolysis, 124–125
 anaerobic, 265
 effect of cyclic 3', 5'-AMP on, 127–128
Glycolytic enzymes, serotonin effects, 124–125
Gonadotropins, 45, 47
Growth:
 malignant, 230–232
 propanolol effects, 99
 reparative, 227–228
 reserpine effects, 97–98
 theophylline effects, 99
Growth hormones, 8, 279–280, 282, 283

Heart:
 adenyl cyclase in, 37, 39, 43, 46–47
 anoxic, 149
 inotropic response in, 13–14
 ischemic, 149
 poikilothermic, 149
Hemoglobin, 96
 three-dimensional structure of, 5
Hepatic glycogenolytic effect of epinephrine, 11
Hirudo medicinalis:
 histochemical fluorescence studies, 85–86
 Retzius cells of, 88–91
Histamine, 8, 28, 30–31, 38, 39
 distribution of, 82–83
 exocrine organs, 208, 212–220
  bacterial origin in stomach, 214–218
 exogenous, 246
 extracellular, 244
 formation, 223–238
  collagen, 227–228
  in foetal tissues, 224–227
  inhibition of, 228–230
  malignant growth, 230–232
  protein synthesis and, 232–237
  reparative growth, 227–228
  tissue origin, 224–237
 intracellular, 244

Histamine (*cont.*):
 loosely bound, 244
 microcirculation effects
  pharmacological, 240
  physiological, 241
 theory of microcirculatory regulation, 243–246
^{14}C-Histamine, 228
Histamine forming capacity (HFC), 225
Histidine, 227, 229, 230
 γ-aminobutyryl, 256
^{14}C-Histidine, 228
D-Histidine, 215
L-Histidine, 213, 215, 247
Histidine decarboxylase, 244, 246–248
Histochemical fluorescence studies, 83–88
 annelids, 84–86
 arthropods, 86–87
 coelenterates, 83–84
 flatworms, 84
 molluscs, 86
 vertebrates, 88
Histofluorescence technique, 332
Homeostasis, microcirculatory, 239–251
 control, 241–242
 dilator metabolites, 242–243
 histamine theory, 243–246
 microvascular regulatory function, 246–250
 pharmacological effects, 240
 physiological effects, 240–241
Homocarnosine, 256
Homogenates, 48
Honeybee, histamine in, 82
Hormones, 7, 11, 36, 43; *see also* names of hormones
 adenyl cyclase activations, 43–45, 48
 adrenocorticotropic (ACTH), 7, 11, 12, 38, 43, 44
 glycogen synthetase effects, 161–180
 growth, 81, 279–280, 282, 283
 in the level of cyclic AMP, 6
 luteinizing, 7, 38, 39, 44, 47
 melanocyte-stimulating (MSH), 7, 20–21, 30
 parathyroid, 7, 40, 45, 47
 polypeptide, 36, 44, 45
 thyroid, 47
 thyroid stimulating, 38, 40, 45, 47
Hornets:
 histamine in, 82–83
 5-hydroxytryptamine in, 79
Hydrochloride, pyridoxine, 316, 318
Hydrocortisone, 172–174
5-Hydroxyindolacetaldehyde, 329, 337
Hydroxyindole-O-methyltransferase (HIOMT), 57
β-Hydroxylation, 315–316

Subject Index

5-Hydroxytryptamine, *see* Serotonin
5-Hydroxytryptophan, 132, 211, 270, 325
Hyperemia:
 post-exercise, 242, 245
 reactive, 242, 245
Hyperthyroidism, 63–64, 100
Hypobranchial glands, 5-hydroxytryptamine in, 78
Hypoglycemia, 134, 175–176
Hypophysectomy, 66
Hypothalamus, 5-hydroxytryptamine in, 79
Hypothyroidism, 63–64, 100

Imidazole, 13
Immunosympathectomy, 62, 63
Indolamine, 124
Innervation in adipose tissue, 191–193
Insects, 5-hydroxytryptamine in, 80
Institut National de la Santé et de la Recherche Médicale, 321
Insulin, 7, 38, 39, 40, 44, 46, 189
 in adenyl cyclase activity, 56
 adipose tissue inhibition, 56
 in fat mobilization, 184, 186
 glycogen synthetase effects, 161, 163, 165, 166, 171, 173, 175–179
 inhibition, 24–26
 release, 20
Invertebrates:
 catecholamines in, 80–82
 histamine in, 82
 5-hydroxytryptamine in, 78, 79
 metabolic regulation in, 119–138
  amine effects, 121–124
  cyclic 3′, 5′-AMP effects, 127–128
  phosphofructokinase control mechanisms, 128–131, 135
  serotonin effects, 121–136
Iproniazid, 325
Ischemia, 148
Ischemic heart, 149
Isocitrate lyase, 103–104, 115
Isopropyl arterenols, 190
Isopropylmethoxamine, 13
Isoproterenol, 10, 14–15, 21, 27, 38, 39, 47

Kidney, 41
 adenyl cyclase in, 37, 39, 47
Kinetosomes, 95
Krebs cycle, 256–257

Lactic acid, 122, 124, 125
Lactobacillus, 215–217, 220

Leeches:
 histochemical fluorescence studies, 85–86
 Retzius cells of, 88–91
Leiurus quinquestriatus, 5-hydroxytryptamine in, 79
[14]C-Leucine, 233–236
Lighting, environmental, 57–59
Lipase:
 activation, 186–187
 diglyceride, 187
 lipoprotein, 197
 monoglyceride, 187
 pancreatic, 197
Lipolysis, 11, 63, 67, 183–186, 187, 188, 191, 194, 196, 197, 200–201
 activators, 183
 stimulation, 23–26
Lipolytic response, 45
Lipoprotein lipase, 197
Liver, 27, 29, 45
 adenyl cyclase in, 37, 39, 45, 47
 cyclic AMP in, 44
 insulin inhibitory action, 56
 regeneration, 277
Liver glycogen, 96
Liver fluke, serotonin effect on metabolism, 121–131
Lumbricus terrestris, histochemical fluorescence studies, 84
Lungs, adenyl cyclase in, 37, 39
Luteinizing hormones, 7, 38, 39, 44, 47
Lymnaea stagnalis, serotonin effects, 133
Lysergic acid diethylamide (LSD), 121, 127

McArdle's disease, 154
Macromolecular complex, 4–5
Macromolecules, 4–5
 synthesis of, 275
Magnesium, 42, 51, 53
Magnesium ions, 6
Malate synthase, 103
Mammals, 5-hydroxytryptamine in, 80
Mammary glands, 41, 233
Manganese, 42
Mast cells, 212
Meganyctophanes, serotonin effects, 133–134
Melanocyte condensation, 24
Melanocytes, 21
Melanocyte-stimulating hormone (MSH), 7, 20–21, 30
Melatonin, 7, 28, 29–30, 64, 199
Metabolic regulation in invertebrates, 119–138
 amine effects, 121–124
 cyclic 3′, 5′-AMP effects, 127–128

Metabolic regulation in
  invertebrates (cont.):
  phosphofructokinase control mechanisms, 128–131, 135
  serotonin effects, 121–136
Metabolism:
  carbohydrate, 121–124
  epinephrine in, 134–135
  serotonin in, 134–135
  cyclic AMP, 6–8
  glucose, 256–258
  glycogen, 104–109, 139–159
    cardiac, 148–152
    epinephrine in, 140–143, 149–156
    norepinephrine in, 155–156
    skeletal muscle, 152–157
Metabolites, dilator, 242–243
Metarterioles, 241
Metazoa, 95–118
^{14}C-Methionine, 277–278, 279–280
α-Methylhistidine, 235
α-Methyl-para-tyrosine, 306
O-Methyltransferase, 260
Methylxanthines, 8, 13, 16, 183–184
Metridium senile, histochemical fluorescence studies, 83–84
Microcirculatory homeostasis, 239–251
  control, 241–242
  dilator metabolites, 242–243
  histamine theory, 243–246
  microvascular regulatory function, 246–250
  pharmacological effects, 240
  physiological effects, 240–241
Microspectrofluorometry, 76, 90
Midbrain, 5-hydroxytryptamine in, 79
Mitochondria, 95
Modiolus demissus, serotonin effects on the gills of, 131–132
Molluscs:
  dopamine in, 81, 82
  histochemical fluorescence studies, 86
  5-hydroxytryptamine in, 78, 79, 80, 82
  norepinephrine in, 81, 82
  serotonin effects, 132–133
Monoamine, 76, 77
  drugs producing decreases in brain, 325
  drugs producing increases in brain, 325–326
Monoamine oxidase, 260, 270
Monoamine oxidase (MAO) inhibitors, 325–326
Monoglyceride lipase, 187
2-Monoglyceride, 187
Muscles:
  adrenergic amine effects, 139–157
  cardiac, 143–145, 146
  glycogen metabolism, 148–152

Muscles (cont.):
  glycogen synthetase in, 161–180
  skeletal, 41, 143–144, 146
  adenyl cyclase in, 37, 40, 47
  cyclic AMP in, 44
  glycogen metabolism, 152–157
  smooth, 29
  adenyl cyclase in, 37, 40
  contraction, 9
  serotonin effects, 132
Mussels, serotonin effects on the gills of, 131–132
Myoepithelial cells, 208
Myometrium, 41
Mytilus muscle, serotonin effects, 132

National Institute of Allergy and Infectious Diseases, 119
National Institute of Mental Health, 3
National Institutes of Health, 9, 95, 214, 216
National Research Council for Medical Sciences, 275
National Science Foundation, 95
Nerves, sympathetic, 209–210
Nervous system:
  catecholamines in, 80–82
  5-hydroxytryptamine in, 78–80
Neurohormones, 37
  adenyl cyclase activations, 43–45
Neurotransmitters, adenyl cyclase responsiveness to, 64
Nialamide, 325, 326
Nicotinic acid, 38
Norepinephrine, 8, 13, 14, 20, 31, 38, 39, 48, 269, 332–336, 338
  in adenyl cyclase activity, 65–67
  adipose tissue role, 187
  circadian rhythm and, 301–320
  distribution of, 80–82
  environmental lighting, 58–59
  exocrine organs, 208
  in free fatty acid release, 198, 199
  in glycogen metabolism, 155–156
  insulin inhibition, 26
  in neuronal activity, 59–61
  tyrosine transaminase interaction, 311–316
Nucleic acid, 279, 282
Nucleotides, 46, 50

Occlusion, 259–262
Octopuses, 5-hydroxytryptamine in, 78
Ornithine decarboxylase (OCD), 278, 279, 281, 282, 283, 285, 286
^{14}C-Orotic acid, 285, 288
Ouabain, 260, 261, 262

Subject Index

Ovaries, adenyl cyclase in, 37, 47
α-Oxoglutarate, 256–257
Oxytocin, 28–29

Pancreas, adenyl cyclase in, 37, 39
Pancreatic *beta* cells, 20
Pancreatic islets, 20, 24
Pancreatic lipase, 197
Paradoxical Sleep (PS), 321–323, 324, 325–327, 329, 332, 335, 337
Paraochlorophenylalanine, 211
Parathyroid hormone, 7, 40, 45, 47
Parathyroidectomy, 66
Parotid glands, 37, 47
Pentobarbital, 265, 266
Pentylenetetrazol, 265
Periplaneta americana, histochemical fluorescence studies, 87
Peroxisomal enzymes, 105
Peroxisomes, in *Tetrahymena pyriformis*, 103–104
Phagocata oregonensis, histochemical fluorescence studies, 84, 85
Pharmacology, ultimate goal of, 5
Pheniprazine, 325
Phenothiazines, in adenyl cyclase activity, 52–56
Phentolamine, 20, 30, 191
Phenylalanine hydroxylating system, 96
β-Phenylisopropylhydrazine, 309
Phoneutria fera, 5-hydroxytryptamine in, 79
Phosphatase, 162
Phosphate:
 adenosine 3′, 5′, 125–126
 pyridoxal, 258, 270
 pyridoxal-5, 312–316, 317, 319
Phosphodiesterase, 7–8, 13, 16, 27, 31, 36, 48, 102
 effects on *Tetrahymena pyriformis*, 101, 102–103
 inhibition of, 56, 102–103
Phosphofructokinase, 29, 124–125, 126
 activation of, 189
 control mechanisms, 128–131, 135
 effect of cyclic 3′, 5′-AMP on, 127–128
Phosphorylase, 99–100, 115, 146
 heat stability of, 114
Phosphorylase deficient myopathy, 154
Phosphorylase kinase, 155
 activation of, 146–147, 148–150, 151–153
Picrotoxin, 255, 265
Pineal adenyl cyclase, 49, 52–55, 64
Pineal enzymes, 43
Pineal glands, 37, 41, 45
 adenyl cyclase activity, 37, 42, 47, 48, 50, 51–52, 59, 63
 denervation, 59–63

Pineal glands (*cont.*):
 environmental lighting, 57–59
Planarians:
 dopamine in, 81
 histochemical fluorescence studies, 84, 85
 norepinephrine in, 81
Planktonic crustaceans, serotonin effects, 133–134
Plasma membrane, 7
Platelets, 23
 adenyl cyclase in, 37, 47
 aggregation, 24–25
 5-hydroxytryptamine in, 78
Poikilothermic heart, 149
Polyamine acid, 279
Polyamines:
 aliphatic, 275
 exogenous, 285
 as metabolic regulators, 275–300
  accumulation in, 277–295
  correlation in, 277–284
Polymerase, 281, 282, 286, 290, 291, 292
Polypeptide hormones, 36, 44, 45
Pontine tegmentum, 324, 332
Potassium, 42
 adenyl cyclase in, 42
Potassium ion, 152
Prenylamine, 116
Procaryotic cells, 113
Progesterone-estrogen balance, 26–27
Proline, 265
Pronethalol, 11, 13, 50
 cyclic AMP level effect, 12
Propanolol, 14–15, 27, 98–99
 effects on *Tetrahymena pyriformis*, 99
Prostaglandin E_1, 21–23, 38, 39, 40, 44, 47
 in adenyl cyclase activity, 50
 cyclic AMP in, 44
Prostaglandins, 7
Protein kinase, 152
Protein synthesis, histamine formation and, 232–237
Protoveratrine, 260, 261, 262
Protozoa, 95–96
Purines, 47
 derivatives in adenyl cyclase activity, 50–52
Puromycin, 248, 306, 307
Putrescine, 275, 278–279, 281, 293, 295, 297
 endogenous, 278
Pyridoxal hydrochloride, 316, 318
Pyridoxal phosphate, 258, 270
Pyridoxal-5-phosphate, 312–316, 317, 319
Pyrophosphates, 7

Rana pipiens, 20, 21
Raphe system, 332, 335–339

Rate-limiting enzymes, 135
Rats:
  liver, 27
  environmental lighting, 57–59
Receptors, 3–34
  adipose tissue, 190–191
  alpha, 9–28
  beta, 9–28
  concept of specific, 3
  definitions, 3, 4–5
  evidence for existence of, 4
  isolation of, 5
  nature of, 4–5
Regulatory subunits, 19
REM sleep, 323
Renal cortex, adenyl cyclase activity, 47
Renal medulla, adenyl cyclase activity, 47
Renshaw cells, 254–255
Reptiles, 5-hydroxytryptamine in, 80
Reserpine, 107–109, 269, 305–306, 307
  effects on *Tetrahymena pyriformis*, 97–98
  in sleep, 325
Reticulum, sacroplasmic, 101
Retzius cells, 88–91
Ribonucleic acid (RNA), 276–297
  accumulation, 277–295
  correlation, 277–284
Ribosomes, stability of, 275–276
RNA, see Ribonucleic acid (RNA)
Rous virus sarcoma, 235–237

Sacroplasmic reticulum, 101
Salivary glands, 5-hydroxytryptamine in, 78
Schistosoma mansoni, 127
Scorpions, 5-hydroxytryptamine in, 78, 79
Sea anemones:
  histochemical fluorescence studies, 83–84
  5-hydroxytryptamine in, 78
Second messenger, cyclic AMP as, 6
Secretory cells, 207–208
Serine, 265, 268
Serotonin, 28, 29, 38, 39, 45, 46, 115–116, 144, 259, 260, 269–270, 271, 306
  in adenyl 3′, 5′-cyclase activation, 125–127
  adipose tissue role, 198–200
  in carbohydrate metabolism, 134–135
  distribution of, 75–80
    in non-nervous tissues and products, 77–78
  effects in gills of mussels, 131–132
  in embryogenesis, 133
  endogenous, 309
  exocrine organs, 208, 210–212
  glycolitic enzyme effects, 124–125
  in metabolic regulation, 121–136
    carbohydrate of mammals, 134–135
    the liver fluke, 121–131

Serotonin (*cont.*):
  microcirculation effects
    pharmacological, 240
    physiological, 240–241
  mollusc effects, 132–133
  planktonic crustacean effects, 133–134
  in Retzius cells, 88–91
  sleep and, 321–340
    decrease in brain monoamines, 325
    decrease of cerebral serotonin, 327–332, 335–339
    histochemical anatomy of the lower brain stem, 332–335
    increase in brain monoamines, 325–326
    neuropharmacological approach, 325–326
  smooth muscle effects, 132
  synthesis occurrence, 121
Sigrid Jus'elius Foundation, 275
Skeletal muscles, 41, 143–144, 146
  adenyl cyclase in, 37, 40, 47
  cyclic AMP in, 44
  glycogen metabolism, 152–157
Sleep:
  Paradoxical (PS), 321–323, 324, 325–327, 329, 332, 335, 337
  REM, 323
  reserpine in, 325
  serotonin and, 321–340
    decrease in brain monoamines, 325
    decrease of cerebral serotonin, 327–332, 335–339
    histochemical anatomy of the lower brain stem, 332–335
    increase in brain monoamines, 325–326
    neuropharmacological approach, 325–326
  slow wave (SWS), 321, 323, 324, 325, 327–329, 335–337
Slow Wave Sleep (SWS), 321, 323, 324, 325, 327–329, 335–337
Smooth muscle, 29
  acenyl cyclase in, 37, 40
  contraction, 9
  serotonin effects, 132
Sodium, 42, 259–261
  transport across frog skin, 29
Sodium chloride, 259, 261
Sodium fluoride
  adenyl cyclase activation, 45–48, 65–66
  in neuronal activity, 59–61
Sodium ions, 259
Spectrofluorimetry, 90
Spermidine, 275, 276, 277–278, 279, 282, 285–287, 288, 289, 291, 292, 293, 294, 296
  exogenous, 285
Spermine, 275, 286–287, 288, 289, 291, 292

Subject Index

Spiders, 5-hydroxytryptamine in, 78, 79
Spinal cord, 5-hydroxytryptamine in, 79
Spleen, adenyl cyclase in, 37, 40
Steroids, 8
Strychnine, 255
Sulfate, ammonium, 292
 fractionation, 110
Sympathetic nerves, 209-210
Sympathin E, 9
Sympathin I, 9
Synoeca surinama, 5-hydroxytryptamine in, 79
Synthetase I kinase, 162, 163, 164, 165, 177, 178
 changes in, 169-179
Systemic stress, 242

Taurine, 256
Telia felina, histochemical fluorescence studies, 83-84
Tenuazonic acid, 248
Testes, 45
 adenyl cyclase in, 37, 47
Testosterones, 281, 291
Tetrahymena pyriformis, 95-118, 270-271
 alpha-adrenergic blocking agents effects, 98-100
 beta-adrenergic blocking agents effects, 98-100
 caffeine effects, 101-102, 103
 gluconeogenesis in, 103-104
 glycogen in, 96, 97-98, 101-102
 glycogen phosphorylase in, 104-109
  properties of, 109-113
 glycogen synthetase in, 104-109
 peroxisomes in, 103-104
 phosphodiesterase activity, 101, 102-103
 reserpine effects, 97-98
 theophylline effects, 99, 101-102, 103
 triiodothyronine effects, 100-101
Tetrodotoxin, 260, 261, 262
Theophylline, 8, 13, 14, 15, 31, 115, 191, 200
 effects on *Tetrahymena pyriformis*, 99, 101-102, 103
 in free fatty acid release, 198, 199
 in glycogen phosphorylase, 104-109
 in glycogen synthetase, 104-109
 potentiation by, 16
Thioacetamide, 282, 286
Thrombocytin, 77
^{14}C-Thymidine, 288
Thyroid glands, 45
 adenyl cyclase in, 37, 40, 47
Thyroid hormones, 47
Thyroid stimulating hormones, 38, 40, 45, 47
Thyroidectomy, 66
Thyroxin, 8, 46, 63-64, 100

Tissue:
 adipose, 11, 42, 45, 181-206
  adenyl cyclase in, 37, 38, 46, 50
  calcium role, 196-197
  cyclic AMP role, 44, 193-198, 200
  fat mobilization, 182-187
  innervation, 191-193
  insulin inhibitory action, 56
  receptors, 190-191
  serotonin role, 198-200
 cerebellar, 5-hydroxytryptamine in, 79
 coelenteric, 5-hydroxytryptamine in, 78
 cyclic AMP concentrations in, 43-44
 foetal, histamine formation in, 224-227
 non-nervous, 5-hydroxytryptamine in, 77-78
 origin of histamine, 224-237
Toad bladders, 20, 24
Tricarboxylic acid cycle, 103
Trifluoperazine, potency of, 54-55
Triglycerides, 182, 186-187
Triiodothyronine, 107-109
 effects on *Tetrahymena pyriformis*, 100-101
Tryosine, synthesis of, 96
Trypanosomatids, 116
Tryptophan:
 hydroxylation, 211
 synthesis, 115
^{3}H-Tryptophan, 327
L-Tryptophan, 213
Tryptophan hydroxylase, 211, 328
Tryptophan pyrolase, 280
Tunicates, 5-hydroxytryptamine in, 79, 80
L-Tyrosine, 306, 313
Tyrosine aminotransferase, 280
Tyrosine hydroxylase, 57, 306
Tyrosine transaminase, 302, 306-311, 317, 319
 norepinephrine interaction, 311-316
 rhythm, 303-304, 305

United States Public Health Service, 3, 75, 139, 239
^{14}C-Uracil, 287
^{14}C-Uridine, 288
^{3}H-Uridine, 288
Uterine relaxation, 14-18
Uterus, 26
 adenyl cyclase in, 37, 40

Vasodilatation, slowly-developing, 242
Vasomotion, 242, 244-245
Vasopressin, 7, 38, 39, 45, 47
Venoms:
 histamine in, 82
 5-hydroxytryptamine in, 78, 79

Subject Index

Vertebrates:
  histamine in, 83
  histochemical fluorescence studies, 88
  5-hydroxytryptamine in, 78, 79, 80
Vespa crabro:
  histamine in, 82–83
  5-hydroxytryptamine in, 79

Vespa vulgaris, histamine in, 82

Wasps:
  histamine in, 82
  5-hydroxytryptamine in, 78, 79

Zinc, 42

DATE DUE

574.192   B52   67855

BLUM

BIOGENIC AMINES AS PHYSIOLOGICAL
REGULATORS

College Misericordia Library
Dallas, Pennsylvania